Handbook of Machine Learning for Computational Optimization

Demystifying Technologies for Computational Excellence: Moving Towards Society 5.0

Series Editors
Vikram Bali and Vishal Bhatnagar

This series encompasses research work in the field of Data Science, Edge Computing, Deep Learning, Distributed Ledger Technology, Extended Reality, Quantum Computing, Artificial Intelligence, and various other related areas, such as natural language processing and technologies, high-level computer vision, cognitive robotics, automated reasoning, multivalent systems, symbolic learning theories and practice, knowledge representation and the semantic web, intelligent tutoring systems, AI, and education.

The prime reason for developing and growing out this new book series is to focus on the latest technological advancements – their impact on the society, the challenges faced in implementation, and the drawbacks or reverse impact on the society due to technological innovations. With the technological advancements, every individual has personalized access to all the services, all devices connected with each other communicating among themselves, thanks to the technology for making our life simpler and easier. These aspects will help us to overcome the drawbacks of the existing systems and help in building new systems with latest technologies that will help the society in various ways, proving Society 5.0 as one of the biggest revolutions in this era.

Industry 4.0, AI, and Data Science
Research Trends and Challenges
Edited by Vikram Bali, Kakoli Banerjee, Narendra Kumar, Sanjay Gour, and Sunil Kumar Chawla

Handbook of Machine Learning for Computational Optimization
Applications and Case Studies
Edited by Vishal Jain, Sapna Juneja, Abhinav Juneja, and Ramani Kannan

Data Science and Innovations for Intelligent Systems
Computational Excellence and Society 5.0
Edited by Kavita Taneja, Harmunish Taneja, Kuldeep Kumar, Arvind Selwal, and Ouh Lieh

Artificial Intelligence, Machine Learning, and Data Science Technologies
Future Impact and Well-Being for Society 5.0
Edited by Neeraj Mohan, Ruchi Singla, Priyanka Kaushal, and Seifedine Kadry

For more information on this series, please visit: https://www.routledge.com/Demystifying-Technologies-for-Computational-Excellence-Moving-Towards-Society-5.0/book-series/CRCDTCEMTS

Handbook of Machine Learning for Computational Optimization
Applications and Case Studies

Edited by
Vishal Jain, Sapna Juneja, Abhinav Juneja, and
Ramani Kannan

CRC Press
Taylor & Francis Group
Boca Raton London New York

CRC Press is an imprint of the
Taylor & Francis Group, an **informa** business

First edition published 2022
by CRC Press
6000 Broken Sound Parkway NW, Suite 300, Boca Raton, FL 33487-2742

and by CRC Press
2 Park Square, Milton Park, Abingdon, Oxon, OX14 4RN

© 2022 selection and editorial matter, Vishal Jain, Sapna Juneja, Abhinav Juneja, and Ramani Kannan; individual chapters, the contributors

CRC Press is an imprint of Taylor & Francis Group, LLC

Library of Congress Cataloging-in-Publication Data
Names: Jain, Vishal, 1983- editor.
Title: Handbook of machine learning for computational optimization :
applications and case studies / Vishal Jain, Sapna Juneja, Abhinav Juneja, Ramani Kannan.
Description: Boca Raton : CRC Press, 2021. | Series: Demystifying
technologies for computational excellence | Includes bibliographical references and index.
Identifiers: LCCN 2021017098 (print) | LCCN 2021017099 (ebook) |
ISBN 9780367685423 (hardback) | ISBN 9780367685454 (paperback) |
ISBN 9781003138020 (ebook)
Subjects: LCSH: Machine learning—Industrial applications. | Mathematical
optimization—Data processing. | Artificial intelligence.
Classification: LCC Q325.5 .H295 2021 (print) | LCC Q325.5 (ebook) | DDC 006.3/1—dc23
LC record available at https://lccn.loc.gov/2021017098
LC ebook record available at https://lccn.loc.gov/2021017099

ISBN: 978-0-367-68542-3 (hbk)
ISBN: 978-0-367-68545-4 (pbk)
ISBN: 978-1-003-13802-0 (ebk)

DOI: 10.1201/9781003138020

Typeset in Times
by codeMantra

Contents

Preface.. vii

Editors... xi

Contributors xiii

Chapter 1   Random Variables in Machine Learning... 1

                 Piratla Srihari

Chapter 2   Analysis of EMG Signals using Extreme Learning Machine
                 with Nature Inspired Feature Selection Techniques 27

                 A. Anitha and A. Bakiya

Chapter 3   Detection of Breast Cancer by Using Various Machine Learning
                 and Deep Learning Algorithms 51

                 Yogesh Jadhav and Harsh Mathur

Chapter 4   Assessing the Radial Efficiency Performance of Bus Transport
                 Sector Using Data Envelopment Analysis..71

                 Swati Goyal, Shivi Agarwal, Trilok Mathur, and Nirbhay Mathur

Chapter 5   Weight-Based Codes—A Binary Error Control Coding
                 Scheme—A Machine Learning Approach.. 89

                 Piratla Srihari

Chapter 6   Massive Data Classification of Brain Tumors Using DNN:
                 Opportunity in Medical Healthcare 4.0 through Sensors 95

                 Rohit Rastogi, Akshit Rajan Rastogi, D.K. Chaturvedi,
                 Sheelu Sagar, and Neeti Tandon

Chapter 7   Deep Learning Approach for Traffic Sign Recognition on
                 Embedded Systems 113

                 A. Shivankit, Gurminder Kaur, Sapna Juneja, and Abhinav Juneja

Chapter 8   Lung Cancer Risk Stratification Using ML and AI on Sensor-
Based IoT: An Increasing Technological Trend for Health of
Humanity...137

*Rohit Rastogi, Mukund Rastogi, D.K. Chaturvedi,
Sheelu Sagar, and Neeti Tandon*

Chapter 9   Statistical Feedback Evaluation System...........................153

Alok Kumar and Renu Jain

Chapter 10   Emission of Herbal Woods to Deal with Pollution and Diseases:
Pandemic-Based Threats...183

*Rohit Rastogi, Mamta Saxena, D. K. Chaturvedi,
and Sheelu Sagar*

Chapter 11   Artificial Neural Networks: A Comprehensive Review..................203

Neelam Nehra, Pardeep Sangwan, and Divya Kumar

Chapter 12   A Case Study on Machine Learning to Predict the Students'
Result in Higher Education 229

Tejashree U. Sawant and Urmila R. Pol

Chapter 13   Data Analytic Approach for Assessment Status of Awareness of
Tuberculosis in Nigeria...243

*Ishola Dada Muraina, Rafeeah Rufai Madaki, and
Aisha Umar Suleiman*

Chapter 14   Active Learning from an Imbalanced Dataset: A Study
Conducted on the Depression, Anxiety, and Stress Dataset.............251

Umme Salma M. and Amala Ann K. A.

Chapter 15   Classification of the Magnetic Resonance Imaging of the Brain
Tumor Using the Residual Neural Network Framework..................267

Tina and Sanjay Kumar Dubey

Index..279

Preface

Machine learning is a trusted technology over decades and has flourished on a global scale touching the lives of each one of us. The modern-day decision making and processes are all dependent on machine learning technology to make matured short-term and long-term decisions. Machine learning is blessed to have phenomenal support from the research community, and have landmark contributions, which is enabling machine learning to find new applications every day. The dependency of human processes on machine learning-driven systems is encompassing all spheres of current state-of-the-art systems with the level of reliability it offers. There is a huge potential in this domain to make the best use of machines in order to ensure the optimal prediction, execution, and decision making. Although machine learning is not a new field, it has evolved with ages and the research community round the globe have made remarkable contribution for the growth and trust of applications to incorporate it. The predictive and futuristic approach, which is associated with machine learning, makes it a promising tool for business processes as a sustainable solution. There is an ample scope in the technology to propose and devise newer algorithms, which are more efficient and reliable to give machine learning an entirely new dimension in discovering certain latent domains of applications, it may support. This book will look forward to addressing the issues, which can resolve the modern-day computational bottom lines which need smarter and optimal machine learning-based intervention to make processes even more efficient. This book presents innovative and improvised machine learning techniques which can complement, enrich, and optimize the existing glossary of machine learning methods. This book also has contributions focusing on the application-based innovative optimized machine learning solutions, which will give the readers a vision of how innovation using machine learning may aid in the optimization of human and business processes.

We have tried to knit this book as a read for all books wherein the learners and researchers shall get insights about the possible dimensions to explore in their specific areas of interest. The chapter-wise description is as follows:

Chapter 1 explores the basic concepts of random variables (single and multiple), their role and applications in the specified areas of machine learning.

Chapter 2 demonstrates Wigner-Ville transformation technique to extract the time-frequency domain features from typical and atypical EMG signals – myopathy (muscle disorder) and amyotrophic lateral sclerosis (neuro disorder). Nature inspired feature selection algorithms, whale optimization algorithm (WOA), genetic algorithm (GA), bat algorithm (BA), fire-fly optimization (FA), and particle swarm optimization (PSO) are utilized to determine the relevant features from the constructed features.

Chapter 3 presents various algorithms of machine learning (ML), which can be used for breast cancer detection. Since these techniques are commonly used in many areas, they are also used for making decisions regarding diagnostic diagnosis and clinical studies.

Chapter 4 measures the efficiency and thoroughly explores the scope for optimal utilization of the input resources owned by depots of the RSRTC. The new slack model (NSM) of DEA is used as it enumerates the slacks for input and output variables. The model satisfies the radial properties, unit invariance, and translation invariance. This study enables policy-makers to evaluate inputs for consistent output up to the optimum level and improve the performance of the inefficient depots.

Chapter 5 presents a binary error control coding scheme using weight-based codes. This method is quite used for classification and employs the K nearest neighbor algorithms. The paper also discussed the role of distance matrix with hamming code evaluation.

Chapter 6 exhibits MRI images of the framed brain to create deep neural system models that can be isolated between different types of heart tumors. To perform this task, deep learning is used. It is a type of instrument-based learning where the lower levels responsible for many types of higher-level definitions appear above the different levels of the screen.

Chapter 7 focuses on creating an affordable and effective warning system for drivers that is able to detect the warning sign boards and speed limits in front of the moving vehicle, and prompt the driver to lower to safer speeds if required. The software internally works on a deep learning-based modern neural network YOLO (You Only Look Once) with certain modifications, which allows it to detect the road signs really quickly and accurately on low-powered ARM CPUs.

Chapter 8 presents an approach for the classification of lung cancer based on the associated risks (high risk, low risk, high risk). The study was conducted using a lung cancer classification scheme by studying micrographs and classifying them into a deep neural network using machine learning (ML) framework.

Chapter 9 presents a statistical feedback evaluation system that allows to design an effective questionnaire using statistical knowledge of the text. In this questionnaire, questions and their weight are not pre-decided. It is established that questionnaire-based feedback systems are traditional and quite straightforward, but these systems are very static and restrictive. The proposed statistical feedback evaluation system is helpful to the users and manufacturers in finding the appropriate item as per their choices.

Chapter 10 presents an experimental work based on the data collected on various parameters on the scientific measuring analytical software tools Air Veda instrument and IoT-based sensors capturing the humidity and temperature data from atmospheric air in certain interval of time to know the patterns of pollution increment or decrement in atmosphere of nearby area.

Chapter 11 concerns with neural network representations and defining suitable problems for neural network learning. It covers numerous substitute designs for the primitive units making up an artificial neural network, such as perceptron units, sigmoid unit, and linear units. This chapter also covers the learning algorithms for training single units. Backpropagation algorithm for multilayer perceptron training is described in detail. Also, the general issues such as the representational capabilities of ANNs, overfitting problems, and substitutes to the backpropagation algorithm are also explained.

Chapter 12 proposes a system which will make use of the machine learning approach to predict a student's performance. Based on student's current performance and some measurable past attributes, the end result can be predicted to classify them among good or bad performers. The predictive models will make students aware who are likely to struggle during the final examinations.

Chapter 13 presents a study that assists in assessing the awareness status of people on the TB towards its mitigation and serves as contribution to the field of health informatics. Indeed, the majority of participants claimed that they had low awareness on the TB and its associated issues in their communities. Though, the participants were from Kano state, a strategic location in the northern part of Nigeria, which means that the result of the experiment can represent major opinions of northern residents.

Chapter 14 deals with psychological data related to depression, anxiety, and stress data to study how the classification and analysis is carried out on imbalanced data. The proposed work not only contributes on providing practical information about the balancing techniques like SMOTE, but also reveals the strategy for dealing with working of many existing classification algorithms like SVM, Random Forest, XGBoost etc. on imbalanced dataset.

Chapter 15 proposes the construction of segmented mask of MRI (magnetic resonance image) using CNN approach with the implementation of ResNet framework. The understanding of ResNet framework using layered approach will provide the extensive anatomical information of higher-dimensional image for precise clinical analysis for curative treatment of patients.

Editors

Vishal Jain is an Associate Professor in the Department of CSE at Sharda University, Greater Noida, India. He has earlier worked with Bharati Vidyapeeth's Institute of Computer Applications and Management (BVICAM), New Delhi, India (affiliated with Guru Gobind Singh Indraprastha University, and accredited by the All India Council for Technical Education). He first joined BVICAM as an Assistant Professor. Before that, he has worked for several years at the Guru Premsukh Memorial College of Engineering, Delhi, India. He has more than 350 research citation indices with Google scholar (*h*-index score 9 and *i*-10 index 9). He has authored more than 70 research papers in reputed conferences and journals, including Web of Science and Scopus. He has authored and edited more than 10 books with various reputed publishers, including Springer, Apple Academic Press, Scrivener, Emerald, and IGI-Global. His research areas include information retrieval, semantic web, ontology engineering, data mining, ad hoc networks, and sensor networks. He was recipient of a Young Active Member Award for the year 2012–2013 from the Computer Society of India, Best Faculty Award for the year 2017, and Best Researcher Award for the year 2019 from BVICAM, New Delhi.

Sapna Juneja is a Professor in IMS, Ghaziabad, India. Earlier, she has worked as a Professor in the Department of CSE at IITM Group of Institutions and BMIET, Sonepat. She has more than 16 years of teaching experience. She completed her doctorate and masters in Computer Science and Engineering from M.D. University, Rohtak, in 2018 and 2010, respectively. Her broad area of research is Software Reliability of Embedded System. Her areas of interest include Software Engineering, Computer Networks, Operating System, Database Management Systems, and Artificial Intelligence etc. She has guided several research theses of UG and PG students in Computer Science and Engineering. She is editing book on recent technological developments.

Abhinav Juneja is currently working as a Professor in the Department of IT at KIET Group of Institutions, Delhi-NCR, Ghaziabad, India. Earlier, he has worked as an Associate Director and a Professor in the Department of CSE at BMIET, Sonepat. He has more than 19 years of teaching experience for postgraduate and undergraduate engineering students. He completed his doctorate in Computer Science and Engineering from M.D. University, Rohtak, in 2018 and has done masters in Information Technology from GGSIPU, Delhi. He has research interests in the field of Software Reliability, IoT, Machine Learning, and Soft Computing. He has published several papers in reputed national and international journals. He has been a reviewer of several journals of repute and has been in various committees of international conferences.

Ramani Kannan is currently working as a Senior Lecturer, Center for Smart Grid Energy Research, Institute of Autonomous system, University Teknologi PETRONAS (UTP), Malaysia. Dr. Kannan completed Ph.D. (Power Electronics and Drives) from Anna University, India, in 2012; M.E. (Power Electronics and Drives) from Anna University, India, in 2006; B.E. (Electronics and Communication) from Bharathiyar University, India, in 2004. He has more than 15 years of experience in prestigious educational institutes. Dr. Kannan has published more than 130 papers in various reputed national and international journals and conferences. He is the editor, co-editor, guest editor, and reviewer of various books, including Springer Nature, Elsevier etc. He has received award for best presenter in CENCON 2019, IEEE Conference on Energy Conversion (CENCON 2019), Indonesia.

Contributors

Shivi Agarwal
Department of Mathematics
BITS Pilani
Pilani, Rajasthan, India

Amala Ann K. A.
Data Science Department
CHRIST (Deemed to be University)
Bangalore, India

A. Anitha
D.G. Vaishnav College
Chennai, India

A. Bakiya
MIT Campus, Anna University
Chennai, India

D. K. Chaturvedi
Department of Electrical Engineering
DEI, Agra, India

Sanjay Kumar Dubey
Department of Computer Science and
  Engineering
Amity University
Noida, Uttar Pradesh, India

Ayushi Ghosh
Maulana Abul Kalam Azad University
  of Technology
Kolkata, West Bengal, India

Swati Goyal
Department of Mathematics
BITS Pilani
Pilani, Rajasthan, India

Yogesh Jadhav
Research Scholar
Madhyanchal Professional University
Bhopal, Madhya Pradesh, India

Renu Jain
University Institute of Engineering and
  Technology
CSJM University
Kanpur, Uttar Pradesh, India

Abhinav Juneja
KIET Group of Institutions
Ghaziabad, Uttar Pradesh, India

Sapna Juneja
Department of Computer science
IMS Engineering College
Ghaziabad, Uttar Pradesh, India

Gurminder Kaur
Department of Computer Science and
  Engineering
BM Institute of Engineering &
  Technology
Sonepat, India

Alok Kumar
University Institute of Engineering and
  Technology
CSJM University
Kanpur, Uttar Pradesh, India

Divya Kumar
Department of ECE
IFTMU
Moradabad, Uttar Pradesh, India

Rafeeah Rufai Madaki
Department of Computer Science
Yusuf Maitama Sule University
  (Formerly, Northwest University)
Kano, Nigeria

Harsh Mathur
Department of Computer Science
Madhyanchal Professional University
Bhopal, Madhya Pradesh, India

Trilok Mathur
Department of Mathematics
BITS Pilani
Pilani, Rajasthan, India

Nirbhay Mathur
Department of Electrical & Electronics
Universiti Teknologi PETRONAS
Perak, Malaysia

Ishola D. Muraina
Department of Computer Science
Yusuf Maitama Sule University
    (Formerly, Northwest University)
Kano, Nigeria

Neelam Nehra
Department of ECE
MSIT
Delhi, India

Urmila R. Pol
Department of Computer Science
Shivaji University
Kolhapur, Maharashtra, India

Akshit Rajan Rastogi
Department of CSE
ABES Engg. College
Ghaziabad, Uttar Pradesh, India

Mukund Rastogi
Department of CSE
ABES Engg. College
Ghaziabad, Uttar Pradesh, India

Rohit Rastogi
Department of CSE
ABES Engg. College
Ghaziabad, Uttar Pradesh, India

Sheelu Sagar
Amity International Business School
Noida, Uttar Pradesh, India

Pardeep Sangwan
Department of ECE
MSIT
Delhi, India

Tejashree U. Sawant
Department of Computer Science
Shivaji University
Kolhapur, Maharashtra, India

Mamta Saxena
Ministry of Statistics
Govt. of India
Delhi, India

A. Shivankit
Department of CSE
BM Institute of Engineering &
    Technology
Sonepat, India

Piratla Srihari
Department of ECE
Geethanjali College of Engineering and
    Technology
Hyderabad, Telangana, India

Aisha Umar Suleiman
Department of ECE
Yusuf Maitama Sule University
    (Formerly, Northwest University)
Kano, Nigeria

Neeti Tandon
Research Scholar
Vikram University
Ujjain, Madhya Pradesh, India

Tina
Department of Computer Science
    Engineering
Amity University
Noida, Uttar Pradesh, India

Umme Salma M.
Department of Computer Science
CHRIST (Deemed to be University)
Bangalore, India

1 Random Variables in Machine Learning

Piratla Srihari
Geethanjali College of Engineering and Technology

CONTENTS

1.1 Introduction 2
1.2 Random Variable 3
    1.2.1 Definition and Classification... 3
        1.2.1.1 Applications in Machine Learning 4
    1.2.2 Describing a Random Variable in Terms of Probabilities 4
        1.2.2.1 Ambiguity with Reference to Continuous
                Random Variable 5
    1.2.3 Probability Density Function... 6
        1.2.3.1 Properties of pdf 6
        1.2.3.2 Applications in Machine Learning 7
1.3 Various Random Variables Used in Machine Learning................................. 7
    1.3.1 Continuous Random Variables 7
        1.3.1.1 Uniform Random Variable.. 7
        1.3.1.2 Gaussian (Normal) Random Variable................................... 8
    1.3.2 Discrete Random Variables 10
        1.3.2.1 Bernoulli Random Variable 10
        1.3.2.2 Binomial Random Variable 11
        1.3.2.3 Poisson Random Variable 12
1.4 Moments of Random Variable... 13
    1.4.1 Moments about Origin... 13
        1.4.1.1 Applications in Machine Learning 13
    1.4.2 Moments about Mean 14
        1.4.2.1 Applications in Machine Learning 14
1.5 Standardized Random Variable.. 15
    1.5.1 Applications in Machine Learning.. 15
1.6 Multiple Random Variables 16
    1.6.1 Joint Random Variables 17
        1.6.1.1 Joint Cumulative Distribution Function (Joint CDF).......... 17
        1.6.1.2 Joint Probability Density Function (Joint pdf) 17
        1.6.1.3 Statistically Independent Random Variables 18
        1.6.1.4 Density of Sum of Independent Random Variables............. 18
        1.6.1.5 Central Limit Theorem 19

DOI: 10.1201/9781003138020-1

      1.6.1.6   Joint Moments of Random Variables.................................. 19
      1.6.1.7   Conditional Probability and Conditional Density
                 Function of Random Variables 22
1.7   Transformation of Random Variables..23
    1.7.1   Applications in Machine Learning.......................................23
1.8   Conclusion 24
References...24

1.1   INTRODUCTION

Predicting the future using the knowledge about the past is the fundamental objective of machine learning.

In a digital communication system, a binary data generation scheme referred to as differential pulse code modulation (DPCM) works on the similar principle, where, based on the past behaviour of the signal, its future value will be predicted, using a predictor. A tapped delay line filter serves the purpose. More is the order of the predictor, better is the prediction, i.e. less is the prediction error.[1]

Thus, machine learning, even though not being referred to by this name earlier, was/is an integral part of technical world.

The same prediction error with reference to a DPCM system is now being addressed as confidence interval in connection with machine learning. Less prediction error implies a better prediction, and as far as machine learning is concerned, the probability of the predicted value to be within the tolerable limits of error (which is the confidence interval) should be large, which is a metric for the accuracy of prediction.

The machine learning methodology involves the process of building a statistical model for a particular task, based on the knowledge of the past data. This collected past data with reference to a task is referred to as data set.

This way of developing the models to predict about 'what is going to happen', based on the 'happened', is predictive modelling.

In detective analysis also, 'Happened', i.e. past data, is used, but there is no necessity of predicting about 'Going to happen'.

For example, 30–35 years back, Reynolds-045 Pen ruled the market for a long time, specifically in South India. Presently, the sales are not that much significant. If it is required to study the journey of the pen from past to present, detective analysis is to be performed, since there is no necessity of predicting its future sales.

Similarly, a study of 'Why the sales of a particular model of an automobile vehicle came down?' also belongs to the same category.

The data set referred above is used by the machine to learn, and hence, it is also referred to as training data set. After learning, the machine faces the test data. Using the knowledge the machine gained through learning, it should act on the test data to resolve the task assigned.

In predictive modelling, if the learning mechanism of the machine is supervised by somebody, then the mode of learning is referred to as supervised learning. That supervising 'somebody' is the training data set, also referred to as labelled training data, where each labelled data element such as D_i is mapped to a data element D_0. Such many pairs of elements are the learning resources for the machine, and are used to build the model, using which the machine predicts, i.e. this knowledge about the mapped will help the machine to map the test data pairs (input-output pair).

It can be inferred about the supervised learning that there is a target variable which is to be predicted.

Example: Based on the symptoms of a patient, it is to be predicted whether he/she is suffering from a particular disease. To enable this prediction, the past history or statistics such as patients with what symptoms (similar) were categorized under what disease. This past data (both symptoms and categorization) is the training data set that supervises the machine in its process of prediction. Here, the target variable is the disease of the patient, which is to be predicted.

In unsupervised learning mechanism, the training data is considered to be unlabelled, i.e. only D_i. Major functionality of unsupervised learning is pattern identification.

Some of the tasks under unsupervised learning are:

Clustering: Group all the people wearing white (near white) shirts.
Density Estimation: If points are randomly distributed along an axis, the regions along the axis with minimum/moderate/maximum number points need to be estimated.

It can be inferred about the unsupervised learning that there is no target variable which is to be predicted.

With reference to previous example of patient with ill-health, it can be told that all the people with a particular symptom of ill-health need to be grouped; however, disease of the patient need not to be predicted, which is the target variable, with reference to supervised learning.

1.2   RANDOM VARIABLE

1.2.1   DEFINITION AND CLASSIFICATION

For an experiment E to be performed, let S be the set of all possible outcomes of the experiment (sample space) and ξ be the outcome defined on S. The domain of $X = f(\xi)$ is S.

The range of the function depends on the mapping between the outcomes of the experiment to a numerical value, specifically real number.

This X is referred to as random variable, and thus, the random variable is a function defined on S and is real-valued.

Example: For the experiment $E = $ 'simultaneous throw of two dice', $S = \{(1,1),(1,2),\dots(1,6),(2,1),(2,2)\dots(2,6),\dots(6,6)\}$, where each number of each pair of this set indicates the face of the particular die.

Each element of S is mapped to a real value by the function

$$X = f(\xi) = \text{sum of the two faces}$$

Table 1.1 gives the pairs of all possible outcomes with the corresponding real values mapped.

TABLE 1.1

Sample Space and Mapped Values

Pair in the Sample Space	Real Value
(1,1)	2
(1,2), (2,1)	3
(1,3),(2,2),(3,1)	4
(1,4),(2,3),(3,2),(4,1)	5
(1,5),(2,4),(3,3),(4,2),(5,1)	6
(1,6),(2,5),(3,4),(4,3),(5,2),(6,1)	7
(2,6),(3,5),(4,4),(5,3),(6,2)	8
(3,6),(4,5),(5,4),(6,3)	9
(4,6),(5,5),(6,4)	10
(5,6),(6,5)	11
(6,6)	12

This X is referred to as random variable taking all the real values as mentioned.

Thus, a random variable can be considered as a rule by which a real value is assigned to each outcome of the experiment.

If $X = f(\xi)$ possesses countably infinite range (points in the range are large in number, but can be counted), X is referred to as discrete random variable (categorical with reference to machine learning).

On the other hand, if the range of the function is uncountably infinite (large in number, which can't be counted), X is referred to as continuous random variable[2,3] (similar terminology is used with reference to machine learning also).

1.2.1.1   Applications in Machine Learning

In the process of prediction, which is the major functionality for which the machine learning is used for, the variables involved are predictor variable and target variable. The very fundamental task of a machine learning methodology is to identify these variables with reference to the given task.

A predictor is an independent variable, which is used for prediction, and target variable is that being predicted and is dependent.

In machine learning terminology, variables are categorical (discrete in nature, e.g. number of people survived in an accident) and continuous (can have an infinite number of values between the maximum and the minimum, e.g. age of a person). These variables are nothing but the discrete and continuous random variables, and are task specific. Identification of these variables is the primary stage of machine learning.[4]

1.2.2   Describing a Random Variable in Terms of Probabilities

With reference to the above example, it can be stated that the value mapped (which is considered as the value taken by the random variable X) is not always 2 or 3 or 4 etc., but there is some certainty associated with the mapping, i.e. X will take a value of 7, only under certain outcomes, with a certainty of $\dfrac{7}{36}$.

Thus, a definition of a random variable is not complete simply by specifying its value, without its probabilistic description that deals with the probabilities that X takes on a specific value or values.

This description is done by the probability mass function (pmf), which assigns a probability will for each value of X.

The probability for $X = x$ (value taken by X) is $P(X = x)$ and will be assigned by the corresponding pmf.[2]

Example: Consider the case of tossing of an unbiased coin and let this process of tossing be repeated till a head occurs for the first time. X = Number of times of tossing the coin is the random variable. The corresponding sample space can be concluded as follows:

Since the coin is a fair coin, the event of getting head (H) or tail (T) in each flip is with equal likelihood, i.e. $P(H) = P(T) = \frac{1}{2}$.

In the first flip, if a head occurs, there will be no second toss. Then, $X = 1$ with probability $= \frac{1}{2}$. On the other hand, if it is a tail, then the user will go to the second flip. If the outcome is a head in the second flip, there will be no third toss. Now, $X = -2$ with a probability $= \frac{1}{4}$. This is recurrent.

Table 1.2 expresses various values taken by X, with the corresponding probabilities. The function that assigns the probability for each value taken by X, i.e. the pmf, is

$$P(X = n) = \left(\frac{1}{2}\right)^n, \quad n = 1, 2, 3 \ldots$$

The properties possessed by pmf are:

$$\text{(i) } 0 \leq P(X = n) \leq 1 \quad \text{(ii) } \sum_n P(X = n) = 1$$

1.2.2.1 Ambiguity with Reference to Continuous Random Variable

Let a discrete random variable X takes values from a set of N values, i.e. $\{0, 1, 2, 3, \ldots N-1\}$, where the values are with equal likelihood. The corresponding pmf is $P(X = k) = \frac{1}{N}$, $k = 0, 1, 2, \ldots (N-1)$ and $\lim_{N \to \infty} P(X = k) = \lim_{N \to \infty} \frac{1}{N} = 0$.

Then, the random variable X is considered to be continuous and $P(X = k)$ (where k is a value among N) is found to be zero.

Thus, it can be concluded that the probability of a continuous random variable to take a specified value is typically zero.[2]

Under such condition, pmf is not suitable to describe a random variable.

TABLE 1.2
Probability Distribution

x_i (value taken by X)	1	2	3	4	–
Probability	$\frac{1}{2}$	$\frac{1}{4}$	$\frac{1}{8}$	$\frac{1}{16}$	–

1.2.2.1.1 Cumulative Distribution Function

Instead of X taking a specific value, consider the case of X to lie in a range, i.e. $X \leq k$, and the corresponding probability $P(X \leq k)$ is referred to as cumulative distribution function (CDF).[1]

Thus, CDF of a random variable X is $F_X(x) = P(X \leq x)$, where 'x' is the value, and instead of pmf, this is used to explain a continuous random variable.

Since CDF is also probability, it is bounded as $0 \leq F_X(x) \leq 1$.

Properties of CDF are:

- $F_X(-\infty) = 0$, since the event $X \leq -\infty$ will never happen.
- $F_X(\infty) = 1$, since the event $X \leq \infty$ is always true.
- $F_X(x_1) \leq F_X(x_2)$, if $x_1 < x_2$, since $X \leq x_2$ is a super set of $X \leq x_1$. Thus, CDF is a nondecreasing function of X

$$P(x_1 < X \leq x_2) = F_X(x_2) - F_X(x_1)$$

1.2.3 PROBABILITY DENSITY FUNCTION

Even though CDF is a substitute for pmf, to explain a continuous random variable, for all the random variables, it may not be in a closed form.

Example: For a Gaussian random variable Y, the CDF $P(Y \leq k) = 1 - Q\left(\dfrac{k-m}{\sigma}\right)$, and the function $Q(\alpha)$ cannot be expressed in a closed form.

Here, m and σ are, respectively, the mean and standard deviation of X.

Under such circumstances, probability density function (pdf) is an alternative tool to describe a random variable statistically. It is defined for a random variable X at x as

$$f_X(x) = \lim_{\delta \to 0} \frac{P(x \leq X < x + \delta)}{\delta}$$

The certainty or the probability with which X is in the interval $(x, x+\delta)$ is $P(x \leq X < x+\delta)$, and the denominator δ is the width (length) of the interval.

Thus, $f_X(x)$ is the probability normalized by the width of the interval and can be interpreted as probability divided by width.

From the properties of CDF, $P(x \leq X < x+\delta) = F_X(x+\delta) - F_X(x)$.

Hence, $f_X(x) = \lim_{\delta \to 0} \dfrac{F_X(x+\delta) - F_X(x)}{\delta} = \dfrac{d}{dx} F_X(x)$, i.e. change in CDF is referred to as pdf.[2]

1.2.3.1 Properties of pdf

$$f(x) \geq 0$$

$$\int_{-\infty}^{\infty} f(x)dx = 1$$

$$F_X(x) = \int_{-\infty}^{x} f(\alpha)d\alpha$$

$$\int_a^b f(x)dx = P(a < X < b)$$

1.2.3.2 Applications in Machine Learning

- Confidence interval is an estimate of a parameter computed based on the statistical observations of it.
  - This specifies a range of reasonable values of being estimated, and the accuracy of estimation is expressed in terms of confidence level.
  - With a given confidence interval, if the number of observations of a parameter 'p' is p_1, p_2, ... p_n, with the confidence level ε, it can be interpreted that the estimated value of 'p' lies in the given confidence interval with a probability ϵ.
  - This is expressed as $P(a < X < b)$, where X is the parameter being estimated, (a, b) is the confidence interval and $P(a < X < b)$ is the confidence level, which can be computed from the probability density of the parameter being estimated.
  - The confidence level $P(\alpha < X < \beta)$ can be computed as $\int_\alpha^\beta f(x)dx$.

- In evaluating a model based on the predicted probabilities (in connection with binary classification), one of the evaluation metrics is area under curve-receiver-operating characteristic (AUC-ROC) metric.
  - This ROC is also referred to as false-positive rate-true-positive rate curve, which can be obtained from the predicted properties, with reference to binary classification.
  - Once this distribution curve is known, area under that curve, i.e. $\int f(x)dx$, will be a measure of the accuracy of the predicted.

- Ideally, the area is assumed as 1 (which is the area enclosed by any valid density function).
- More close to 1 is the area measured, i.e. much better is the performance of the predicted model.[4,5]

1.3 VARIOUS RANDOM VARIABLES USED IN MACHINE LEARNING

1.3.1 CONTINUOUS RANDOM VARIABLES

1.3.1.1 Uniform Random Variable

- The density function of a random variable X (continuous) uniform over (α, β) is $f(x) = \begin{cases} \dfrac{1}{\beta - \alpha} & \text{for} \quad \alpha < x < \beta \\ 0 & \text{elsewhere} \end{cases}$ with constant density within its specified limits is referred to as uniform random variable (continuous).
- A uniform discrete X takes all the possible values between (α, β) with equal likelihood.

The physical significance of uniform density is that the random variable X can lie in any interval within the limits (α, β) with the same probability, i.e. $\dfrac{1}{\beta - \alpha}$, for any confidence interval $(k, k + \delta)$, where $\alpha < k < \beta$, the confidence level $P(\alpha < X < x + \delta) = \dfrac{1}{\beta - \alpha}$.

- Any confidence level $P(\alpha_1 < X < \beta_1)$ for the given confidence interval (α, β) can be obtained as $\displaystyle\int_{\alpha_1}^{\beta_1} \dfrac{1}{\beta - \alpha} \cdot dx$
- The CDF of uniformly distributed random variable X (continuous) is

$$F_X(x) = \begin{cases} 0 & \text{for } x < \alpha \\ \dfrac{x - \alpha}{\beta - \alpha} & \text{for } \alpha < x \leq \beta \\ 1 & \text{for } x > \beta \end{cases}$$

- A uniform random variable is said to be symmetric about its mean $\left(= \dfrac{\alpha + \beta}{2} \right)$.[6]

1.3.1.1.1 Applications in Machine Learning

- When there is no prior realization of the distribution of the variable being predicted, the variable is considered to lie anywhere in the interval under consideration with the same probability, i.e. the variable under consideration is treated as uniform.
- In classification, decision tree learning is the most widely used algorithm. The objective of decision tree is to have pure node. The purity of a node and the information gain are related as:
  - More impure is the node, more is the information required to describe.
  - More is the information gain, more homogeneous or pure is the node.
  - Information gain $= 1 -$ entropy
  - More is the entropy, less pure is the node, and vice versa.
  - A uniform random variable will have the maximum entropy.
  - For example, in a decision tree, in a node if there are two equiprobable classes, then the corresponding entropy is maximum and is the indication for more impure node.[5,6]

1.3.1.2 Gaussian (Normal) Random Variable

- The density of normally distributed X denoted as $N(m, \sigma^2)$ is

$$f(x) = \frac{1}{\sqrt{2\pi\sigma^2}} \exp\left[-\frac{(x - m)^2}{2\sigma^2} \right], \text{ where '}m\text{' and '}\sigma^2\text{' are the mean and}$$
variance of X, respectively.

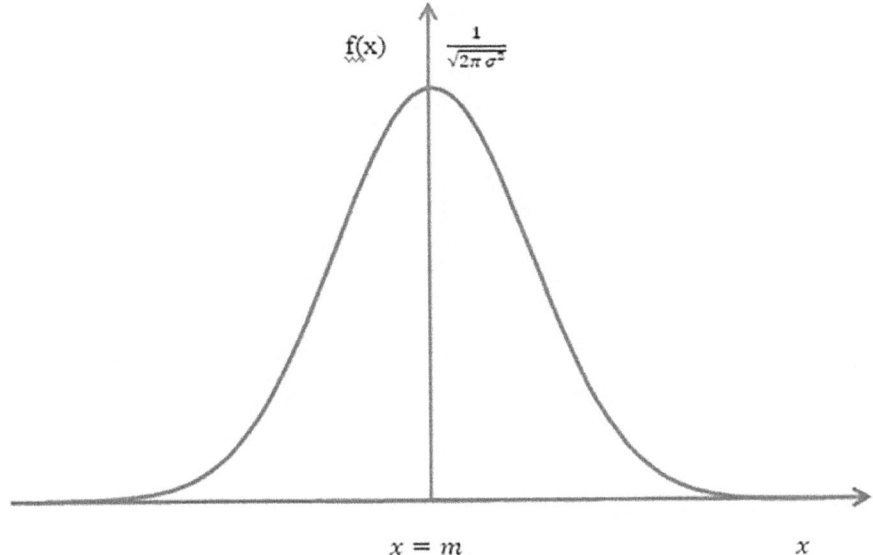

FIGURE 1.1 Gaussian density function.

- This density is a bell-shaped curve, with a point of symmetry at $x = m$, and will have its maximum value at $x = m$ (Figure 1.1).
- The CDF of normal variable X is $F_X(x) = 1 - Q\left(\dfrac{x - m}{\sigma}\right)$, where

$Q(k) = \dfrac{1}{\sqrt{2\pi}} \displaystyle\int_k^\infty \exp\left(-\dfrac{x^2}{2}\right) dx$, which is not having a closed form of solution.

- The density curve is symmetrical about its mean, i.e. equal distribution about its mean.
- If the distribution is more oriented to the right of its mean, it is said to be right-skewed.
- Similarly, left-skewed distribution can also be identified.
- Generally, it is preferred to have zero coefficient of skewness (it is a measure of symmetry of the given density function).[1]

1.3.1.2.1 Applications in Machine Learning

- Experimental observations in many case studies can be fitted with Gaussian density.
  - Marks of all the students in a class in a particular subject
  - Variations in the share price of a particular company
- With reference to machine learning, to make more predictions, more data is to be added to the model. The following are the ways of adding the data:
  - Add external data
  - Use existing data more effectively

- Feature engineering refers to the technique of generating new features using existing features. No new data is added.
- Feature pre-processing is one of the primary steps in feature engineering. This involves updating or transforming the existing features. This is referred to as feature transformation.
- Feature transformation involves replacing a variable by some mathematical functions such as 'log', 'square', 'square root', 'cube', 'cube root' etc.
- If any distribution is right-skewed or left-skewed, it is made normally distributed using nth root or log and nth power or exponential, respectively.
- As per central limit theorem, the density of sum of 'n' number independent random variables approaches Gaussian density. This is the basis for assuming the channel is normally distributed with reference to a communication system, which facilitates the study of the noise performance of a communication system.[7]

1.3.2 DISCRETE RANDOM VARIABLES

1.3.2.1 Bernoulli Random Variable

When an experiment with only two outcomes is repeated independently multiple number of times, such repetitions are referred to as Bernoulli trials.

Example: The experiment can be

- Single flip of a coin, where the possible outcomes are head and tail
- Identifying about the survival of a person in an accident: survived or not

The pmf of Bernoulli random variable is $P(X = m) = (p)^m (q)^{1-m}$, where m assumes only two values either 0 or 1. It can assume any one value at a time. p and $q(= 1 - p)$ are the probability of success and failure, respectively. Success is the required whose probability is to be computed.

For Example: when an unbiased coin is tossed, if it is required to compute the probability of getting a head (represented as $m = 1$), then $P(X = 1) = \left(\dfrac{1}{2}\right)^1 \left(\dfrac{1}{2}\right)^{1-1} = \dfrac{1}{2}$. Here, success is getting a head.

$$\text{The CDF of Bernoulli random variable is } F_X(x) = \begin{cases} 0 & \text{for } x < 0 \\ q & \text{for } 0 \le x < 1^{[6]} \\ p + q = 1 & \text{for } x \ge 1 \end{cases}$$

1.3.2.1.1 Applications in Machine Learning

- Classification is used to predict categorical variables through machine learning algorithms.
- The test data is to be assigned to a category based on some classification criteria.

- When the variable to be predicted is binary valued, i.e. the test data is to be categorized into any one of the two classes available, such classification is binary classification and the performance of such algorithms can be analysed using Bernoulli process.[5]

1.3.2.2 Binomial Random Variable

When multiple number of independent Bernoulli trials are repeated, such sequence of trials is referred to as Bernoulli process. The output of a Bernoulli process is a binomial random variable/distribution.

Example: Let a fair coin be thrown. Since the outcome can be either head or tail, it comes under Bernoulli trial. When the experiment is repeated for a number of times, such sequence of trials is Bernoulli process.

When the experiment is performed 'r' times (each experiment being a Bernoulli trial), the probability of getting the success for 'k' times is given as $p(X = k) = r_{c_k} (p)^k (q)^{r-k}$, where p and q are the probabilities of success and failure, respectively, such that $p + q = 1$. Here, X is the number of times of having the success in the experiment.

Here, X is the binomial random variable, since the above probability is the coefficient of kth term in the binomial expansion of $(p + q)^r$.

If it is required to find the probability for tail to occur four times, when a fair coin is tossed ten times, then such probability is $p(X = 4) = 10_4 \left(\dfrac{1}{2}\right)^4 \left(\dfrac{1}{2}\right)^6$.

The probability of having success for 'k' times in 'r' number of trials of the experiment in a random order is $p(X = k) = \begin{cases} r_{c_k} (p)^k (q)^{r-k} & \text{for } k = 0,1,2,...r \\ 0 & \text{otherwise} \end{cases}$, and this is the pmf of X.

The corresponding CDF is $F_X(m) = P(X \leq m) = \sum_{k=0}^{m} r_{c_k} (p)^k (q)^{r-k}$.

Binomial distribution summarizes the number of successes in a series of Bernoulli experiments, with success probability $= p$.

Bernoulli distribution is the binomial distribution with single trial.[8]

1.3.2.2.1 Multinoulli Distribution

Similar to Bernoulli distribution used for binary classification, multinoulli distribution also deals with categorical variables. It is a generalization of Bernoulli distribution (binary classification) to a multiclass classification, where k is the number of classes.

Example: In the case of throwing a die, let the sample space be $S = \{1,2,3,4,5,6\}$. Thus, the number of classes can be considered as 6.

Thus, Bernoulli distribution is multinoulli distribution with the number of classes $= 2$.[8]

1.3.2.2.2   Multinomial Distribution

Multiple number of independent multinoulli trials (e.g. throwing a die multiple number of times) follows multinomial distribution, which is a generalized binomial distribution for discrete (categorical) variables/experiments, where each experiment is with k number of outputs. Here, in n number of trials of an experiment, each experiment has k number of outcomes, which are with the probability of occurrence given as $p_1, p_2, \ldots p_k$.[8]

1.3.2.2.3   Applications in Machine Learning

All the algorithms in machine learning may not be able to deal with categorical variables. In feature pre-processing for categorical variables, to enable this handling, the categorical variables will be converted to numerical values and this process of conversion is referred to as variable encoding.

Example: Consider a supermarket having a chain of outlets. Different outlets in a city are of different size and are graded as small size, medium size, and big size. A machine learning algorithm cannot deal with such categorical values, i.e. small, medium and big. In this example, it is the case of multiclass classification with $k = 3$.

Table 1.3 represents the conversion of the above categorical values into numerical values.

This process of converting the categorical variables into numeric values is referred to as one hot encoding and is an example of multinoulli distribution.[6]

1.3.2.3   Poisson Random Variable

When a Bernoulli trial (with a binary outcome, i.e. success or failure) is repeated independently for multiple number of times (n), the probability of getting the success (p) for a defined number (m) of times (no restriction on the sequence/order of getting the success) will be dealt by binomial distribution.

In the limit $n \rightarrow \infty$ and $p \rightarrow 0$, i.e. probability of success is infinitesimal, binomial random variable can be approximated as Poisson random variable.

Example: In digital data transmission, when a large number of data bits are being transmitted, the computation of the probability of bit error will be dealt by this random variable.

TABLE 1.3
Variable Encoding

Outlet	Size	Small	Medium	Big
1	Big	0	0	1
2	Big	0	0	1
3	Medium	0	1	0
4	Small	1	0	0
5	Medium	0	1	0

Its pmf is $p(X = m) = e^{-\lambda} \dfrac{(\lambda)^m}{m!}$, where $\lambda = np$ is a constant and the corresponding

CDF is $F_X(x) = p(X \le x) = \displaystyle\sum_{k=0}^{x} e^{-\lambda} \dfrac{(\lambda)^k}{k!}$.[3]

1.4 MOMENTS OF RANDOM VARIABLE

Moments of a random variable are also referred to as its statistical averages. A random variable can have two types of moments: moments about origin and moments about mean or central moments.

1.4.1 MOMENTS ABOUT ORIGIN

Expected value [$E(x)$] or mean (m) or average value or expectation of a random value is referred to as its first moment about origin and is given as $E(X) = \displaystyle\sum_{i} x_i\, p(x_i)$,

where $p(x_i)$ is the certainty with which the random variable $X = x_i$ and $\displaystyle\int_{-\infty}^{\infty} f(x)dx$,

respectively, for discrete (categorical) and continuous cases.

For a random variable, the second moment about origin is its mean square value $E[X^2]$.

Its nth moment $E[X^n]$.[1]

1.4.1.1 Applications in Machine Learning

- The averages only are used in the study of random variable, as the certainty with which takes different values is not unique.
- Linear models are used in regression, when the dependent and independent random variables are linearly related.
- The preliminary modelling is referred to as benchmark model, where the mean of the variable will be the solution for the prediction problem.
  - Prediction of the relation between the experience of a person and the salary.
  - The initial linear modelling will be the benchmark model, taking the mean, i.e. average of all salaries, i.e. dependent variable as the solution for the model.
- This may not be the accepted one, since the people with different experiences may have the same salary.
- The model can be improved by introducing the curves of the form $Y = mX + C$, i.e. salary $= m$ (Experience) $+ C$, which is a linear model.
- The values of 'm' and 'C' for which the model gives the best prediction can be obtained from the cost function, which is given as

$$\text{Cost Function} = \text{Mean Square Error (MSE)} = \dfrac{\displaystyle\sum_{i=1}^{n} \left(\hat{s_i} - s_i\right)^2}{n}, \text{ where } \hat{s_i}$$

and s_i are the predicted and actual ith value, respectively.

- This can be referred to as the second moment of the variable $S_i^{\wedge} - s_i$.
- Better model results in lower value of MSE.[7,9]

1.4.2 Moments about Mean

Let $p(X = 2) = p(X = 6) = 0.5$, where X is the random variable. Its first moment about the origin, i.e. mean, is $m = \sum_i x_i \, p(x_i) = 2 \cdot \dfrac{1}{2} + 6 \cdot \dfrac{1}{2} = 4$.

To find the average amount by which the values taken by X differ from its mean (the answer is 2), the first moment about origin or the first central moment $E[(X - m)] = \sum_i (x_i - m) p(x_i)$ is defined. But its value $(2 - 4)\dfrac{1}{2} + (6 - 4)\dfrac{1}{2} = 0$. Thus, the very purpose of defining the first central moment is not served.

To find the same for X, the second central moment $E[(X - m)^2] = \sum_i (x_i - m)^2 \, p(x_i)$ is defined. Its value for the above X is $(2 - 4)^2 \dfrac{1}{2} + (4 - 6)^2 \dfrac{1}{2} = 4$. Its positive square root is 2, which is the value required.

Thus, $E[(X - m)^2]$ is an indication of the average amount of variation of the values taken by the random variable with reference to its mean, and hence, its variance (σ^2), and standard deviation (σ) is its positive square root.

The third central moment of X is $E[(X - m)^3]$ and is referred to as its skew. The normalized skew, i.e. coefficient of skewness, is given as $\dfrac{E[(X - m)^3]}{\sigma^3}$ and is a measure of symmetry of the density of X.

A random variable with symmetric density about its mean will have zero coefficient of skewness.

If more values taken by the random variable are to the right of its mean, the corresponding density function is said to be right-skewed and the respective coefficient of skewness will be positive (>0). Similarly, the left-skewed density function can also be specified and the respective coefficient of skewness will be negative (<0).[1]

** Identically distributed random variables will have identical moments.

1.4.2.1 Applications in Machine Learning

- Variance is used to address the concepts of underfitting and overfitting of a machine learning model. When the proposed model is trained using different data sets that differ significantly, at the time of performing on the test data, the model may not result in the required accuracy. This can be referred to the error due to variance and is due to the significant difference in various training data models.
- Standard deviation can be used to measure the spread of the data, i.e. it gives the average distance of a point in a data set from the mean of the data set.

- Learning through decision trees is a widely adopted algorithm in classification problems of machine learning.
  - In a decision tree, root node represents the entire data.
  - Nodes of the tree are divided into sub nodes using the process of splitting.
  - Leaf nodes cannot be further splitted.
  - The best split among the available splits is that it results in the most homogeneous subnodes.
  - A higher homogenous node is said to be more purity.
  - To decide the best spilt, its variance is used as a metric.
  - In this tree, the variance of each child node is computed.
  - Weighted average variance of each child node is the variance of the split.
  - The split with less variance is considered to be the best split.
- In feature pre-processing of feature engineering, feature transformation is applied to a variable through some mathematical functions (as mentioned in Section 1.3.1.2.1). This transformation is used to make the net distribution of the variable to be symmetric about its mean, thereby aiming to get zero coefficient of skewness.[5,7]

1.5 STANDARDIZED RANDOM VARIABLE

Let X be a random variable with mean m and variance σ^2. Let $X' = \dfrac{X - m}{\sigma}$ be the new random variable. Irrespective of the nature of X, X' always will be of zero mean and unity variance. This X' is referred to as standardized random variable associated with X.[3]

1.5.1 APPLICATIONS IN MACHINE LEARNING

- Distance-based machine learning algorithms require all the variables (data) in the same scale.[8,10]

  Table 1.4 gives an example where all the data used is not in the same scale.

TABLE 1.4
Data in Nonuniform Scaling

Loan amount (Rs. in lakhs)	EMI (Rs. in thousands)	Income (Rs. in hundreds)
30	50	1600
40	40	1200
50	30	800

TABLE 1.5

Statistical Parameters of the Data

Variable	Mean (m)	Standard deviation (σ) (amount by which value taken by the variable differs from its mean)
Loan amount	40	10
EMI	40	10
Income	1200	400

TABLE 1.6

Scaled Data

Loan Amount (Rs. in lakhs)	EMI (Rs. in thousands)	Income (Rs. in hundreds)
$\dfrac{30-40}{10}=-1$	$\dfrac{50-40}{10}=1$	$\dfrac{1600-1200}{400}=1$
$\dfrac{40-40}{10}=0$	$\dfrac{40-40}{-10}=0$	$\dfrac{1200-1200}{400}=0$
$\dfrac{50-40}{10}=1$	$\dfrac{30-40}{-10}=-1$	$\dfrac{800-1200}{400}=-1$

- All these data are subjected to scaling to make them to be on the same scale, such that they are comparable. Thus, feature pre-processing requires scaling of the variables.
- Standard scaling is a scaling method, where the scaling of variables is done as per the formula $X' = \dfrac{X - m}{\sigma}$.
- Table 1.5 gives the computations of mean and standard deviation for different variables of Table 1.4.
- Table 1.6 represents the above data subjected to scaling.
- It appears that all the scaled variables appear on the same reference scale.
- Thus, the concept of standardized random variable is used in feature scaling.

1.6 MULTIPLE RANDOM VARIABLES

Data exploration and variable identification is one of the constituent stages of machine learning life cycle.

In data exploration, the process of exploring only one variable at a time and its study is referred to as univariate analysis.

In bivariate analysis, two variables are studied together. In such contexts, the concept of multiple random variables finds its place.

1.6.1   JOINT RANDOM VARIABLES

Let E be an experiment and S be the corresponding sample space. Let X and Y be two variables defined as real functions of S. Then, this pair of variables is referred to as two-dimensional random variable or joint random variables.

1.6.1.1   Joint Cumulative Distribution Function (Joint CDF)

$F_{XY}(x,y) = P(X \leq x, Y \leq y)$ is the joint CDF of the random variables X and Y, where x and y are the values taken by them, respectively.[1,11]

1.6.1.1.1   Properties

1. $F_{XY}(-\infty, -\infty) = 0$

2. $F_{XY}(\infty, \infty) = 1$

3. $F_{XY}(-\infty, y) = 0$

4. $F_{XY}(x, -\infty) = 0$

5. $F_{XY}(x, \infty) = F_X(x)$, which is marginal CDF of X

6. $F_{XY}(\infty, y) = F_Y(y)$, which is marginal CDF of Y

7. $P(x_1 < X \leq x_2,\ y_1 < Y \leq y_2) = F_{XY}(x_2, y_2) - F_{XY}(x_1, y_2) - F_{XY}(x_2, y_1)$

$$+ F_{XY}(x_1, y_1)$$

8. $0 \leq F_{XY}(x,y) \leq 1$

1.6.1.2   Joint Probability Density Function (Joint pdf)

The joint pdf of two random variables X and Y is given as $f_{XY}(x,y) = \dfrac{\partial^2}{\partial x \cdot \partial y} F_{XY}(x,y).$

1.6.1.2.1   Properties

1. $\displaystyle \int_{-\infty}^{\infty} \int_{-\infty}^{\infty} f_{XY}(x,y)\, dx\, dy = 1$

2. $\displaystyle \int_{y=-\infty}^{\infty} f_{XY}(x,y)\, dy = f(x),$ which is the marginal density of X

3. $\displaystyle \int_{x=-\infty}^{\infty} f_{XY}(x,y)\, dx = f(y),$ which is the marginal density of Y

4. $\displaystyle \int_{-\infty}^{x} \int_{-\infty}^{y} f_{XY}(x,y)\, dx\, dy = F_{XY}(x,y)$

5. $P(x_1 < X < x_2, y_1 < Y < y_2) = \displaystyle \int_{x_1}^{x_2} \int_{y_1}^{y_2} f_{XY}(x,y)\, dx\ dy$

6. $F_X(x) = \int_{-\infty}^{x} \int_{y=-\infty}^{\infty} f_{XY}(x,y)dx\ dy$

7. $F_Y(y) = \int_{x=-\infty}^{\infty} \int_{-\infty}^{y} f_{XY}(x,y)dx\ dy$[12]

1.6.1.2.2 Joint Occurrence of Random Variables

The joint occurrence of two discrete random variables X and Y can be represented by a matrix referred to as joint probability matrix $P(X,Y)$, which is given as

$$
P(X,Y) = \begin{array}{c|cccc}
X/Y & \beta_1 & \beta_2 & \cdots & \beta_n \\
\alpha_1 & p(\alpha_1,\beta_1) & p(\alpha_1,\beta_2) & \cdots & p(\alpha_1,\beta_n) \\
\alpha_2 & p(\alpha_2,\beta_1) & \vdots & \cdots & p(\alpha_2,\beta_n) \\
\vdots & \vdots & \vdots & \cdots & \vdots \\
\alpha_n & p(\alpha_n,\beta_1) & p(\alpha_n,\beta_2) & \cdots & p(\alpha_n,\beta_n)
\end{array}
$$

$p(\alpha_i, \beta_j)$ is the joint probability of occurrence of the pair $(X = \alpha, Y = \beta_j)$.[1,3]

1.6.1.2.2.1 Properties of Joint Probability Matrix

1. Each element of the matrix is non-negative

2. $\sum_i \sum_j p(\alpha_i, \beta_j) = 1$

3. $\sum_i p(\alpha_i, \beta_j) = p(\beta_j)$

4. $\sum_j p(\alpha_i, \beta_j) = p(\alpha_i)$

1.6.1.3 Statistically Independent Random Variables

For such X and Y

- $F_{XY}(x,y) = F_X(x)F_Y(y)$

- $f_{XY}(x,y) = f_X(x)f_Y(y)$

- $P(x_1 < X < x_2, y_1 < Y < y_2) = P(x_1 < X < x_2)P(y_1 < Y < y_2)$

1.6.1.4 Density of Sum of Independent Random Variables

Let $Z = X + Y$, where X and Y are the random variables with individual density functions $f_X(x)$ and $f_Y(y)$, respectively, and are independent.[3] Then,

$$f_Z(z) = f_X(x) * f_Y(y) = \int_{-\infty}^{\infty} f_X(x) f_Y(z-x) dx = \int_{-\infty}^{\infty} f_X(z-y) f_Y(y) dy, \quad \text{which is}$$

convolution of their individual density functions.

This principle can be extended to multiple number of independent random variables also.

1.6.1.5   Central Limit Theorem

This theorem deals with the density of sum of independent random variables.

Central limit theorem can be stated as 'density of sum of n number of independent identically distributed random variables approaches Gaussian density, in the limit $n \to \infty$'.

This is true in the case of distinct distributions also.[3]

1.6.1.6   Joint Moments of Random Variables

1.6.1.6.1   Joint Moments about Origin

For two random variables X and Y, $m_{nk} = E\left[X^n Y^k\right]$ is the $(n+k)$th-order joint moment o about origin.

For discrete X and Y, $m_{nk} = \sum_i \sum_j x_i^n y_j^k \cdot p(x_i, y_j)$ and for continuous

X and Y, $m_{nk} = \int_{-\infty}^{\infty} \int_{-\infty}^{\infty} x^n y^k f(x,y) dx\, dy$.

For two random variables X and Y, there will be three number of second-order joint moments.

They are

$$m_{20} = E\left[X^2\right] = \text{Mean Square value of } X \quad m_{02} = E\left[Y^2\right] = \text{Mean Square value of } Y$$

$$m_{11} = E[XY] = R_{XY}, \quad \text{Correlation between } X \text{ and } Y$$

For two independent random variables, $E\left[X^n Y^k\right] = E\left[X^n\right] E\left[Y^k\right]$.

If the correlation $R_{XY} = 0$, X and Y are said to be orthogonal.

Independently, X and Y with zero individual means are orthogonal.

1.6.1.6.2   Joint Central Moments

$m_{nk} = E\left[(X - m_x)^n (Y - m_y)^k\right]$ is referred to as $(n+k)$th-order joint central moments or joint moment about mean of two random variables X and Y.

For two random variables X and Y, there will be three number of second-order joint moments. They are

$$m_{20} = E\left[(X - m_x)^2\right] = \sigma_x^2, \quad \text{variance of } X \quad m_{02} = E\left[(Y - m_y)^2\right] = \sigma_y^2, \quad \text{variance of } Y$$

$$m_{11} = E\left[(X - m_x)(Y - m_y)\right] = \sigma_{XY}, \quad \text{covariance of } X \text{ and } Y, \quad \text{Cov}(X,Y),$$

which is zero for two independent random variables.

Let X' and Y' be two standardized variables associated with X and Y.

The second-order joint central moment m_{11} of X' and Y' is referred to as correlation coefficient (ρ_{XY}) of X and Y.

$$\rho_{XY} = E[X'Y'] = E\left[\left(\frac{X - m_x}{\sigma_x}\right)\left(\left(\frac{Y - m_y}{\sigma_y}\right)\right)\right] = \frac{E[(X - m_x)(Y - m_y)]}{\sigma_x \sigma_y} = \frac{\sigma_{XY}}{\sigma_x \sigma_y}$$

It is an indication of similarity between random variables and is bounded as $-1 \leq \rho \leq 1$.

If the $X = +\beta Y$, $(\beta$ being a constant$)$, irrespective of the value of β, $\rho_{XY} = +1$, which indicates that X and Y are similar.

If $X = -\beta Y$, irrespective of the value of β, $\rho_{XY} = -1$, which indicates that X and Y are dissimilar.

A value of $\rho_{XY} = +1$ indicates the maximum similarity between X and Y, while $\rho_{XY} = -1$ is an indication of maximum dissimilarity.

Correlation coefficient is used to quantify the relation between two random variables.

A value of $\rho_{XY} = 0$ is an indication of the uncorrelated nature of X and Y.

For two jointly Gaussian random variables, independence implies uncorrelated nature, and vice versa.[3]

1.6.1.6.3 Applications in Machine Learning

- Covariance is the simultaneous variation for two random variables.
  - Linear regression involves only one predictor, where a linear relationship is assumed between the dependent variable and the predictor variable (independent).
  - In linear regression models, slope-intercept format $Y = p + qX$ is used.
  - Here, X, Y, p, and q are the independent (predictor) variable, dependent (being predicted) variable, intercept and the slope of the linear model, respectively.
  - The test data regarding X and Y is fed to the model, and the corresponding p and q are to be identified that relate X and Y.
  - The error in identifying the values of p and q is the marginal error or residual error.
  - This modifies the slope-intercept format as $Y = p + qX + \delta$.
  - Ordinary least-squares technique is used to estimate the straight line (linear model) that minimizes the difference between the actual and predicted values of Y, which is the error.
  - For each value of the predicted, this error can be calculated and the sum of the squares of these errors (SSE) is $\sum_i \delta_i^2$, which is found to be least for $q = \dfrac{\text{Cov}(X,Y)}{\text{Var}(X)}$.
  - From this value of q, the corresponding value of p can be computed.
- The linear relationship and direction of that relation between two random variables can be measured using the correlation coefficient.

- In a passenger ticketing system, the variables can be the age of a person and the ticket fare.
- To develop a model for the relation between these two variables, their correlation coefficient is a metric used, with age being the independent and fare of the ticket being the dependent variable.
- The possibilities in the linear modelling can be

$$\text{Fare} = K(\text{Age}) \rightarrow \text{fare of the ticket increases with age}$$

$$\text{Fare} = -K(\text{Age}) \rightarrow \text{fare of the ticket decreases with age}$$

  – Fare vs. age variation is constant → fare is independent of age
- Similar inference can be made from the correlation coefficient between these variables (Figure 1.2).

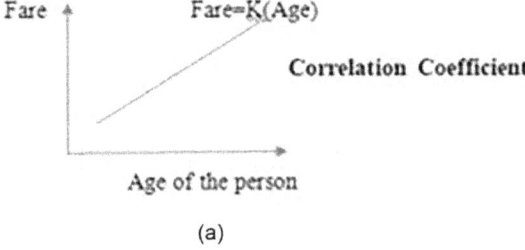

(a)

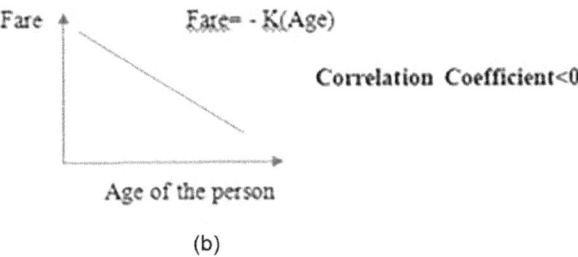

(b)

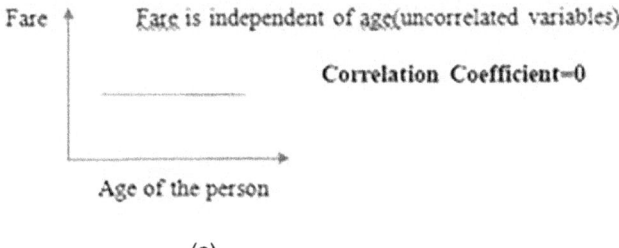

(c)

FIGURE 1.2 Correlation coefficient between the variables.

- Thus, correlation coefficient can be used as the metric for the measure of the linear relation between the variables.
- Variance and standard deviation are the measures of the spread of the data set around its mean and are one-dimensional measure. When dealing with data sets with two dimensions, the relation between these two variables in the data set (e.g. number of hours spent by a tailor in stitching and the number of shirts stitched can be the variables) i.e. the statistical analysis between the two variables will be studied using covariance. Covariance for one random variable is nothing but its variance. In the case of n variables, covariance matrix $[n \ X \ n]$ is used for the statistical analysis of all the possible pairs of the variables.[9,13,14]

1.6.1.7 Conditional Probability and Conditional Density Function of Random Variables

During the study of bivariate random variables, conditional probability is also of good importance.

This deals with the conditional occurrence of an event, while the occurrence of other event being the condition is denoted as $p\left(A/B\right) = \dfrac{p\left(A \bigcap B\right)}{P(B)} = \dfrac{p(A,\, B)}{p(B)}$, which is the probability of event A, under the occurrence of the event B.

For categorical variables $X\left(\text{taking values } x_1, x_2, \ldots x_n\right)$ and $Y\left(\text{taking values } y_1, y_2, \ldots y_n\right)$, the probability $p\left(X = x_i \middle/ Y = y_j\right) = \dfrac{p\left(X = x_i,\, Y = y_j\right)}{p\left(Y = y_j\right)}$.

On similar lines, for continuous X and Y, density function is used in the place of probabilities.

1.6.1.7.1 Properties of Conditional Density Function

1. $f_{\frac{X}{Y}}\left(x\middle/y\right)$ is always non-negative

2. $\int_{-\infty}^{\infty} f_{\frac{X}{Y}}\left(x\middle/y\right)dx = 1$

Similar properties hold good for discrete variables also, but defined under discrete summation.[15]

1.6.1.7.2 Applications in Machine Learning

- Baye's theorem is stated as $p\left(A/B\right) = \dfrac{p\left(B/A\right)p(A)}{p(B)}$.
- Let the data set consist of various symptoms leading to corona/malaria/typhoid.

- The prior knowledge about the certainty (probability) of such various hypotheses is referred to as prior, and such probability is a priori probability.
- Based on the data set, the conclusion about the patient is to be made.
- Using above such hypothesis, based on the symptoms of the patient, conclusions are to be drawn whether the patient is suffering from corona/malaria/typhoid etc., i.e. which of these hypotheses is applicable for the patient and with what probability. Such probability is a posteriori probability.
- Then, the probability $p\left(\dfrac{\text{having a specific disease}}{\text{Symptom } S}\right)$, i.e. having a specific symptom S, the probability of suffering from a specific disease is the conditional probability.
- If all the hypotheses are of equal a priori probability, then the above conditional probability can be obtained from the probability of having those symptoms, knowing the disease, i.e. $p\left(\dfrac{\text{Symptom } S}{\text{having the specific disease}}\right)$. This probability is referred to as maximum likelihood (ML) of the specific hypothesis.[7,9]
- Then, the required conditional probability is

$$p\left(\frac{\text{having a specific disease}}{\text{Symptom } S}\right)$$

$$= \frac{p\left(\dfrac{\text{Symptom } S}{\text{having the specific disease}}\right) \cdot p(\text{having the specific disease})}{p(\text{Symptom } S)}$$

1.7 TRANSFORMATION OF RANDOM VARIABLES

Any system, depending on its performance, transforms the input variable X into the output variable Y, with the transformation being the system's performance, i.e. $Y = g(X)$, where $g()$ is the transformation.

This transformation can be (1) monotonic and (2) non-monotonic.

The unknown density of the resulting variable Y is given as $f(y) = f(x)|_{x=x_1} \cdot \left|\dfrac{dx_1}{dy}\right|$, where x_1 is the real solution of x from the transformation.

If the transformation results in multiple number of real solutions for X,

$$f(y) = \sum_i f(x)|_{x=x_i} \cdot \left|\frac{dx_i}{dy}\right| \quad [3]$$

1.7.1 APPLICATIONS IN MACHINE LEARNING

- Regression models are used to identify a functional relation between the dependent (target) variable and the predictor variable.
- Based on this relation, future values of the target variable are to be predicted as a function of predictor variable.
- In linear regression, this functional relation is assumed to be linear and of the form $\beta = p + q\alpha$.

TABLE 1.7

Linear Regression Transformation of Variables[10]

Nonlinear Relations	Reduced to Linear Law
$\beta = p\alpha^n$	$\log(\beta) = \log(p) + n \cdot \log(\alpha) \rightarrow Y = nX + C$, with $Y = \log(\beta), X = \log(\alpha), C = \log(x)$
$\beta = m\alpha^n + C$	$Y = mX + C$, with $X = \alpha^n, Y = \beta$
$\beta = p\alpha^n + q.\log(\alpha)$	$Y = aX + b$, with $Y = \dfrac{\beta}{\log(\alpha)}, X = \dfrac{\alpha^n}{\log(\alpha)}, a = p, b = q$
$\beta = pe^{q\alpha}$	$Y = m\alpha + c$, with $Y = \log(\beta), m = q \cdot \log(e), c = \log(p)$

- If the existing relationship is not linear, the dependent variable is subjected to some transformations to convert the existing nonlinear relation to linear.[8]
- Table 1.7 specifies some of the transformations.

1.8   CONCLUSION

Thus, random variables are playing a vital role in the fields of machine learning and artificial intelligence. Since prediction about the future values of a variable involves some amount of uncertainty, theory of probability and random variables are essential constituent building blocks of the algorithms used to teach a machine to perform certain tasks that are dealing with the principles of learning based on the experience. These random variables are very much specific in the theory of signal estimation too.[11]

REFERENCES

1. Bhagwandas P. Lathi and Zhi Ding – *Modern Digital and Analog Communication Systems*, Oxford University Press, New York, International Fourth Edition, 2010.
2. Scott L. Miller and Donald G. Childers – *Probability and Random Processes with Applications to Signal Processing and Communications*, Academic Press, Elsevier Inc., Boston, MA, 2004.
3. Henry Stark and John W. Woods – *Probability and Random Processes with Applications to Signal Processing*, Pearson, Upper Saddle River, NJ, Third Edition, 2002.
4. Kevin P. Murphy – *Machine Learning: A Probabilistic Perspective*, The MIT Press, Cambridge, MA, 2012.
5. Jose Unpingco – *Python for Probability, Statistics, and Machine Learning*, Springer, Cham, 2016.
6. Steven M. Kay – *Intuitive Probability and Random Processes using MATLAB*, Springer, New York, 2006.
7. Peter D. Hoff – *A First Course in Bayesian Statistical Methods*, Springer, New York, 2009.
8. Shai Shalev-Shwartz and Shai Ben-David – *Understanding Machine Learning: From Theory to Algorithms*, Cambridge University Press, New York, 2014.
9. Bernard C. Levy – *Principles of Signal Detection and Parameter Estimation*, Springer, Cham, 2008.

10. Michael Paluszek and Stephanie Thomas – *MATLAB Machine Learning*, Apress, New York, 2017.
11. Rober M. Gray and Lee D. Davisson – *An Introduction to Statistical Signal Processing*, Cambridge University Press, Cambridge, 2004.
12. Friedrich Liese and Klaus-J. Miescke – *Statistical Decision Theory – Estimation, Testing and Selection-Springer Series in Statisitcs*, Springer, New York, 2008.
13. James O. Berger – *Statistical Decision Theory and Bayesian Analysis*, Springer-Verlag, New York Inc., New York, Second Edition, 2013.
14. Ruise He and Zhiguo Ding (Eds.) – *Applications of Machine Learning in Wireless communications, IET Telecommunication Series 81*, IET The Institution of Engineering and Technology, London, 2019.
15. Robert M. Fano – *Transmission of Information: A Statistical Theory of Communications*, The MIT Press, Cambridge, MA, 1961.

2 Analysis of EMG Signals using Extreme Learning Machine with Nature Inspired Feature Selection Techniques

A. Anitha
D.G. Vaishnav College

A. Bakiya
MIT Campus, Anna University

CONTENTS

2.1 Introduction 27
2.2 Data Set 30
2.3 Feature Extraction 30
2.4 Nature Inspired Feature Selection Methods 32
    2.4.1 Particle Swarm Optimization Algorithm (PSO) 32
    2.4.2 Genetic Algorithm (GA) 33
    2.4.3 Fire-Fly Optimization Algorithm (FA) 34
    2.4.4 Bat Algorithm (BA) 36
    2.4.5 Whale Optimization Algorithm (WOA) 37
        2.4.5.1 Exploitation Phase 37
        2.4.5.2 Exploration Phase 38
2.5 Extreme Learning Machine (ELM) 39
2.6 Results and Discussion 41
2.7 Conclusion 45
References 47

2.1 INTRODUCTION

Neuromuscular Impairment (NMI) is an ailment that affects the neuromuscular system by breaking in the communication path between the muscles and the nervous system (Reed et al. 2017). The symptoms of NMI include muscle numbness, fatigue muscles, abnormal pain sensation, atrophy in muscles, and fasciculation in

DOI: 10.1201/9781003138020-2

muscles. Amyotrophic Lateral Sclerosis (ALS) is a progressive degenerative NMI and more dreadful condition, which becomes severe when it is not treated properly (Blottner & Salanova 2015). Myopathy is a NMI, which majorly affects the muscles. Electromyogram (EMG) is a galvanic activity used by medical practitioners for recording the muscle activity, which assists in identifying the NMI (Sadikoglu et al. 2017). EMG helps in indicating the symptoms of muscles and nerves. However, to identify the abnormalities, timely intervention is required to prevent further deterioration of neuromuscular system.

Various computer-assisted methods are available in the literature (Tuncer et al. 2020; Khan et al. 2019; Too et al. 2017, Subasi et al. 2018) in identifying normal and NMI impairments from the signals. However, this chapter focuses on identifying and analyzing the abnormalities that exist in EMG (ALS and myopathy) signals using the Extreme Learning machine (ELM) with Nature Inspired Feature Selection (NIFS) techniques.

Feature Construction (FC) is a crucial process in extracting the original and novel features from the EMG signals. FC elevates the prediction accuracy of the classifier by deriving features from the given input data (Phinyomark et al. 2012). FC techniques also reduce dimensionality reduction combining the original features to form a novel feature. FC techniques for signals are accumulated as time-domain (TD) (Phinyomark et al. 2012), frequency-domain (FD) (Subasi et al. 2018), and time-frequency (TF) (Subasi et al. 2018; Phinyomark et al. 2012) methods. Extraction of TD features uses mathematical properties and does not involve any transformation. In contrast, extraction of frequency features requires Fourier transform. However, both time and frequency features assume stationary signals for precise results (Zawawi et al. 2018). The combined features of TD and FD are referred as time-frequency (TF) features. The TF features provide highly nonstationary information of signals (Oskoei & Hu 2007). Extraction of TF features is more appropriate, since EMG signals are nonstationary signals. However, the TF transformation techniques are mandatory to convert single-dimensional time series of EMG signals into two-dimensional TF images. Many transformation techniques are available to transform the EMG signals into TF features (Bakiya et al. 2020; Ambikapathy et al. 2018). In this chapter, Wigner-Ville transformation is employed in transforming the TF features.

In machine learning applications, real-life problems demand enormous number of features. From huge number of features, identifying essential features after eliminating redundant and irrelevant features is crucial for machine learning task. Feature Selection (FS) is an indispensable task in enhancing the capabilities of machine learning and data mining problems by reducing data dimensionality, computational time, and assists in constructing simplified learning model (Kira & Rendell 1992). FS techniques focus on selecting feature subset from the original feature space without any transformations. In contrast, feature extraction constructs the features from the dataset. The best feature subset accommodates reduced dimensions, concurrently enhancing the degree of accuracy by discarding the irrelevant features. In addition, the complexity of FS techniques relies on the extent of search space. Figure 2.1 depicts the essential steps of FS process, which includes the generation of feature subset, evaluation of feature subset, termination criterion and validation of performance (Dash & Liu 1997).

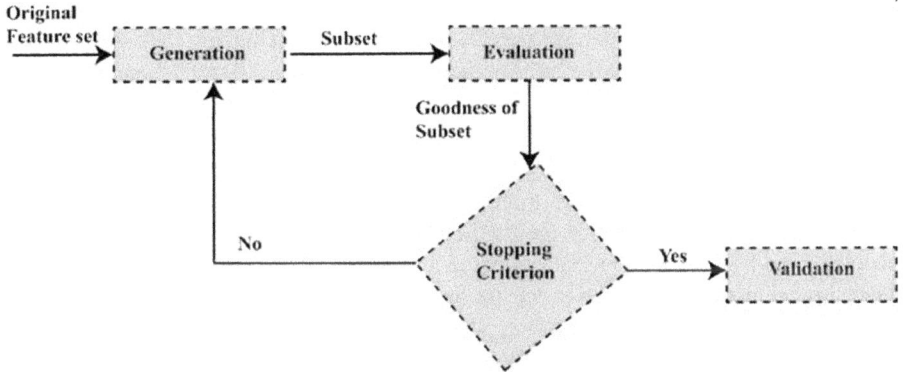

FIGURE 2.1 Feature selection process.

The subset evaluation is a pivotal step in estimating the effectiveness of the feature subset using evaluation criteria. Examining the evaluation of features, FS techniques can be categorized as filter methods (Guyon & Elisseeff, 2003), wrapper methods (Chandrashekar & Sahin 2014), and embedded methods (Stańczyk 2015). Filter model depends on the common data characteristics, and learning model is not used for evaluating feature subsets. In filter model, features are ranked, and it is evaluated independently (univariate) or in batch (multivariate). The chosen feature subsets contain features with highest scores. The features with highest score are grouped as feature subset. Filter model takes an advantage of less computational complexity (Duch 2006); alternatively during classification task, it excludes the performance of selected features. Wrapper model utilizes a learning algorithm (Chandrashekar & Sahin 2014) to evaluate the efficiency of selected feature subset and thus subjugating the disadvantage of filter model. General framework of wrapper model in FS is shown in Figure 2.2. Embedded model merges FS and classification into one process (Stańczyk 2015).

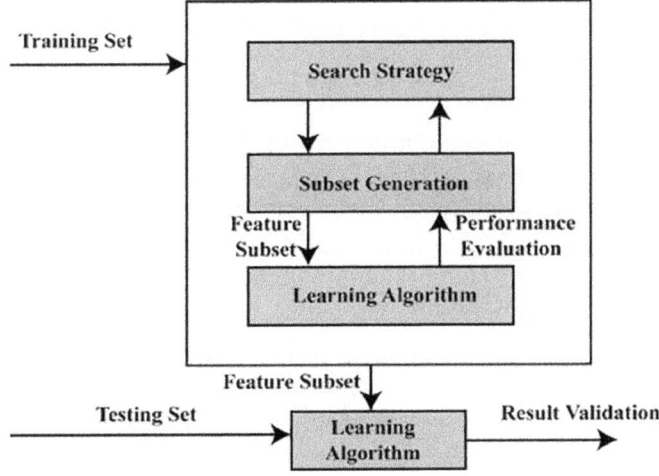

FIGURE 2.2 Framework of wrapper model for feature selection.

Recent advancements of enormous data collections mechanism had further intricated the FS technique more complex. Searching the data set for choosing optimal feature subset exhaustively is pragmatically not feasible. Many researchers have explored a variety of search mechanisms for FS algorithms (Lee et al. 2017) broadly categorized as optimal, heuristic and randomized. Predominant FS techniques suffer from huge computational cost and local optimum problems. However, NIFS algorithms have received special attention among research community due to their global search prospect. Due to their efficiency and performance, NIFS algorithms are utilized for various applications in machine learning, including classification, clustering, regression, pattern recognition and image processing (Abualigah et al. 2018; Gupta et al. 2019; Gokulnath & Shantharajah 2019; Mafarja & Mirjalili 2018; Zhang et al. 2018; Tran et al. 2017). Recent literature using NIFS for the classification of EMG signals gained a prominent attention (Wu et al. 2016; Oskoei & Hu 2006; Sharma et al. 2019). This chapter essentially concentrates on analyzing the NIFS performance efficiency of whale optimization algorithm (WOA), genetic algorithm (GA), bat algorithm (BA), firefly optimization algorithm (FA) and particle swarm optimization (PSO) in classifying normal and abnormal EMG signals. The considered NIFS algorithms fall under wrapper model since these algorithms use learning model to assess the performance of selected feature subset, and further they exploit the advantages of the wrapper models effectively.

Finally, in this chapter, ELM is constructed for evaluating the performance of reduced subset of extracted TF features from EMG signals using NIFS techniques. The traditional neural network uses backpropagation for learning; instead, single-layer feed-forward neural network (SLFN) is employed for learning in ELM, necessitating many user-defined parameters leading to over-fitting, and can stick to local optimum (Huang et al. 2015). ELM also takes advantage of rapid learning and better generalization performance.

2.2   DATA SET

The ALS, myopathy and normal signals were collected from the standard EMGLAB database (Nikolic 2001). In this chapter, 50 normal EMG signals, 50 EMG signals (myopathy) and 50 EMG signals (ALS) were chosen for assessing NIFS performance. The brachial biceps muscle region was preferred for this work, and 23437.5 Hz frequency sampling rate of EMG signals was acquired. Figure 2.3a–c shows the sample EMG signals (ALS, myopathy and normal).

2.3   FEATURE EXTRACTION

In this chapter, Wigner-Ville-transformed TF are used for the extraction of efficient features of abnormal (ALS and myopathy) and normal EMG signals. The Wigner-Ville transformation technique distributes instantaneous power and energy spectrum with respect to time and frequency. Further, it provides multivariate high resolution of frequency as well as time for transformed images (Qian & Chen 1994). The Wigner-Ville transformation is expressed in equation (2.1),

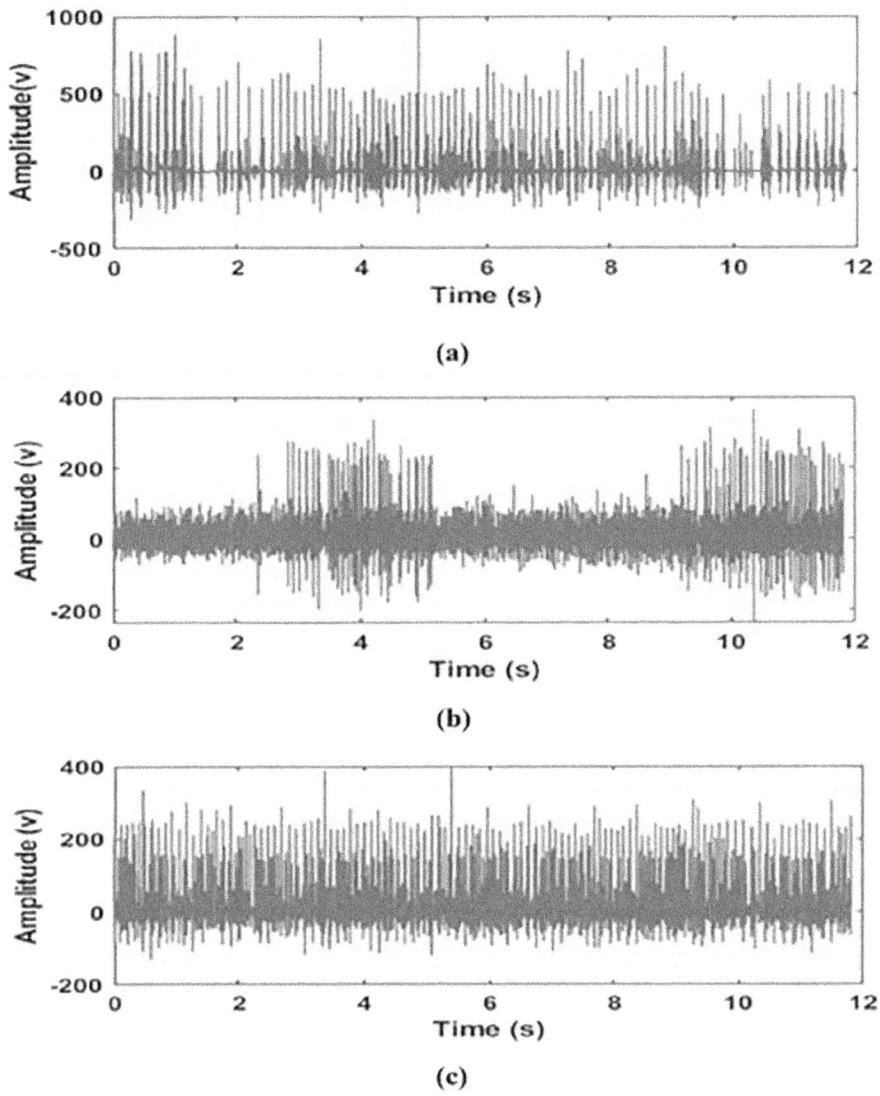

FIGURE 2.3 Illustrative EMG signals. (a) ALS. (b) Myopathy. (c) Normal.

$$Y_{wvt}(t,f) = \int_{-\omega}^{\omega} e^{-j2\pi fx} H^*\left(t - \frac{1}{2}\tau\right) H\left(t + \frac{1}{2}\tau\right) d\tau \qquad (2.1)$$

where
    $H(t)$ denotes the real signals, and $H^*(t)$ indicates the complex signals.

Wigner-Ville-transformed TF images of abnormal and normal EMG signals are shown in Figure 2.4a–c.

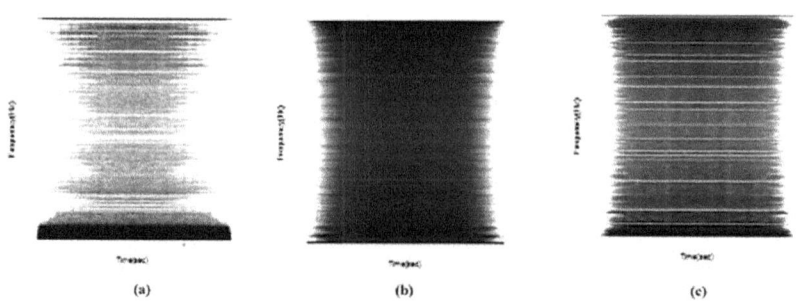

FIGURE 2.4 Wigner-Ville-transformed images. (a) ALS. (b) Myopathy. (c) Normal signals.

2.4 NATURE INSPIRED FEATURE SELECTION METHODS

In this chapter, nature inspired optimization algorithms such as PSO, GA, FA, BA and WOA were used for picking the best substantial features for distinguishing the EMG. The comparison of different FS methods was made using ELM.

2.4.1 PARTICLE SWARM OPTIMIZATION ALGORITHM (PSO)

The PSO computational technique devised by Kennedy and Eberhart (1995) is a randomized procedure influenced from intelligent social behavior of bird's flocking or fish gathering. Generally, the fundamental phenomenon of well-known PSO algorithm is that information has been optimized by social interface in the population (Xue et al. 2012; Too et al. 2019). The selection process of PSO algorithm is simple, and it can converge faster with less computational time. Hence, the PSO algorithm has been effectively used for many applications such as FS, segmentation, classification etc. (Too et al. 2019).

PSO procedure assumes i as the particle in the swarm, and its position with the dimensionality (dim) in the search space is $X_i = \left(x_{i1}, x_{i2}, \ldots, x_{i\,\mathrm{dim}}\right)$ (Xue et al. 2012). Particles are moving randomly in the search space for searching the optimal solution (Too et al. 2019; Xue et al. 2012). Each and every particle has the velocity $V_i = \left(v_{i1}, v_{i2}, \ldots, v_{i\,\mathrm{dim}}\right)$. Each particle updates its velocity as well as position in the course of motion of each particle in the exploration (problem) space. Based on p_{best} (best position of the particle) and g_{best} (best position obtained by population), PSO limits for the optimal solution. The updated position and the velocity of each particle are expressed in equations (2.2) and (2.3) (Xue et al. 2014).

$$x_{i\,\mathrm{dim}} = x_{i\,\mathrm{dim}} + v_{i\,\mathrm{dim}} \tag{2.2}$$

$$v_{i\,\mathrm{dim}} = \omega * v_{i\,\mathrm{dim}} + c_1 * \left(p_{\mathrm{best}} - x_{i\,\mathrm{dim}}\right) + c2 * \left(g_{\mathrm{best}} - x_{i\,\mathrm{dim}}\right) \tag{2.3}$$

where
   ω represents the inertia weight,
   t denotes the iteration and
   $c1$ and $c2$ represents the acceleration coefficients.

Particle Swarm Optimization Algorithm

Begin

Step1 : Initialize a population of particles with random positions and velocities on dimensions (dim) in the problem space

Step2 : For each particle evaluate the desired optimization fitness function in dim variables

Step3 : Compare particle's fitness evaluation with particle's pbest. If current value is better than pbest, then set pbest value equal to current value, and pbest location equal to the current location in dimensional space

Step4 : Compare fitness evaluation in population's overall previous best. If current value is better than gbest, then reset gbest to the current particle's array index and value

Step5 : Change the velocity and position of the particle according to the update equations:

$$V_{idim} = w_{idim}V_{idim} + c_1(b_{best}-x_{idim}) + c_2(g_{best}-p_{idim})$$
$$x_{idim} = x_{idim} + v_{idim}$$

Step6 : Loop to step 2 until a stopping criterion (maximum number of iterations of good fitness) is reached

End

FIGURE 2.5   PSO algorithm.

Figure 2.5 shows the algorithm for PSO. The parameters used for PSO FS are as follows: particles size: 100; maximum iteration: 100; acceleration coefficients: 2; and inertia weight: 1.

2.4.2   GENETIC ALGORITHM (GA)

GA is a conventional, heuristic nature inspired optimization search algorithm devised to emulate the natural genetics and selection process (Katoch et al. 2020). GA is devised by John Holland for detailing the adaption processes of natural genetic systems, which are imitated for developing new artificial systems. In GA, transmutation of one generation of chromosomes (solution candidates) to the next generation is performed through crossovers and mutations iteratively (Abuiziah & Nidal 2013). The GA starts with the selection of individuals (fittest) from the population, for producing the offspring by inheriting the characteristics and features from their parents with better fitness. The selection is iterated till the fittest individuals are discovered. Fitness of the solution candidates is chosen based on the best fitness score generated by the objective function or fitness function (Katoch et al. 2020; Abuiziah & Nidal, 2013). The algorithm converges if the population offspring produced were not significantly different from previous generations.

Figure 2.6 manifests the flowchart of steps involved in GA. The parameters used for GA FS are as follows: number of chromosomes: 100; crossover rate: 0.6; iteration: 100; and mutation rate: 0.001.

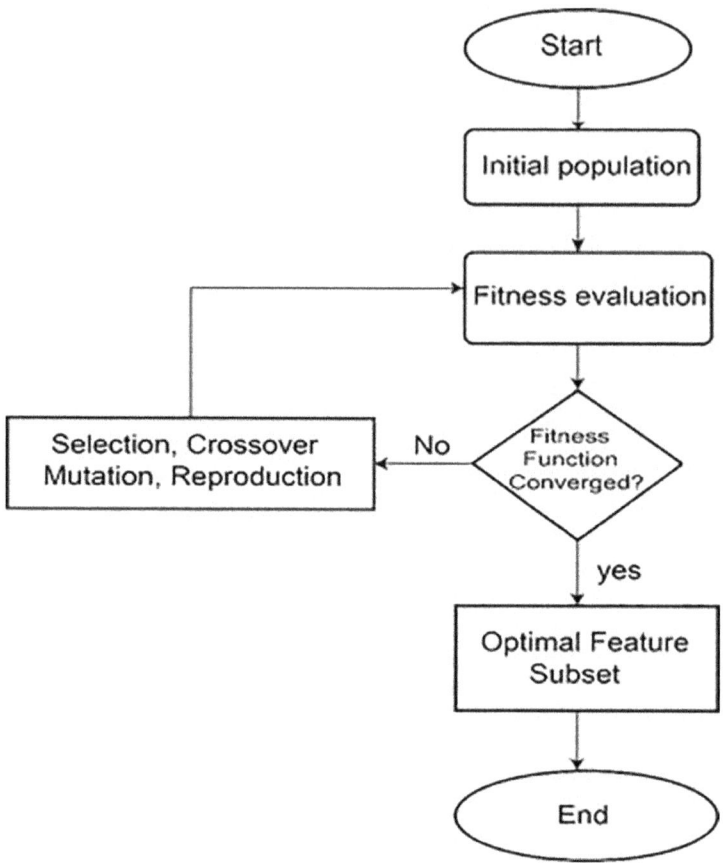

FIGURE 2.6   Steps of FS in genetic algorithm.

2.4.3   FIRE-FLY OPTIMIZATION ALGORITHM (FA)

Xin-She Yang has initially developed and introduced FA in 2008, which is extensively used in many applications such as FS, image segmentation, classification etc. (Yang & He 2013; Yang 2010a). There are two key factors that influence in this optimization algorithm, namely, intensity and position of the fireflies. The FA follows some generalized rules described below:

- The attractiveness and intensity of the firefly are inversely proportional to the distance among the two fireflies. The fireflies will move randomly in the absence of intensity between the fireflies. Suppose the fireflies have higher intensity and then it will attract to each other.
- Fireflies do not belong to any gender so it will attract to each other based on their intensity.
- The intensity of firefly is assessed by means of cost function.

When distance $r = 0$, the attractiveness of the firefly is denoted as β_0, and in general, it is represented as (β), which is given in equation (2.4) (Yang & He 2013):

$$\beta = \beta_0 e^{-\gamma r^2} \qquad (2.4)$$

Equation (2.5) defines the position of firefly (i) attracted by the position of another firefly (j) (Yang 2010a):

$$x_i(t+1) = x_i(t) + \beta_0 e^{-\gamma r_{ij}^2} (x_j(t) - x_i(t)) + \phi_t \varepsilon_i(t) \qquad (2.5)$$

Where the attraction of fireflies is represented as $\beta_0 e^{-\gamma r_{ij}^2} \left(x_j(t) - x_i(t) \right)$ and the randomized component is ϕ_t. Figure 2.7 shows the FA. The parameters used for FA-based FS are as follows: the maximum number of iteration: 100; number of fireflies: 100; light absorption coefficient: 1; and mutation coefficient: 1.

Firefly Optimization Algorithm

Begin
Step 1: Initialize the n number of fireflies in random manner
Step 2: Initialize the absorption coefficient (β) and
        light intensity(γ)
Step 3: Number of fitness evaluation = 0
Step 4: Repeat
Step 5: For i= 1: n
Step 6: For j =1:n
Step 7: If light intensity of i is less than light intensity of j
Step 8: Transfer fireflies from i to j in all dimensions
Step 9: Update the position using the equation
        $x_i(t+1) = x_i(t) + \beta_0 e^{\gamma r_{ij}^2}(x_j(t) - x_i(t) + \Phi_t \varepsilon_i(t)$
Step 11: Update the light intensity of the firefly
Step 12: Increment the number of fitness evaluation
Step 13: Stop the iteration j and i
Step 14: Based on the best cost, list the fireflies in the best
         order, Until stopping criteria.
End

FIGURE 2.7   FA algorithm.

2.4.4   Bat Algorithm (BA)

BA was introduced by Xin-She Yang in 2010, discerning the micro-bat's character-istics and behaviors. The fundamental structure of BA is formulated by imitating the three principle characteristics of the micro-bat (Yang 2010b).

The first fundamental characteristic of BA procedure is echolocation behavior. Several bats trace their prey using echolocation; however, not all bats stick to the same thing. The second characteristic is the predetermined frequency (f_{min}), used by micro-bat for hunting the prey accompanied with loudness (L_0) and wavelength (λ). The third characteristic is imitating the variations in pulse emission rates and bat's loudness dur-ing search process by automated zooming, which balances the exploration (Wang et al. 2015; Yang 2010b). To balance the parameter loudness (L_0), the values are ranged in the interval $[L_0 \ L_{min}]$ (positive maximum constant to minimum constant). The virtual bat motion is simulated with equations (2.6) and (2.7) (Yang 2010b):

$$v_k^t = v_k^{t-1} + \left(p_k^t - p_{best} \right) \cdot f_k \tag{2.6}$$

$$p_k^t = p_k^{t-1} + v_k^t \tag{2.7}$$

where f_k is the frequency-generated random values from maximum to minimum $[f_{max}, f_{min}]$ given in equation (2.8) as (Wang et al. 2015); v_k: bat's velocity; β: uni-formly distributed random vector in the range [0, 1]; p_k: position of the kth bat; p_{best}: global best solution among the entire population; and k: iteration.

$$f_k = f_{min} + \left(f_{max} - f_{min} \right) \cdot \beta \tag{2.8}$$

Certain bats will move with the following local search manner and then the updated position is given by equation (2.9) (Wang et al. 2015; Yang 2010b):

$$p_k^{t+1} = p^* + \varepsilon \times \overline{L} \tag{2.9}$$

where
   $\varepsilon \in [1,1]$ is the uniformly distributed random number and
   $\overline{L}$ is the mean loudness of all bats.

Once the position of bat is updated, the emission rate and loudness are updated by the equations (2.10) and (2.11) (Wang et al. 2015):

$$q_k^{t+1} = q_k^0 \left(1 - e^{-\gamma t} \right) \tag{2.10}$$

$$L_k^{t+1} = \sigma L_k^t \tag{2.11}$$

where
   σ is the predefined factor, which is greater than 0.

Figure 2.8 shows the algorithm of BA for FS. The parameters used for BA algorithm are as follows: maximum bat size: 100 and iteration: 100.

Bat Optimization Algorithm

Begin

Step1 : Initialize positions, velocities, bat size and iteration for
each bat in the problem space

Step2 : Randomly generate the frequency for each bat with equation (8)

Step3 : Update the velocity with equation (6)

Step4 : Update the position with equation (7)

Step5 : If rand is greater than the emission rate (q) then update
the position with equation (9) until a stopping criterion.

Step6 : Calculate the fittness

Step7 : If rand is less than the loudness (L) and ($f(p_k^1)<f(P^*)$) then
replace the position with the new one. Also, update the
emission rate and loudness with the equation (10) and (11)
until the stopping criterion.

Step6 : Select the current global best position

Step7 : Output the best position.

End

FIGURE 2.8   BA algorithm.

2.4.5 WHALE OPTIMIZATION ALGORITHM (WOA)

The WOA was initiated by Mirjalili and Lewis for solving the complex optimization problem and motivated by chasing quality of humpback whales (Mirjalili & Lewis 2016). There are two mechanisms involved in this optimization algorithm: chasing the target with best or random search and inspiring the bubble net hunting approach (Mirjalili & Lewis 2016; Mohammed et al. 2019; Rana et al. 2020). Hunting whales have an extraordinary hunting technique, namely, bubble net feeing method. WOA comprises two phases, namely, exploitation phase and exploration phase. In the exploitation phase, surrounding prey as well as spiral update position are employed, whereas randomized searching of target is achieved in the exploration phase (Mohammed et al. 2019; Mirjalili & Lewis 2016).

2.4.5.1 Exploitation Phase

In this stage, the target or prey's position is decided, and it is surrounded by humpback whales. Initially, WOA assumes the present candidate as the prey or target solution since the optimal solution is not determined currently. This whale optimization behavior is represented by equations (2.12) and (2.13) (Mirjalili & Lewis 2016):

$$\vec{P}_{(t+1)} = \vec{P}^*_{(t)} - \vec{S} \cdot \vec{D} \tag{2.12}$$

$$\vec{D} = \left| \vec{Q} \cdot \vec{P}^*_{(t)} - \vec{P}_{(t)} \right| \tag{2.13}$$

where

$\vec{P}(t+1)$: present whale's position whale,

$\vec{D}$: distance vector between the target and whale,

$\vec{P}^*(t)$: preceding whale's position in the iteration,

t: iteration and

S, Q: vector coefficients represented in equations (2.14) and (2.15) (Mirjalili & Lewis 2016),

$$\vec{S} = 2 \cdot \vec{a} \cdot \vec{r} + \vec{a} \qquad (2.14)$$

$$\vec{Q} = 2 \cdot \vec{r} \qquad (2.15)$$

The value of $\vec{a}$ is decreased when shrinking is applied in equation (2.14), and then the vector coefficient A is also reduced. The value of A is fixed in the span $[-a, a]$; similarly, parameter a is lowered in the range 2–0 in the iteration. Equations (2.16) and (2.17) are the spiral equations produced while updating the spiral position along the target location and whales (Mirjalili & Lewis 2016).

$$\vec{P}_{(t+1)} = e^{bk} \cdot \cos(2\pi k) \cdot \vec{D}^* + \vec{P}_{(t)}^* \qquad (2.16)$$

$$\vec{D}^* = \left| \vec{P}_{(t)}^* - \vec{P}_{(t)} \right| \qquad (2.17)$$

where

b signifies the constant value and

k denotes the random number scaled as $[-1, 1]$.

In this stage, the WOA has the option to switch whale's position in the course of optimization. The complete exploitation phase includes target surrounding mechanism and spiral model in equation (2.18) (Mirjalili & Lewis 2016):

$$\vec{P}_{(t+1)} = \begin{cases} \vec{P}_{(t)}^* - \vec{S} \cdot \vec{D} & \text{if } h < 0.5 \\ e^{bk} \cdot \cos(2\pi k) \cdot \vec{D}^* + \vec{P}_{(t)}^* & \text{if } h \geq 0.5 \end{cases} \qquad (2.18)$$

2.4.5.2 Exploration Phase

In the exploration stage, the whale practices the random search to determine their target, subject to the position of each other. Random search is accomplished to identify the target's location. The WOA is expressed in equations (2.19) and (2.20) (Mirjalili & Lewis 2016):

$$\vec{P}_{(t+1)} = \vec{P}_{\text{rand}} - \vec{S} \cdot \vec{D} \qquad (2.19)$$

$$\vec{D} = \left| \vec{C} \cdot \vec{P}_{\text{rand}} - \vec{P}_{(t)} \right| \qquad (2.20)$$

$\vec{P}_{\text{rand}}$ indicates the current population random position vector. The WOA is shown in Figure 2.9.

```
                        Whale Optimization Algorithm
Begin
Step1 : Initialize the population size n, the parameter a, coefficients
        A and C and the maximum number of iteration.
Step2 : Initialize the iteration counter t.
Step3 : Initialize the population n which is generated randomly and each
        search agent P₁ in the population is evaluated by calculating
        its fitness function f(P₁).
Step4 : Assign the best search agent P.
Step5 : The following steps are repeated until a stopping
        criterion statisfied.
        Step5.1 : The iteration counter is incremented t=t+1.
        Step5.2 : All the parameters a,S,Q,r and D are updated.
        Step5.3 : The exploration and exploitations are applied
                  according to the value of D and |S|.
Step6 : The best search agent P is updated.
Step7 : The overall process is repeated until termination
        criterion is statisfied.
Step8: Produce the best found search agent (P) so far.
End
```

FIGURE 2.9 WOA algorithm.

2.5 EXTREME LEARNING MACHINE (ELM)

ELM was instigated during 2006 by G. Haung and was initially built for the SLFN, which is then broadened for generalized SLFN. ELM is different from other networks since it does not require backpropagation to set its weights and it takes lesser training error. ELM learns extremely fast and produces more generalized performance (Huang et al. 2015). ELM can be implemented by applying computational nodes arbitrary of the hidden layers, which are self-determined by the training data and are fixed without tuning iteratively. The hidden layers in the ELM need not take the same number of neurons (Huang et al. 2011). The weights involved in hidden and output layers is the only variable that needs to be trained. The result produced by ELM is calculated with equation (2.21) (Huang et al. 2015):

$$f_l(y) = \sum_{i=1}^{l} \alpha_i g_i(y) = \sum_{i=1}^{l} \alpha_i g_i(w_i * y_k + b_i), \quad k = 1,\dots,n \qquad (2.21)$$

where
  l: number of hidden neuron units,
  n: number of training samples,
  g: activation function,
  w, b: weights and bias of the input and hidden layers and
  y: input samples.

The variable alpha is used to evaluate the weights involved in the output and hidden layers; then the training is denoted using equation (2.22), where D and α are defined in equations (2.23) and (2.24) (Huang et al. 2015):

$$TE = D\alpha \tag{2.22}$$

$$D = \begin{bmatrix} g(w_1 * y_1 + b_1) \dots g(w_l * y_l + b_l) \\ \cdot \qquad\qquad \cdot \\ \cdot \qquad \dots \qquad \cdot \\ g(w_1 * y_n + b_1) \dots g(w_l * y_n + b_l) \end{bmatrix}_{n \times l} \tag{2.23}$$

$$\alpha = \begin{bmatrix} \alpha_1^{TE} \\ \cdot \\ \cdot \\ \cdot \\ \alpha_l^{TE} \end{bmatrix}_{l \times m} \quad \text{and} \quad TE = \begin{bmatrix} t_1^{TE} \\ \cdot \\ \cdot \\ \cdot \\ t_n^{TE} \end{bmatrix}_{n \times m} \tag{2.24}$$

where
   m: output units,
   TE: training data label matrix and
   D: output matrix of the hidden layer (Huang et al. 2015).

The classic SLFN used for ELM is shown in Figure 2.10. The training parameters used for this network are as follows: the maximum number of iterations is 100, and Tan sigmoid activation function and hidden neurons are manifold in the range of 1–20.

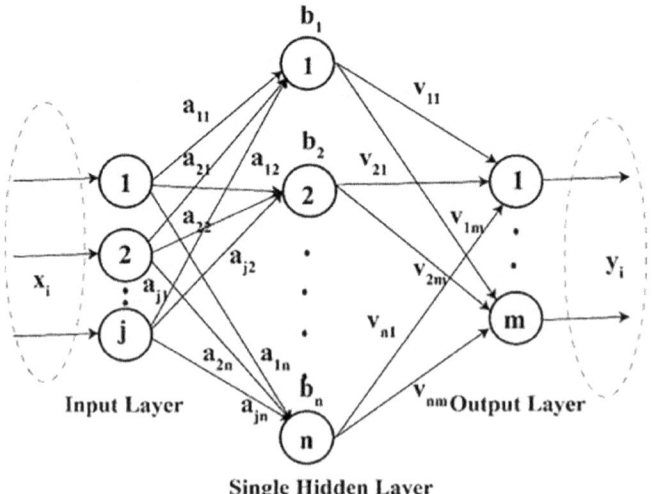

Single Hidden Layer

FIGURE 2.10   Single-hidden-layer feed-forward network (SLFN) for ELM classifier.

2.6   RESULTS AND DISCUSSION

This segment analyzes the results of ELM classification model using NIFS techniques for EMG signals. From acquired EMG wave signals (typical and abnormal), totally, 19 Wigner-Ville-transformed TF features were extracted. Figure 2.11 shows the extracted Wigner-Ville-transformed TF features from ALS, myopathy and normal EMG signals.

Figure 2.12a–e shows the convergence plot for NIFS techniques of Wigner-Ville-transformed TF features of EMG signals. From the results, it is perceived that the PSO FS has converged at 37th iteration. In the case of both GA and WOA FS algorithms have converged in the first iteration itself with the fitness value of 0.333 and 0.3 respectively. Similarly, the convergence fitness value of FA and BA is found to be 2.7560 and 362.98 at 33rd iteration. Based on the fitness value, the best feature subsets selected using PSO, GA, FA, BA, and WOA are presented in Figure 2.13.

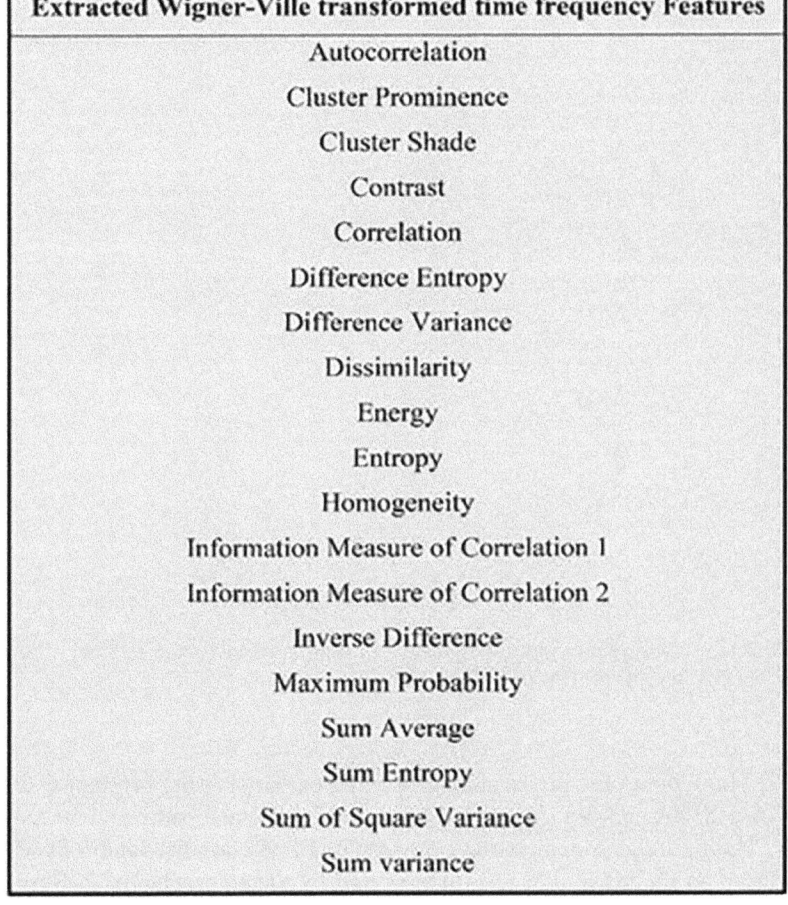

Extracted Wigner-Ville transformed time frequency Features
Autocorrelation
Cluster Prominence
Cluster Shade
Contrast
Correlation
Difference Entropy
Difference Variance
Dissimilarity
Energy
Entropy
Homogeneity
Information Measure of Correlation 1
Information Measure of Correlation 2
Inverse Difference
Maximum Probability
Sum Average
Sum Entropy
Sum of Square Variance
Sum variance

FIGURE 2.11   Extracted Wigner-Ville-transformed TF features.

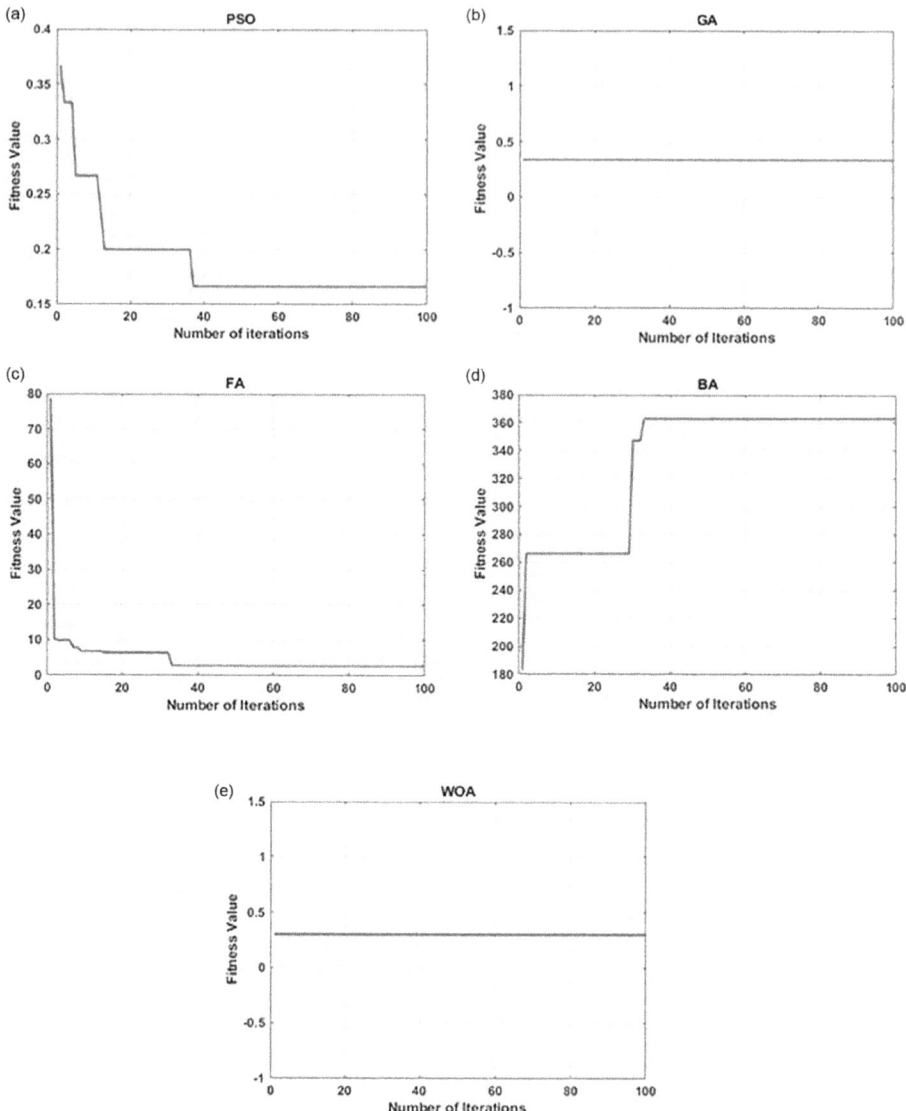

FIGURE 2.12 Convergence plot for NIFS algorithms as a function of number of iterations. (a) PSO. (b) GA. (c) FA. (d) BA. (e) WOA.

Figure 2.14a–f shows the test accuracy of ELM classifier using original feature set, PSO, GA, FA, BA, and WOA selected features of ALS, myopathy and normal EMG signals. The figure also depicts the accuracy of ELM classifier manifold with hidden neurons in the range of 2–20 incremented by the step value of 2. Results discerned that a maximal accuracy of 66.67% has been recorded, considering all the

PSO	GA	FA	BA	WOA
Autocorrelation	Cluster Prominence	Contrast	Cluster Prominence	Contrast
Correlation	Cluster Shade	Correlation	Cluster Shade	Correlation
Difference Entropy	Correlation	Difference Entropy	Correlation	Information Measure of Correlation 2
Energy	Entropy	Difference Variance		Inverse Difference
Entropy	Information Measure of Correlation 1	Energy		Sum Entropy
Information Measure of Correlation 1	Information Measure of Correlation 2	Information Measure of Correlation 1		Sum of Square Variance
Inverse Difference	Maximum Probability	Information Measure of Correlation 2		
Sum Average	Sum Average	Maximum Probability		
Sum Entropy	Sum Entropy			
Sum of Square Variance				
Sum variance				

FIGURE 2.13   Selected Wigner-Ville-transformed time-frequency features of PSO, GA, FA, BA and WOA.

Wigner-Ville-transformed TF features with hidden neuron ($N = 12$) when compared with other hidden neurons using developed ELM classifier. Significantly, it is also observed that in the performance of NIFS techniques, there is a proliferated accuracy of 80% for ELM classifier using PSO and WOA selected features with neurons ($N = 6$ and $N = 16$), respectively, in contrast to the other hidden neurons.

Concurrently, it is found that the ELM classifier has recorded the highest accuracy of 73.3% for classifying EMG signals for the features using GA and FA with hidden neurons ($N = 14$ and $N = 16$), respectively, when compared to the other hidden neurons. With respect to the classification accuracy of ELM classifier, using BA selected features with hidden neurons ($N = 12$) has shown 76.67% accuracy in contrast to the other neurons considered.

Considering the accuracy of the constructed ELM in classifying EMG signals, the AUC plot for the maximum accuracy recorded using NIFS techniques with respective neurons is shown in Figure 2.15a–f. The results reveal that the ELM classifier has generated an AUC value of 0.8 for classification with neurons ($N = 6$).

Table 2.1 depicts the overall performance measures of ELM classifier with hidden neurons using original features and NIFS selected features (PSO, GA, FA, BA and WOA). The results clearly show that the ELM classifiers with PSO as well as WOA feature subsets have performed with the highest accuracy of 80% in classifying the EMG signals of the given data set. Similarly, the AUC values were recorded as 0.8 for both PSO and WOA feature subsets. Further, ELM classifier with BA feature subset has shown a prediction accuracy of 76.67% with an AUC value of 0.9. Significantly, the prediction accuracy of 73.3% has been shown by the ELM

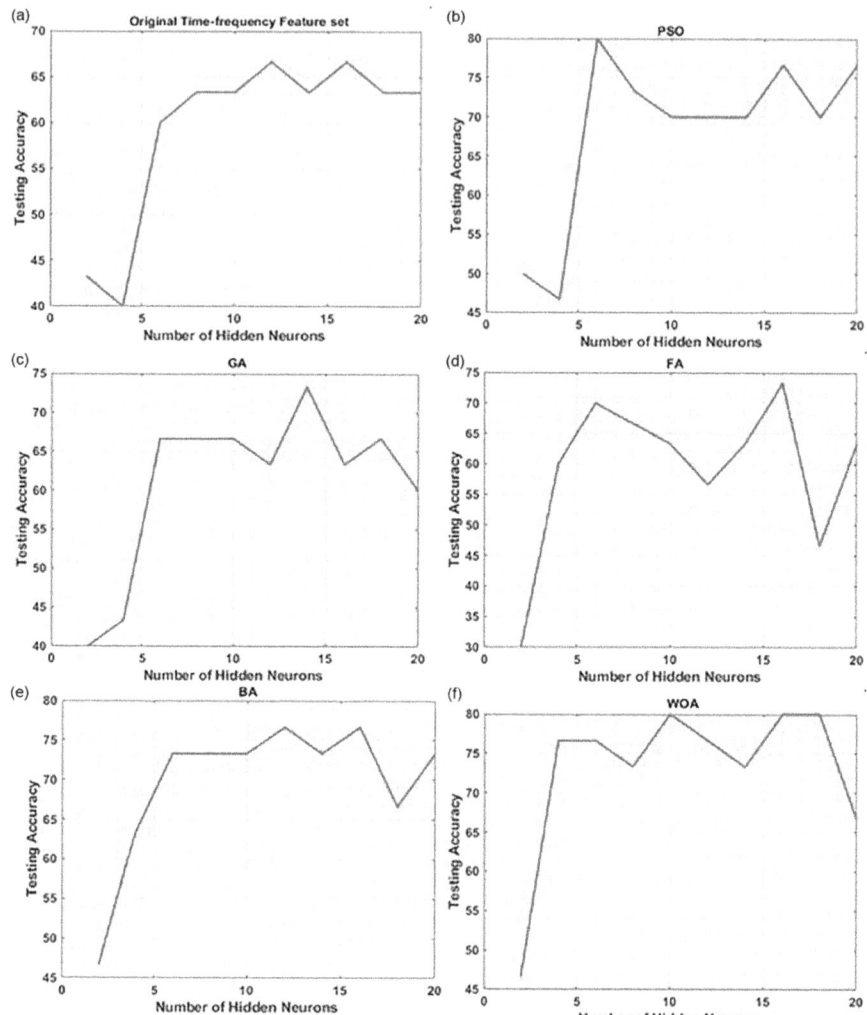

FIGURE 2.14 Testing accuracy of ELM classifier using (a) original feature set, (b) PSO, (c) GA, (d) FA, (e) BA and (f) WOA for EMG signals, for varied hidden neurons ($N=2$–20).

classifiers with GA as well as FA feature subsets. Both have exhibited the same performance: ELM with FA feature subset has recorded an exceptional AUC value of 1, whereas GA has recorded an AUC value of 0.6. Figure 2.16 shows the overall prediction accuracy of ELM classifier using original features and selected features of NIFS algorithms. From the figure, it is obvious that the performance of ELM with NIFS techniques has shown an overall supremacy than that of the ELM with original features.

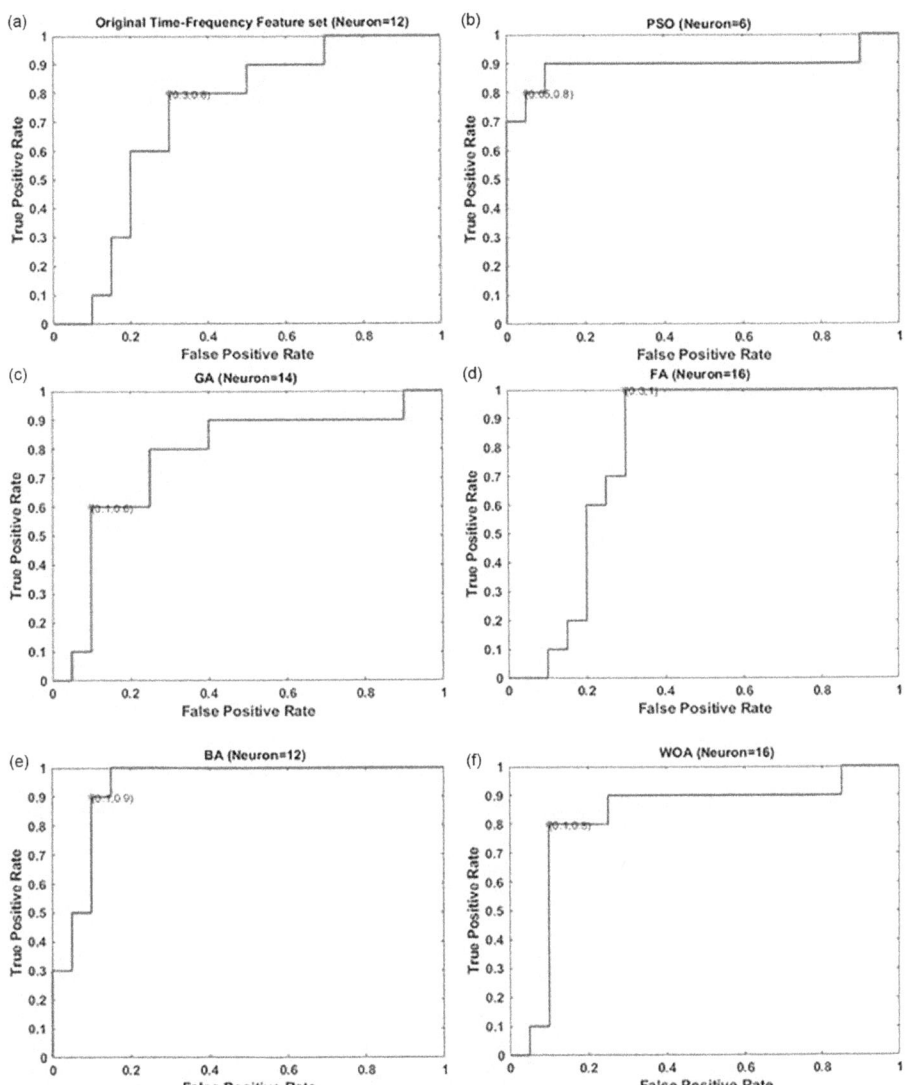

FIGURE 2.15 AUC plot for ELM classifier using selected features of EMG signals. (a) Original feature sets. (b) PSO. (c) GA. (d) FA. (e) BA. (f) WOA.

2.7 CONCLUSION

EMG signals are used to assess the aliments that occur in nerves and muscles of human body. Due to the fact that EMG signals are nonstationary, qualitative analyses cannot be carried out directly. Hence, appropriate feature extraction, selection and efficient classification is inevitable for diagnosing NMI. In constructing an efficient classification model, FS algorithm plays a pivotal role in increasing classifier productivity and

TABLE 2.1

Overall Performance Measures of ELM Classifier with Hidden Neurons Using Original and NIFS Feature Sets

Classifiers	Number of Hidden Neurons	Accuracy (%)	Sensitivity (%)	Specificity (%)	Positive Predictive Value (PPV) (%)	Negative Predictive Value (NPV) (%)	AUC
ELM + Original Features	12	66.67	66.66	83.3	70.15	83.92	0.8
ELM + PSO Features	6	80	79.9	90	80.83	90.27	0.8
ELM + GA Features	14	73.3	73.27	86.66	74	87.20	0.6
ELM + FA Features	16	73.3	73.29	92.5	77.96	88.4	1
ELM + BA Features	12	76.67	76.58	88.33	79.63	88.94	0.9
ELM + WOA Features	16	80	79.97	90	80.83	90.26	0.8

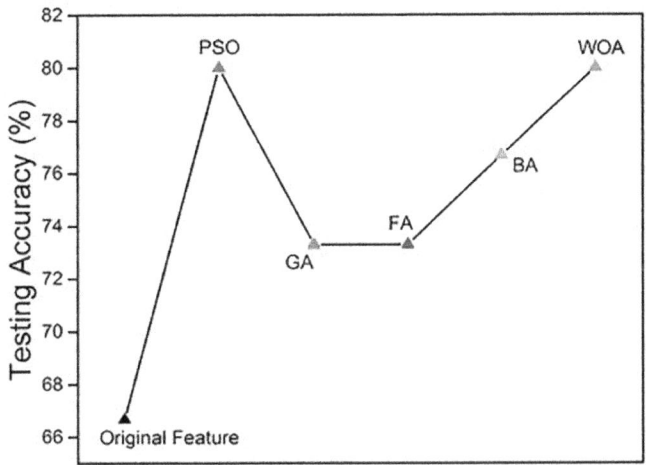

FIGURE 2.16 Overall accuracy of ELM classifier.

reducing the computational time. Recently, NIFS technique has gained predominant attention in building efficient classification systems. In this chapter, 19 Wigner-Ville-transformed TF features of EMG signals (myopathy, ALS and normal) were extracted. Further, WOA, GA, BA, FA and PSO are exploited to select the feature subset from the extracted TF features. The feature subsets generated from NIFS algorithms were fed as inputs for developing a three-class ELM classifier. The constructed ELM classifier using PSO and WOA feature subsets has achieved a highest accuracy of 80% with neurons 6 and 16, respectively, when compared to the other feature subsets considered. Interestingly, ELM with FA feature subset has attained the maximum AUC value of 1. Overall, the results demonstrate that the performance of developed ELM with NIFS feature subsets has shown supremacy than that of ELM with original features.

REFERENCES

Abualigah, L. M., Khader, A. T., & Hanandeh, E. S. (2018). A new feature selection method to improve the document clustering using particle swarm optimization algorithm. *Journal of Computational Science*, 25, 456–466.

Abuiziah, I., & Nidal, S. (2013). A review of genetic algorithm optimization: Operations and applications to water pipeline systems. *International Journal of Physical, Natural Science and Engineering*, 7(12), 341–347.

Ambikapathy, B., Kirshnamurthy, K., & Venkatesan, R. (2018). Assessment of electromyograms using genetic algorithm and artificial neural networks. *Evolutionary Intelligence*, 14(2), 1–11.

Bakiya, A., Kamalanand, K., Rajinikanth, V., Nayak, R. S., & Kadry, S. (2020). Deep neural network assisted diagnosis of time-frequency transformed electromyograms. *Multimedia Tools and Applications*, 79(15), 11051–11067.

Blottner, D., & Salanova, M. (2015). Skeletal muscle. In *The NeuroMuscular System: From Earth to Space Life Science* (pp. 9–62). Springer International Publishing, Berlin.

Chandrashekar, G., & Sahin, F. (2014). A survey on feature selection methods. *Computers & Electrical Engineering*, *40*(1), 16–28.

Dash, M., & Liu, H. (1997). Feature selection for classification. *Intelligent Data Analysis*, *1*(3), 131–156.

Duch, W. (2006). Filter methods. In *Feature Extraction* (pp. 89–117). Springer, Berlin, Heidelberg.

Gokulnath, C. B., & Shantharajah, S. P. (2019). An optimized feature selection based on genetic approach and support vector machine for heart disease. *Cluster Computing*, *22*(6), 14777–14787.

Gupta, D., Arora, J., Agrawal, U., Khanna, A., & de Albuquerque, V. H. C. (2019). Optimized Binary Bat algorithm for classification of white blood cells. *Measurement*, *143*, 180–190.

Guyon, I., & Elisseeff, A. (2003). An introduction to variable and feature selection. *Journal of Machine Learning Research*, *3*(Mar), 1157–1182.

Huang, G., Huang, G. B., Song, S., & You, K. (2015). Trends in extreme learning machines: A review. *Neural Networks*, *61*, 32–48.

Huang, G. B., Wang, D. H., & Lan, Y. (2011). Extreme learning machines: A survey. *International Journal of Machine Learning and Cybernetics*, *2*(2), 107–122.

Katoch, S., Chauhan, S. S., & Kumar, V. (2020). A review on genetic algorithm: Past, present, and future. *Multimedia Tools and Applications*, *80*, 1–36.

Kennedy, J., & Eberhart, R. (1995, November). Particle swarm optimization. In *Proceedings of ICNN '95-International Conference on Neural Networks* (Vol. 4, pp. 1942–1948). IEEE, Perth, Western Australia.

Khan, M. U., Aziz, S., Bilal, M., & Aamir, M. B. (2019, August). Classification of EMG signals for assessment of neuromuscular disorder using empirical mode decomposition and logistic regression. In *2019 International Conference on Applied and Engineering Mathematics (ICAEM)* (pp. 237–243). IEEE, Taxila, Pakistan.

Kira, K., & Rendell, L. A. (1992). A practical approach to feature selection. In *Machine Learning Proceedings 1992* (pp. 249–256). Morgan Kaufmann, San Mateo, CA.

Lee, P. Y., Loh, W. P., & Chin, J. F. (2017). Feature selection in multimedia: The state-of-the-art review. *Image and Vision Computing*, *67*, 29–42.

Mafarja, M., & Mirjalili, S. (2018). Whale optimization approaches for wrapper feature selection. *Applied Soft Computing*, *62*, 441–453.

Mirjalili, S., & Lewis, A. (2016). The whale optimization algorithm. *Advances in Engineering Software*, *95*, 51–67.

Mohammed, H. M., Umar, S. U., & Rashid, T. A. (2019). A systematic and meta-analysis survey of whale optimization algorithm. *Computational intelligence and neuroscience*, *2019*, 25.

Nikolic, M. (2001). *Detailed analysis of clinical electromyography signals emg decomposition, findings and firing pattern analysis in controls and patients with myopathy and amytrophic lateral sclerosis*. PhD Thesis, Faculty of Health Science, University of Copenhagen. [The data are available as dataset N2001 at http://www.emglab.net].

Oskoei, M. A., & Hu, H. (2006, December). GA-based feature subset selection for myoelectric classification. In *2006 IEEE International Conference on Robotics and Biomimetics* (pp. 1465–1470). IEEE, Kunming, China.

Oskoei, M. A., & Hu, H. (2007). Myoelectric control systems – A survey. *Biomedical Signal Processing and Control*, *2*(4), 275–294.

Phinyomark, A., Phukpattaranont, P., & Limsakul, C. (2012). Feature reduction and selection for EMG signal classification. *Expert Systems with Applications*, *39*(8), 7420–7431.

Qian, S., & Chen, D. (1994). Decomposition of the Wigner-Ville distribution and time-frequency distribution series. *IEEE Transactions on Signal Processing*, *42*(10), 2836–2842.

Rana, N., Latiff, M. S. A., Abdulhamid, S. I. M., & Chiroma, H. (2020). Whale optimization algorithm: A systematic review of contemporary applications, modifications and developments. *Neural Computing and Applications*, *32*, 16245–16277.

Reed, S. M., Bayly, W. M., & Sellon, D. C. (2017). *Equine Internal Medicine-E-Book*. Elsevier Health Sciences, Saint Lois, MO.

Sadikoglu, F., Kavalcioglu, C., & Dagman, B. (2017). Electromyogram (EMG) signal detection, classification of EMG signals and diagnosis of neuropathy muscle disease. *Procedia Computer Science*, *120*, 422–429.

Sharma, S., Singh, G., & Singh, D. (2019). Role and performance of different traditional classification and nature-inspired computing techniques in major research areas. *EAI Endorsed Transactions on Scalable Information Systems*, *6*(21), e2.

Stańczyk, U. (2015). Feature evaluation by filter, wrapper, and embedded approaches. In *Feature Selection for Data and Pattern Recognition* (pp. 29–44). Springer, Berlin, Heidelberg.

Subasi, A., Yaman, E., Somaily, Y., Alynabawi, H. A., Alobaidi, F., & Altheibani, S. (2018). Automated EMG signal classification for diagnosis of neuromuscular disorders using DWT and bagging. *Procedia Computer Science*, *140*, 230–237.

Too, J., Abdullah, A. R., & Mohd Saad, N. (2019, June). A new co-evolution binary particle swarm optimization with multiple inertia weight strategy for feature selection. In *Informatics* (Vol. 6, No. 2, p. 21). Multidisciplinary Digital Publishing Institute, Basel.

Too, J., Abdullah, A. R., Zawawi, T. T., Saad, N. M., & Musa, H. (2017). Classification of EMG signal based on time domain and frequency domain features. *International Journal of Human and Technology Interaction (IJHaTI)*, *1*(1), 25–30.

Tran, B., Xue, B., & Zhang, M. (2017). A new representation in PSO for discretization-based feature selection. *IEEE Transactions on Cybernetics*, *48*(6), 1733–1746.

Tuncer, T., Dogan, S., & Subasi, A. (2020). Surface EMG signal classification using ternary pattern and discrete wavelet transform based feature extraction for hand movement recognition. *Biomedical Signal Processing and Control*, *58*, 101872.

Wang, G. G., Chang, B., & Zhang, Z. (2015, May). A multi-swarm bat algorithm for global optimization. In *2015 IEEE Congress on Evolutionary Computation (CEC)* (pp. 480–485). IEEE, Sendai, Japan.

Wu, Q., Mao, J. F., Wei, C. F., Fu, S., Law, R., Ding, L., …Yang, C. H. (2016). Hybrid BF–PSO and fuzzy support vector machine for diagnosis of fatigue status using EMG signal features. *Neurocomputing*, *173*, 483–500.

Xue, B., Zhang, M., & Browne, W. N. (2012). Particle swarm optimization for feature selection in classification: A multi-objective approach. *IEEE Transactions on Cybernetics*, *43*(6), 1656–1671.

Xue, B., Zhang, M., & Browne, W. N. (2014). Particle swarm optimisation for feature selection in classification: Novel initialisation and updating mechanisms. *Applied Soft Computing*, *18*, 261–276.

Yang, X. S. (2010a). *Engineering Optimization: An introduction with Metaheuristic Applications*. John Wiley & Sons, Hoboken, NJ.

Yang, X. S. (2010b). A new metaheuristic bat-inspired algorithm. In *Nature Inspired Cooperative Strategies for Optimization (NICSO 2010)* (pp. 65–74). Springer, Berlin, Heidelberg.

Yang, X. S., & He, X. (2013). Firefly algorithm: Recent advances and applications. *International Journal of Swarm Intelligence*, *1*(1), 36–50.

Zawawi, T. N. S. T., Abdullah, A. R., Jopri, M. H., Sutikno, T., Saad, N. M., & Sudirman, R. (2018). A review of electromyography signal analysis techniques for musculoskeletal disorders. *Indonesian Journal of Electrical Engineering and Computer Science*, *11*(3), 1136–1146.

Zhang, L., Mistry, K., Lim, C. P., & Neoh, S. C. (2018). Feature selection using firefly optimization for classification and regression models. *Decision Support Systems*, *106*, 64–85.

3 Detection of Breast Cancer by Using Various Machine Learning and Deep Learning Algorithms

Yogesh Jadhav and Harsh Mathur
Madhyanchal Professional University

CONTENTS

3.1 Introduction 51
    3.1.1 Risk Factors for Breast Cancer 52
    3.1.2 Screening Guidelines 53
    3.1.3 Consequences of Misidentifying the Tumor 53
    3.1.4 Materials and Methods 54
3.2 Model Selection 55
    3.2.1 Logistic Regression 56
    3.2.2 Nearest Neighbor 58
    3.2.3 Support Vector Machine 60
    3.2.4 Naive Bayes Algorithm 63
    3.2.5 Decision Tree Algorithm 64
    3.2.6 Random Forest Classification 66
3.3 Detection of Breast Cancer by Using Deep Learning 66
3.4 Conclusion 69
References 69

3.1 INTRODUCTION

A lump or development of a cell growth in certain size is considered as tumor [3]. When cancers cells are tiny or unformed, they are benign. It looks like there has been something wrong and has a lump. When the cells are erratic in shape and cannot be regulated, they are cancer cells and the tumor is malignant. To confirm the assumption that a tumor is benign or malignant, a physician can take an interventional diagnostic procedure, such as an interventional biopsy. It is a totally benign disease, meaning that it is not a cancerous tumor. It is unclear whether the virus

DOI: 10.1201/9781003138020-3

can migrate through the body or whether it can be contained within one area. A benign tumor is less dangerous because it is rare and moves and forces adjacent tissue, nerves or blood vessels to hurt. Some benign tumors contain fibroids that are found within the uterus. Benign tumors will need to be removed by surgery. They will become very tall, weighing up to pounds sometimes. They could be dangerous, for example, where in the brain and in the restricted portion of the skull, the normal structures were greatly stretched. Perhaps you can block essential channels in the human body. Intestinal polyps are precancerous and, thus, should be removed in order to avoid malignancy. Benign tumors [2] are usually removed from the exact location in which they develop. However, they will also recur in the exact location in which they arose. Malignant means that the tumor is made up of cancer cells and can end up inside your tissue nearby. Any cancer cells that are present in the body can travel along blood or lymphatic vessels and infect other organs in the body causing further damage. The skin, breast, intestines, lungs, sexual organs, blood and skin are present in all parts of the body. Additionally, tumor cells may reach the breast and then migrate to the liposomal tissue, thereby causing breast cancer. It is delivered directly to the lymph system through the lymph vessels before it is caught and handled early enough. Cancer cells may travel from the breast to other areas of the body, such as the bone or liver, after breast cancer has spread across the body and has entered the lymph nodes. Within the breast cancer tissues, breast cancer cells can develop tumors. If biopsies of the cancerous sections are made, the initial tumor features can be identified. A tumor is not always a simple indication that the tumor has or has not spread. Diseases have many signs, and physicians who treat them rely on many different approaches to determine which one applies. You could be afflicted with some assorted possibilities. Breast cancer is the most common cancer among women, resulting in the majority of new cases and deaths related to cancer around the world. This, in addition to the global spread of the disease, means it has become a huge public health problem today.

3.1.1 RISK FACTORS FOR BREAST CANCER

It is well known that a vast number of risk factors precede the development of breast cancer. The disease may have multiple causes, but most cases can only be due to one cause. The risk of breast cancer increases with the age of a woman although 20% of breast cancers are diagnosed in women over the age of 50; it is also evident that the risk of a breast cancer diagnosis is much higher in older women. In this situation, as with other conditions, a woman with cancer in one breast (not necessarily her own) has a greater chance of the cancer spreading to her other breast. Women who have had breast cancer or who have a family member, such as a mother, sister, or daughter, who had breast cancer are at a higher risk of developing the disease themselves. In addition, if you and your other mothers have breast cancer, you have a higher chance of having it. Women are at an increased risk of contracting breast cancer during their lives with such genetic mutations, including BRCA1 and BRCA2 changes. Other genes involved in tumor formation also raise the chances of breast cancer. As a woman ages, the risk of her having breast cancer first rises as soon as she has her first child. They are at a greater risk of disease. Women who have begun their cycles at an early age, women who started their periods later than average, women who don't have kids, and women who have not yet gone through menopause are at greater risk of breast cancer.

3.1.2 SCREENING GUIDELINES

Breast examination is very common method for the detection of breast cancer. Breast and lymph nodes are supervised by doctors. Doctor will try to identify anomalies in the breast through which they can make conclusion regarding the breast cancer. As the most effective screening test for detecting this disease, the mammogram is very useful. An X-ray can probably be regarded as a mammography. It can detect cancer in the breast, which is good for only two years before you or your treating doctor notices something is wrong. Ultrasound is another tool used to detect the presence of breast cancer. Sound waves used to create the representation of breast. Ultrasound waves can be used to check for breast lump. Biopsy, i.e., a sample from breast cell, is taken and examined. A biopsy is a technique for extracting from the body a piece of tissue or a cell sample for analyses in a laboratory. If you have any signs and symptoms or have a field of interest reported by a doctor, you can perform a biopsy to confirm if you have cancer or some other disease. Magnetic resonance imaging (MRI) is also used for the detection of breast cancer, in which radio waves are used to create images of breast and through which the development of cancer in the breast can identified. Women who are at an elevated risk of breast cancer and who are 40–45 years of age or older should have a mammogram once a year. Women that are at an elevated risk of developing breast cancer should have an annual mammogram and an MRI.

3.1.3 CONSEQUENCES OF MISIDENTIFYING THE TUMOR

Where the doctor misclassifies the tumor as benign instead of malignant, if the tumor is potentially malignant and the patient is not continuing to receive treatment, tumors will over time be metastasized or spread to other parts of the body. This will enable the patient to experience a life-threatening condition. This is known to be the flaw in the scientific terms of the "type 1" doctor in which the patient has malignant tumor but is not marked as having it. Considering this probability, if the doctor suggests conservatively treating a patient with a tumor, irrespective of if they have a benign or malignant form of tumor, then certain people may recover from excessive mental distress and other care expenses. In the event of benign cancers, the patient will usually live their lives without life-threatening complications, even if she may not wish for medication. This form of mistake by a surgeon is typically known as a "type 2" mistake: the patient may not have a malignant tumor, but has it. We must also realize that the doctor is more likely to commit a type 2 mistake than a type 1 error in such a situation. In this case, having type 1 errors leads to life-threatening risks for the patient, while type 2 errors result in needless patient costs and emotional strain. It is also critical that all malignant tumor patients are classified as such. If patients with malignant tumors are deemed to be genuinely optimistic, sensitivity is the fraction of the persons with malignant tumors that are appropriately detected by the examination. If we plan to rescue patients with benign tumors from needless care bills, we need to assess the diagnostic test accuracy. Specificity is the percentage of people who have not received malignant tumor. An optimum tumor diagnostic method will have accuracy and sensitivity of 1.

3.1.4   Materials and Methods

Below are subsections that will address the data that was used for the detection of breast cancer as well as the machine learning (ML) algorithms that were used for cancer detection. Dr. William H. Wolberg was the MD at Madison of Wisconsin Hospital in Madison, Wisconsin, USA. He collected the data used for this simulation. Dr. Wood used a digital scanner to take cell samples from patients with their breast mass, and then used more sophisticated technology on top of that to try to see if there were any issues with their cells. In order to determine 10 of each of the cells in the dataset [16,17], the program uses an algorithm for the calculation of each image element's average, unbelievably high value and standard error, returning 30 real-assessed vectors. In this sample, there are 569 patients' data. Following are the attributes of datasets – ID number, diagnosis ((M = malignant, B = benign) 3–32). Ten real-valued features are also calculated for each nucleus of cell. Following are those parameters: radius, texture, perimeter, area, smoothness, compactness, concavity, concave points, symmetry and fractal dimensions. The study aims to see what models are the most effective in predicting the malignancy or benign existence of cancers as well as to see whether trends can help select a model and set hyperparameters. The main purpose is to conduct a universal breast scan for cancer cells. I have done a disruptive classification with classification methods recommended for ML by matching a feature that could predict the discrete class of new data. For the purposes of counting the number of patients with this type of disease, we will imagine the cells as malignant and noncancerous in the following (Figure 3.1).

The initial phase is data exploration phase in which null data is removed from the dataset. It is also called as data preprocessing. In the next phase, we are going

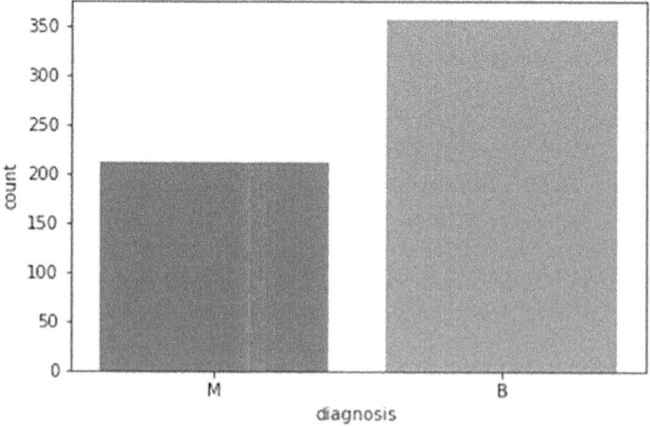

FIGURE 3.1   Bar graph showing M-malignant and B-benign.

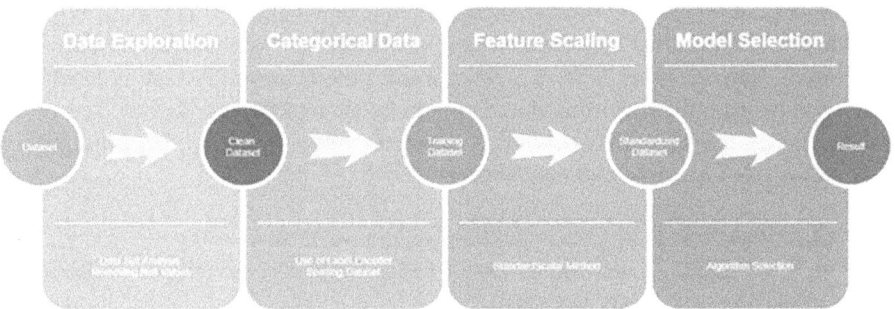

FIGURE 3.2 Stages involved in machine learning.

to consider categorical data. Categorical data means data that contains labels rather than values. As example, users can be categorized based on age group, country, gender etc. We cannot directly give such data to ML algorithm for training purpose. This data needs to be converted into values. We used label encoder to assign values instead of labels. Label encoder will assign 1 to M and 0 to B. Dataset is split into two sections in the next step. One part is training, and the other part is testing. Training portion is used to build a system. However, testing part of the dataset is used to test the system. In a next phase, feature scaling is done where dataset may contain features that are highly varying in magnitude or in range. Euclidian distance to data points is often used for computing ML algorithms. All functionality must be taken to the same standard. The effect is a scaling method. In scaling, all data points are scaled between values such as 0–1 or 0–100. Most commonly standard scalar is used for scaling operation. The next phase is the most important phase in which an appropriate algorithm needs to be selected. We will train our system with the most common algorithms, and these algorithms are selected based on their accuracy in predicting breast cancer. Data researchers typically use different kinds of algorithms for ML for a large volume of data. These different algorithms can therefore be classified into two classes at a high level: supervised and unsupervised learning (Figure 3.2).

3.2 MODEL SELECTION

A ML algorithm is required in this step to have some datasets. Choosing the correct algorithm is the most critical factor in predicting the future. When working with massive datasets, data scientists prefer to use a different kind of ML algorithm. A more general way to characterize the different algorithms in these two groups is that they can be thought of as supervised learning or unsupervised learning.

Supervised Learning:
For the most part, ML [4] algorithms use supervised learning. The supervised learning algorithms used in this study rely on mapping a function that is frequently applied when an input variable (X) is related to an output variable (Y).

$$Y = f(X)$$

By using this mapping function whenever there is new value for *X,* we try to predict the value of *Y.* It is called supervised learning because the system tries to learn from the training dataset. Over here, we can consider the training dataset as teacher. A prediction based on the training data is made by an algorithm and corrected by the teacher. When the algorithm reaches a sufficient output level, the learning process ends. Supervised learning can be categorized into two parts:

1. **Classification**: In the classification, output variable is category or a class such as "disease" and "no disease." Color is "black" or "White" Filtering emails are "spam" or "not spam."
2. **Regression**: In the regression, the output is a real value, such as height, weight.

Unsupervised Learning:
Unsupervised learning is when only input data is given to it, there is no output associated with it at all. Unsupervised learning aims to discover structure inside data. In unsupervised learning, there is no presence of qualified professionals. Algorithms recognize trends. Unsupervised learning is classified into two groups:

1. **Clustering**: In clustering, grouping is done based on some factors. Example is grouping of customers based on their purchasing pattern.
2. **Association**: In the association rule algorithm, the goal is to find new rules by evaluating the current rules. Person who can buy *X* can also buy *Y* is an example worth considering.

The dataset in which we have the result or dependent variable is what we have called here the "our dataset." *Y,* with only two sets of values, may be one of two options: M or B (benign). To achieve this, we will use the learning classification supervised algorithm.

Over here, we discussed seven classification algorithms [7] used in ML [1].

3.2.1 LOGISTIC REGRESSION

Logistic regression [16] is used in the logistic function; therefore, it is called as logistic regression. Sigmoid function is also known as the logistic function. Logistic function provides a curve in the form of *S* that can be used to map the real values between 0 and 1. Following is the expression for logistic function:

$$1/(1 + e^{\wedge}\text{-value})$$

Here, the term e is used to represent the base of natural algorithms (Figure 3.3).

Logistic regression [6] is similar to linear regression in that both equations use a logistic function. An input value is multiplied by a weight and added to another input value to approximate the output value. The difference between linear regression and logistic regression is that when we get a numerical value from logistic regression, we get 0 or 1 in the same manner as with linear regression.

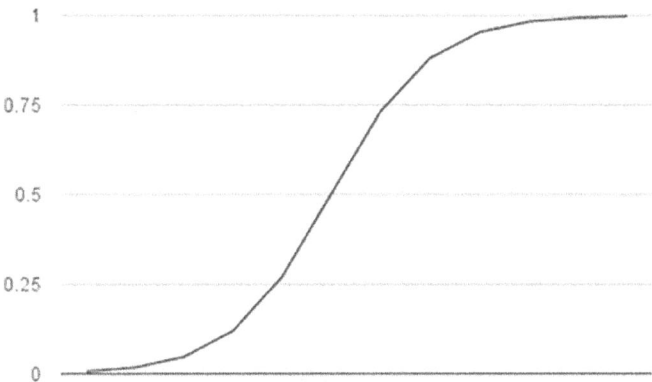

FIGURE 3.3   Representation of logistic regression.

Logistic regression is represented as given example:

$$y = e^{\wedge}(b_0 + b_1 * x)\big/\big(1 + e^{\wedge}(b_0 + b_1 * x)\big)$$

where
   y is the expected output,
   b_0 is the intercept and
   b_1 is the single input value coefficient (x).

There is a corresponding b coefficient in each column in your input data that you would learn from training data. For logistic regression, data preparation is needed.

Binary Output: In the case of binary classification, the logistic regression model is used. It divides class into 0 (zero) or 1 (one).

Noise Removal: No error occurs in the output variable Y in the logistic regression. The outliers must be eliminated.

Gaussian Distribution: A more robust model would allow the modeling of additional input variable data transformations that go deeper in explaining this linear relationship. One clear example of this technique is to reveal the relationship between log, base, and box-cox transformations as well as a few others such as the log, base, and base-expansion transformations.

If multiple strongly correlated inputs are in place, the model will fit over, such as linear regression. Take the coefficients of each input wisely calculated and tightly clustered inputs eliminated.

It is conceivable for the expected form of probability assessment that recognizes that the coefficients do not converge. This is when many highly correlated inputs are included in the data, or the data is very sparse.

In order to conduct a logistic regression analysis on the breast cancer dataset, we used the ML library of Python. By using this library, we are trying to evaluate test set outcomes using each supervised ML algorithm, and testing for each whether they yield an accurate prediction. In order to interpret the data, we will be using an uncertainty matrix. The confusion matrix is a tabular method of calculating how

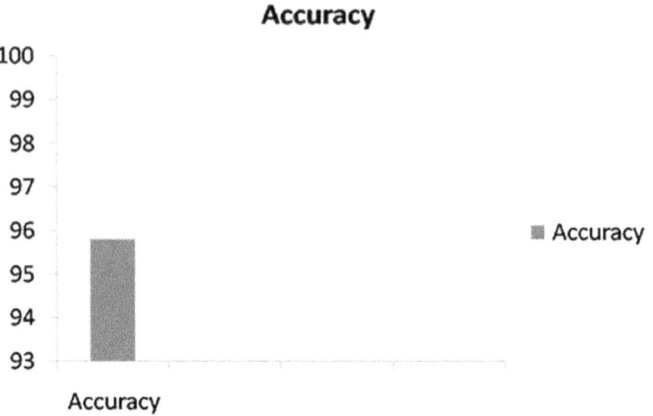

FIGURE 3.4   Accuracy of logistic regression.

many errors in classification there are. The number of separate classes for which an employee was wrongly reclassified, provided that only the classes they are teaching are actually significant. With the aid of a classification accuracy technique, the Random Forest model's accuracy is discovered. The number of accurate predictions as a percentage of the total number of samples taken by the data pipeline is referred to as the accuracy.

Accuracy = Number of Correct Predications/Total Number of Predictions Done

To verify the right prediction, the confusion matrix is used. The results are diagonally predicted and split into complete predictions. The results are correctly predicted.

	0	1
0	87	3
1	3	50

Example: Confusion Matrix

Logistic regression provides 95.8% accuracy (Figure 3.4).

3.2.2 NEAREST NEIGHBOR

The neighbor immediately adjacent is a basic algorithm that keeps track of all of the available cases and classifies new data or events based on a similarity metric. In addition, it can also be used to define a data point based on how its neighbors are classified. In KNN (K nearest neighbor) [5], "k" is a parameter, which is equal to the number of nearest neighbors that will be used for a large portion of the voting process. When it comes to providing better precision, there are procedures that must be performed, which are called parameter tuning. The KNN algorithm has these criteria, and parameter tuning is another one of them. To find the value of k, finding is far from easy. At the moment, there is no set mechanism to discovering the "right" value for "K." We need to decide this by testing, with different values, and then draw the conclusion that the training data is poorly understood. Choosing K values that are smaller is complicated, and that can have a huge impact on the result. The other way to pick K is by cross-validation. The method of cross-validation is achieved by using the training dataset. A small part of the training dataset is called the validation dataset. To do this, a validation dataset can be used, which will provide a whole range of K values. Using this approach, the validation set will take different values of K, such as $K = 1$, 2, and 3, and use these values to predict each case's mark. After we have validated the data, we will use the K value that produces the best results from the validation set. On the whole, most of the time the $K = sqrt(N)$ function is used to select its value. In this case, N is the number of samples in the training dataset. Most of the time, you can strive to apply K in order to avoid conflict between two data classes. When computing the similarity between data points in KNN, you take the distance between them into consideration. One of the most common techniques is the Euclidean distance algorithm. In addition to these types of distance methods, there are other techniques as well, such as Manhattan, Minkowski, and Hamming distance measures. The Hamming distance is the most important algorithm used in categorical variables because if there are categorical variables, then that is where it must be used. Here we used the sklearn.neighbors() function from the KNeighborsClassifier class. 5 is the implementation number of the number of neighbors. The distance-finding technique employed is the Minkowski distance-finding technique, in which distance between two points in normalized vector space is taken into consideration. After training the system with this algorithm, we got an accuracy of 95.1%, which is less than that obtained using the logistic regression (Figure 3.5).

KNN can be introduced very quickly and is versatile to pick features. It is used to manage cases in many grades. But we must specify the parameter K value in KNN, which is the number of closest neighbors. The computational cost for KNN algorithms is very high, since all training samples must be distanced from each other. Data storage is required. You should be aware of distance approaches when using KNN.

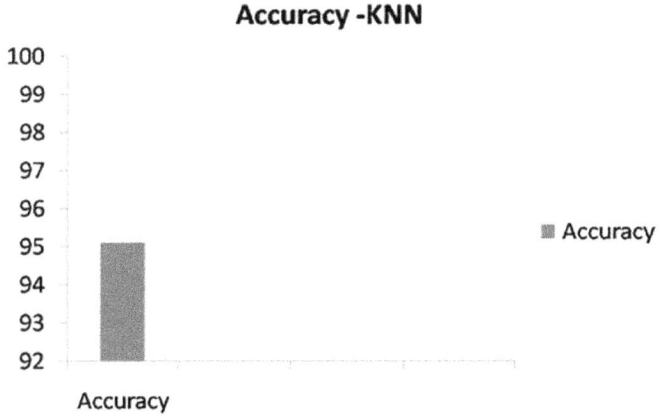

FIGURE 3.5 Accuracy-KNN.

3.2.3   SUPPORT VECTOR MACHINE

A Support Vector Machine [6] is supervised ML algorithm, which is built on top of a hyperplane, which can make predictions on unseen or new samples (a regression problem) or on existing training data (a classification problem). For this product, this is not an important problem for classification. In this linear regression prototype, each point of data is plotted as an input in the nth dimension. This format often assumes a variety of characteristics, but a strong one is n [8,9].

There are several different types of potential ways to group data in order to represent categories of interest. For example, place the terms "sine" and "cosine" in a numerical matrix. Our main objective is to find a plane that is the furthest away from the dimensions of its class based on the difference in data points to do a similar thing with the next class. While research helps us to see what types of threats we may be facing, it can be used to help us further inform the amount of time it will be to produce a consensus and the margin of error that may be found in these findings (Figure 3.6).

Let us explore an example to understand support vector machine.

In the situation above, let us consider that we have two tags black and grey that we want to bear in mind when analyzing our choices. Data is made up of two distinct features: X and Y. The training data is plotted with label. Now support vector machine algorithm will try to find hyperplane which will best to separate these tags. This line that is best to separate out these tags is called as decision boundary. This boundary is decided by the maximum distance between support vector data points. Below figure is showing the same (Figure 3.7).

In the above example, it is very simple to separate out data points because they are linearly separable. We can easily draw line to separate out black and grey points.

But if these points are like below, then it is very difficult to separate them (Figure 3.8).

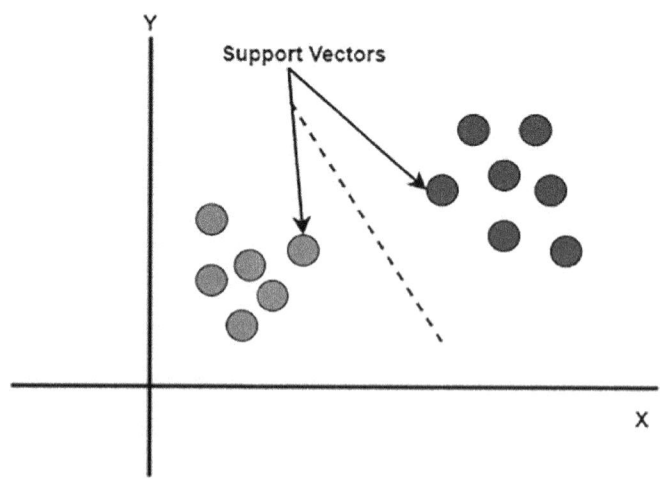

FIGURE 3.6   Support vector machine with hyperplane.

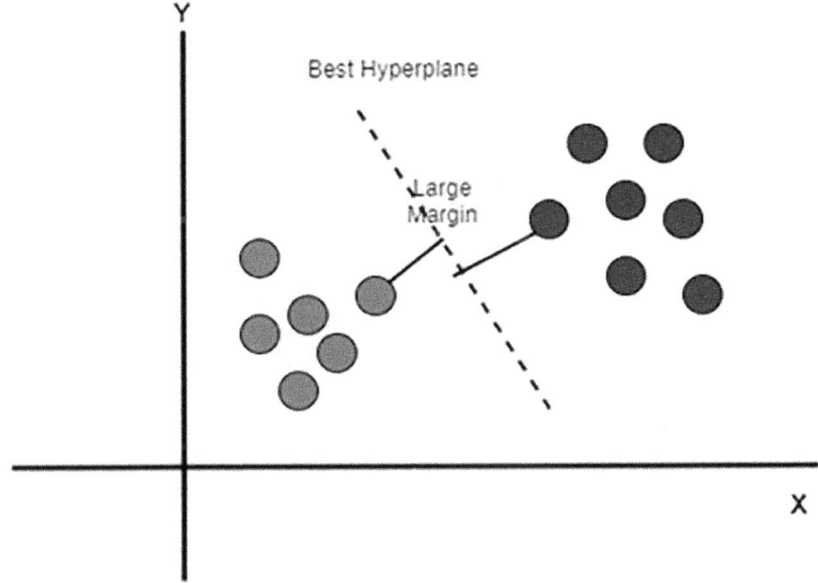

FIGURE 3.7 Best hyperplane selection.

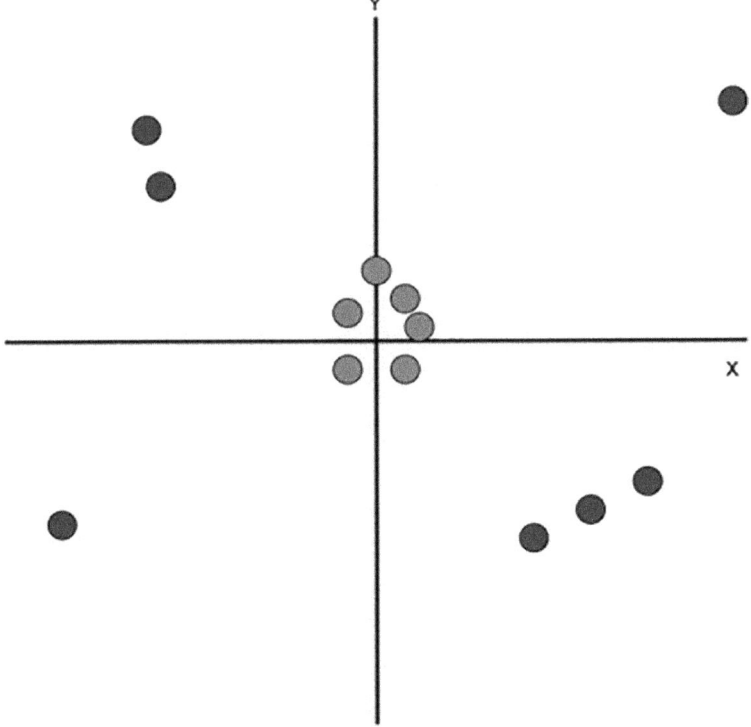

FIGURE 3.8 Complex dataset.

It is very difficult to draw a line over here but the vectors are segregated and it is very easy to separate them. So here we can add the third dimension. Previously, we used only two dimensions X and Y. We will now consider the third dimension that is Z.

This $Z = X^2 + Y^2$, which is the equation of circle. This will give three-dimensional spaces. Now this space will look like (Figure 3.9).

Now the hyperplane is parallel to X-axis and certain at Z (Figure 3.10).

We have identified a way to cleverly map our room to a higher dimension, to recognize nonlinear data. However, it turns out that computing this transformation is extremely expensive in computational terms: several new calculations can be carried out, and each can involve a complex calculation. This can be a lot of work in the dataset for each vector. SVM does not have to do its magic with the actual vectors; it really can only function with the dot products between them. That means that we can miss the expensive new measurement calculations.

$$z = x^2 + y^2$$

$$\_a \cdot b \; = \; xa \cdot xb + ya \cdot yb + za \cdot zb\_$$

$$\_a \cdot b \; = \; xa \cdot xb + ya \cdot yb + \left(xa^2 + ya^2\right) \cdot \left(xb^2 + yb^2\right)\_$$

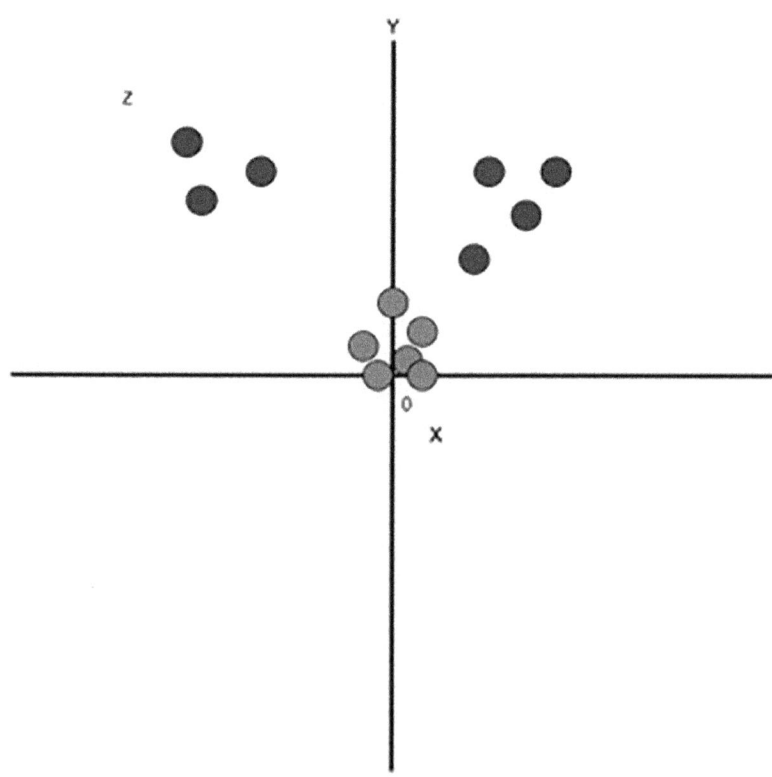

FIGURE 3.9 Two separable groups.

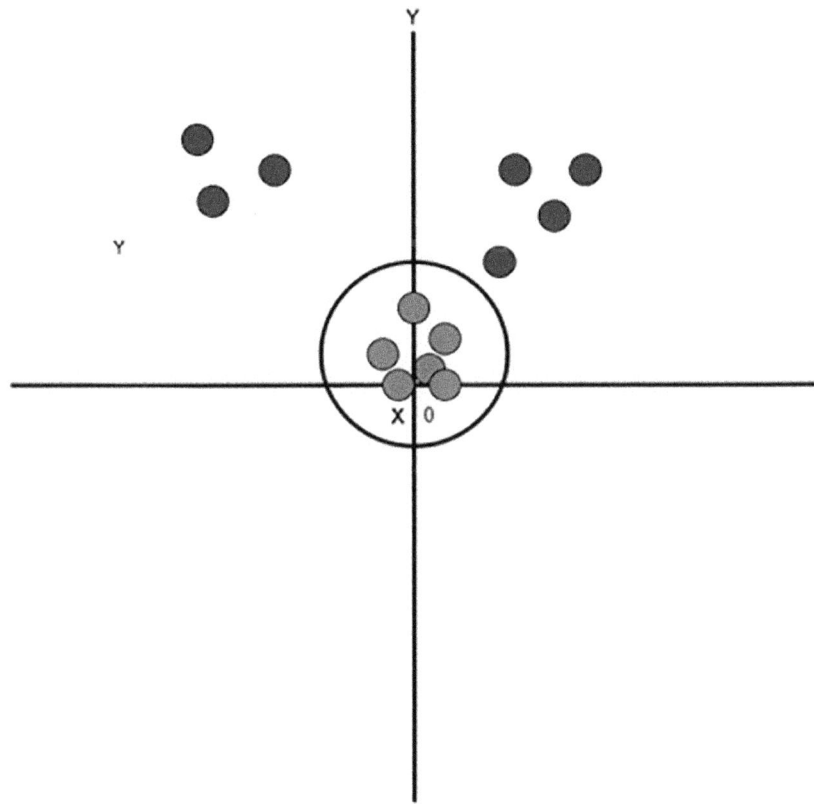

FIGURE 3.10 Hyperplane.

Kernel allows sidestepping a lot of computations. Normally, the kernel is linear but when nonlinear hyperplane is required, then nonlinear kernel is required. Kernel function selection is very important. Every problem is different, and based on data, we need to select the kernel function.

Over here for breast cancer dataset, we used the linear kernel, which is very simple, and data is linearly separable. We used SVC method of SVM class from sklearn. By using this method, we achieve 97.2% accuracy, which is better than previous algorithms such as logistic regression and KNN (Figure 3.11).

We changed the kernel functions of support vector machine. We used the radial basis function kernel, which is called rbf. It is well known that in ML, the radial basis function is very common (Figure 3.12).

$\|x - y\|^2$ is the squared Euclidean distance.

3.2.4 NAIVE BAYES ALGORITHM

A naïve Bayes classifier is a means of estimating the probability of obtaining a positive diagnosis on a computer. Based on Bayes theorem, the classifier has an accuracy that is relatively high. The Bayesian classifiers (naive Bayes and naive Bayes

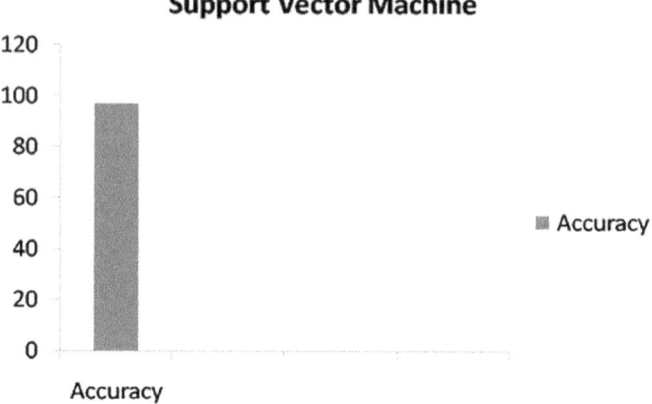

FIGURE 3.11   Accuracy of SVM.

$$k\left(x,y\right)=\exp\left(-\frac{\left\|x-y\right\|^{2}}{2\sigma^{2}}\right)$$

FIGURE 3.12   RBF formula.

$$P(A|B) = \frac{P(B|A)P(A)}{P(B)}$$

FIGURE 3.13   Bayes formula.

classifiers) are algorithms whose classifier form is based on the Bayes theorem. In this learning process, we are not using a single algorithm, but we are employing a family of algorithms, where each one of these algorithms is different from the others, but they all have the same meaning, i.e., every pair of features is fully independent from each other (Figure 3.13).

We may use Bayes theorem to find the likelihood of A's occurring provided B has occurred. In this case, proof is B and the hypothesis is A. It is presumed that the predictors/features are independent. Factors are separate of each other. People refer to this as naive.

When naïve Bayes algorithm is applied on the dataset, the accuracy is 91.6% (Figure 3.14).

3.2.5   DECISION TREE ALGORITHM

Decision tree [4] is a supervised learning technique that can be used for both classifying and regressing problems. This is a tree structure classifier, where internal nodes represent the data attributes, branches represent the classification rules, and leaf nodes represent the correct labels.

In a tree diagram, there are two roots, which are the decision node and the leaf node. Decision nodes are used to make decisions and branch out, while leaf nodes

Accuracy

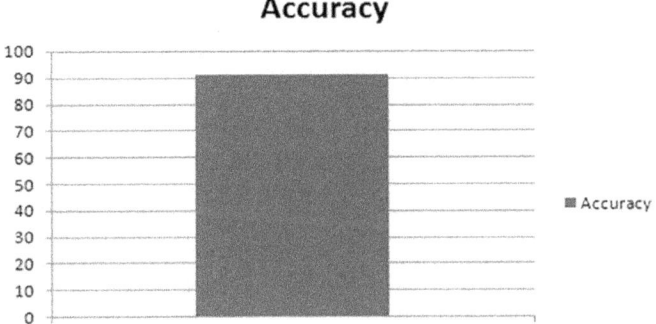

FIGURE 3.14   Accuracy of Naive Bayes.

are the product of those decisions and do not have any more branches. The test is performed according to the features of the given dataset.

It is a graphical way to display all paths to a certain solution when a certain problem occurs.

In order to construct a tree, it is important to use the CART algorithm.

If we were to use a decision tree algorithm, starting at the root node of the tree, we could predict the class of our dataset. An algorithm can leap to another node within the branch in terms of values of parameters. This process continues until it reaches the terminal node of the tree. The advantages of decision tree algorithm are as follows: it is easy to understand how it follows the way a person makes every real-life decision. It can be extremely helpful in helping us solve decision-related problems. It will encourage you to think about all of the potential solutions to a dilemma. Scientists calculated that the algorithm needed less cleaning than the other algorithms.

When decision tree algorithm is applied on breast cancer dataset, the accuracy achieved is 95.8% (Figure 3.15).

accuracy

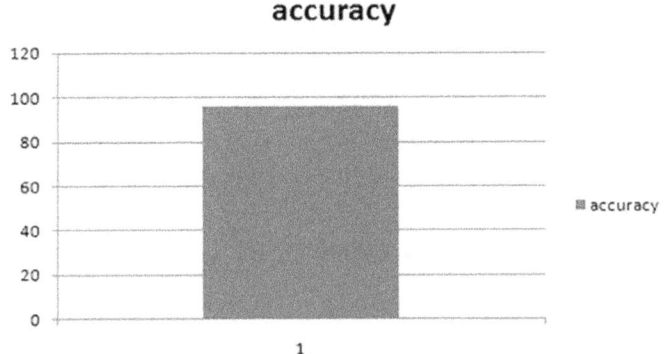

FIGURE 3.15   Accuracy of decision tree.

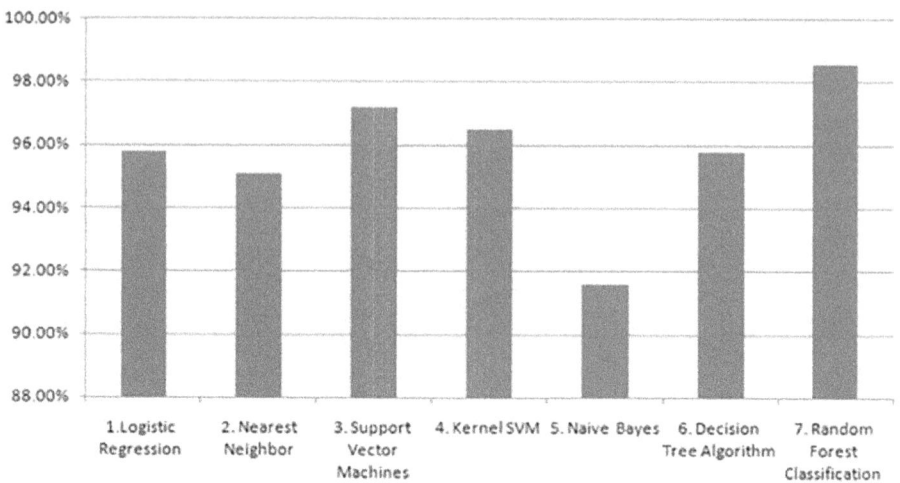

FIGURE 3.16   Summary of algorithms with their accuracy.

3.2.6   RANDOM FOREST CLASSIFICATION

The Random Forest [11] is a method of sampling several different decision trees to make a joint decision. Each tree in the forest makes a prediction when the Random Forest algorithm works and predicts the class that gets the most votes for the best prediction. Although it is recognized as a model, the quality of the crowd is really significant. In data scientists, it is because a large number of relatively uncorrelated forest models (trees) are analytically tested simultaneously by a committee and then coupled with an educated decision-making method to see which one model is the best. A Random Forest algorithm has the following advantages: the idea behind Random Forest is, instead of just a single expert, that the chance was provided by a large number of people. The fact that a large number of relatively unrelated models (trees) are simultaneously evaluated by a committee and then combined with informed decision-making in which the best outcomes of individual models are obtained is in data sciences. When Random Forest algorithm is applied on the breast cancer dataset, the accuracy achieved is 98.6%.

So finally, we have all classification models, and we summarized these accuracies together and then we can see that Random Forest algorithm is having the highest accuracy, i.e., 98.6%. It is not necessary that the same algorithm will give the best accuracy for any other dataset. Selection of model depends on the datasets. According to the datasets, we can select the best ML model (Figure 3.16).

3.3   DETECTION OF BREAST CANCER BY USING DEEP LEARNING

Deep learning [10,12,13,24] is like learning to do something instinctively by watching someone else do it, or how a machine can be programmed to learn how to learn on its own. When it comes to driverless vehicles, the use of deep learning has been important in improving the capabilities of vehicle computers and scanners when used in autonomous mode. The final secret to consumer voice recognition in devices like mobile phones, tablets, TVs, and speakers is what we call latent fingerprinting. Because of the recent upswing in interest in deep learning, there are even greater reasons to embrace it now. Prior to this, our job was limited to building something we had not achieved before. Deep learning, a brain-like computer program, learns to perform its classification tasks by visualizing, listening to, the emotional dimensions of its results. Even the deepest learning models have been developed that can achieve the state-of-the-art precision, and are often faster than human-level performance. All models are first broken down into several smaller parts, and then a set of labeled examples are obtained from the results. The number of layers, meanwhile, varies from a few to a few hundred. Deep learning is a special type of artificial intelligence known as "deep learning." The image is extracted manually first, as it is in the beginning of the feature extraction process. Using this approach, they are able to distinguish between images and actual objects. A deep learning technique operates in combination with features' extraction. Deep learning performs "end-to-end learning" – i.e., a network is given raw data and a task to perform, and then learns the task on its own without human control.

Another important difference is that deep learning algorithms scale with data. Deep learning vulnerabilities affect a number of artificial intelligence models. ML is prone to the problem of overfitting. There are pros and cons of deep learning artificial neural network as the amount of information grows. A convolutional neural network [24] (CNN or ConvNet) is a subset of deep neural networks that explicitly process visual imagery. CNNs are simpler versions of the layered perceptrons. Multilayer perceptrons mean "fully connected networks," which means "all the neurons in one layer are connected to each neuron in the next layer." This makes the networks vulnerable to overfitting the data. These include the approaches such as partial regularization, regularization with metric magnitudes, as well as using hyperbolic loss functions. CNN processes regular structure and analyzes the relationship between it. On the computational power, CNNs are on the lower score.

Similar to how a brain organizes visual information, CNNs organize visual information in the same way. Individual neurons only respond to a restricted area of the visual field known as the receptive field. The receptive fields of various neurons have overlapping coverage of the visual field. CNNs use much less preprocessing than other kinds of classification algorithms. It means the network knows how to create new ones. It is a significant advantage of nonspecific human genetic elements. Dataset we used here contains 250 grayscale images, which are ultrasonic images which are having tumor. Out of 250 images, 100 images are of benign and 150 are malignant (Figure 3.17).

For the purpose of training, a highly accurate deep learning neural network model can successfully identify tumor type as benign or malign by using the following distorted images for training purpose.

These training images can be supplemented by slightly spinning, flipping, stretching, or rendering them and then moved to the network for training. This approach improves neural network's ability to learn and identify correctly unseen data. The original dataset is divided into three parts: training, validation, and testing purposes. The ratio for split-up is maintained as 70:20:10. Here, we have used keras [17] library to build the neural network.

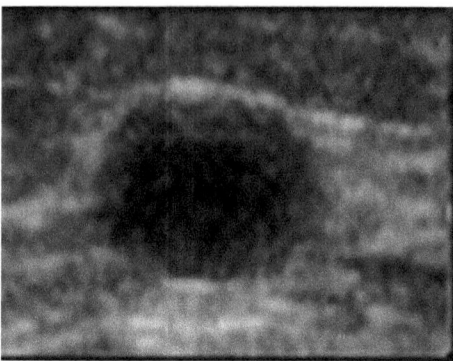

FIGURE 3.17   Sample grayscale image with tumor present in the dataset [23].

CNN has multilayer perceptron at the core and consists of three main types of layers.

Convolutional Layer:
There are layers where filters such as "edges, "shapes," and "objects" are added to layers that precede them. It is important that the hyperparameters of the neural network be explicitly provided in the next layer. The two convolutional layer parameters are stride and padding. The stride will monitor the sum in the change as it determines the next output for the sheet. The padding controls whether to add extra dummy points on the outer boundary of the input layer so that the resulting output after applying the Fourier transforms, if you will, maintains the same size or shrinks from the boundaries as opposed to the preceding layer.

Pooling Layer:
It decreases dimension and prevents loud activations from the following sheet. Either taking maximum or average of inputs for a particular process can apply. Max pooling is a special process.

Flattening Layer:
This transforms 2D or higher dimensions into one-dimensional vectors and better serves to feed the fully connected sheet.

Fully Connected Layer:
In this way, this method is used to learn nonlinear decision boundaries to perform classification tasks with the aid of densely connected layers to previous layer in a simple feed-forward fashion.

Dropout layer:
This model contains a dropout layer between fully linked layers. Output of certain nodes is randomly shunned from input during the training phase and linked to obtain the output later. It is very difficult for the training method to manage every dropout layer. Dropout causes all the nodes to arbitrarily shun all the connections coming out of some fraction of edges during the training process. In order to preserve the

same affect of the effects, the activation from the previous layers is dampened by the same proportion as the fraction of dropout. These networks can operate better without blocking nodes and edges in some cases. This technique helps neural networks apply better to identify unseen inputs during preparation. The hidden layers are transferred to the active layer of the RELU, which only enables the next level to be positively enabled. The output node uses the sigmoid activation function, which fluctuates from 0 to 1 for the negative to positive input. By using the CCN, breast cancer can be detected through the deep learning model. By using this model, the accuracy achieved is 98.3% and the precision achieved is 0.65. Recall value achieved is 0.95. F1 score achieved is 0.77. ROC-AUC achieved is 0.692.

3.4 CONCLUSION

Here, we addressed the breast cancer detection with different ML algorithms and their accuracy. The distinction is made according to their accuracy between these algorithms. A deep learning algorithm, such as the CNN, is also used for grayscale pictures of breast cancer, along with standard ML algorithms.

REFERENCES

1. M. F. Akay, "Support vector machines combined with feature selection for breast cancer diagnosis." *Expert Systems with Applications* 36(2): 3240–3247, 2009.
2. M. Nawaz, A. A. Sewissy and T. H. A. Soliman, "Multi-class breast cancer classification using deep learning convolutional neural network." *International Journal of Advanced Computer Science and Applications* 9(6), 2018, doi:10.14569/IJACSA.2018.090645.
3. L. Hussain, W. Aziz, S. Saeed, S. Rathore and M. Rafique, "Automated breast cancer detection using machine learning techniques by extracting different feature extracting strategies," *2018 17th IEEE International Conference On Trust, Security and Privacy in Computing and Communications/12th IEEE International Conference On Big Data Science And Engineering (TrustCom/BigDataSE)*, New York, NY, 2018, pp. 327–331, doi: 10.1109/TrustCom/BigDataSE.2018.00057.
4. M. Amrane, S. Oukid, I. Gagaoua and T. Ensari, "Breast cancer classification using machine learning," *2018 Electric Electronics, Computer Science, Biomedical Engineerings' Meeting (EBBT)*, Istanbul, 2018, pp. 1–4, doi: 10.1109/EBBT.2018.8391453.
5. M. R. Al-Hadidi, A. Alarabeyyat and M. Alhanahnah, "Breast cancer detection using K-nearest neighbor machine learning algorithm," *2016 9th International Conference on Developments in eSystems Engineering (DeSE)*, Liverpool, 2016, pp. 35–39, doi: 10.1109/DeSE.2016.8.
6. A. Ali and M. Kim, "Amalgamation of SVM based classifiers for prognosis of breast cancer survivability," *2010 Fourth International Conference on Genetic and Evolutionary Computing*, Shenzhen, 2010, pp. 173–176, doi: 10.1109/ICGEC.2010.50.
7. F. Firoozbakht, I. Rezaian, L. Porter and L. Rueda, "Breast cancer subtype identification using machine learning techniques," *2014 IEEE 4th International Conference on Computational Advances in Bio and Medical Sciences (ICCABS)*, Miami, FL, 2014, pp. 1–2, doi: 10.1109/ICCABS.2014.6863912.
8. N. Jafarpisheh, N. Nafisi and M. Teshnehlab, "Breast cancer relapse prognosis by classic and modern structures of machine learning algorithms," *2018 6th Iranian Joint Congress on Fuzzy and Intelligent Systems (CFIS)*, Kerman, 2018, pp. 120–122, doi: 10.1109/CFIS.2018.8336649.

9. A. Mert, N. Kilic and A. Akan, "Breast cancer classification by using support vector machines with reduced dimension," *Proceedings ELMAR-2011*, Zadar, 2011, pp. 37–40.

10. S. Charan, M. J. Khan and K. Khurshid, "Breast cancer detection in mammograms using convolutional neural network," *2018 International Conference on Computing, Mathematics and Engineering Technologies (iCoMET)*, Sukkur, 2018, pp. 1–5, doi: 10.1109/ICOMET.2018.8346384.

11. D. Bazazeh and R. Shubair, "Comparative study of machine learning algorithms for breast cancer detection and diagnosis," *2016 5th International Conference on Electronic Devices, Systems and Applications (ICEDSA)*, Ras Al Khaimah, 2016, pp. 1–4, doi: 10.1109/ICEDSA.2016.7818560.

12. A. Dogra, B. Goyal and K. Kaushik, "A brief review of breast cancer detection via computer aided deep learning methods." *International Journal of Engineering Research & Technology* 8(12), 2019, doi: 10.17577/IJERTV8IS120191.

13. G. Vani, R. Savitha and N. Sundararajan, "Classification of abnormalities in digitized mammograms using extreme learning machine," *2010 11th International Conference on Control Automation Robotics & Vision*, Singapore, 2010, pp. 2114–2117, doi: 10.1109/ICARCV.2010.5707794.

14. Y. Jadhav and H. Mathur, "Comparative Study of breast cancer detection methods." (June 26, 2020). Available at SSRN: https://ssrn.com/abstract=3702283 or doi: 10.2139/ssrn.3702283.

15. "Deep learning for humans." https://keras.io/.

16. D. Selvathi and A. AarthyPoornila, "Performance analysis of various classifiers on deep learning network for breast cancer detection," *2017 International Conference on Signal Processing and Communication (ICSPC)*, Coimbatore, 2017, pp. 359–363, doi: 10.1109/CSPC.2017.8305869.

17. N. Khuriwal and N. Mishra, "Breast cancer diagnosis using deep learning algorithm," *2018 International Conference on Advances in Computing, Communication Control and Networking (ICACCCN)*, Greater Noida, UP, India, 2018, pp. 98–103, doi: 10.1109/ICACCCN.2018.8748777.

4 Assessing the Radial Efficiency Performance of Bus Transport Sector Using Data Envelopment Analysis

Swati Goyal, Shivi Agarwal, and Trilok Mathur
BITS Pilani

Nirbhay Mathur
Universiti Teknologi PETRONAS

CONTENTS

4.1 Introduction 71
    4.1.1 Background Work 73
4.2 Methodology Framework 75
    4.2.1 DEA Background 75
    4.2.2 New Slack Model 75
4.3 Performance Evaluation of Depots 76
    4.3.1 Data Collection 76
    4.3.2 Region-wise Classification of Depots 77
    4.3.3 Input and Output Parameters 77
    4.3.4 Empirical Results 78
    4.3.5 Input Targets for Inefficient Depots 81
4.4 Conclusion 84
Acknowledgement 84
References 84
Appendix (A) 86

4.1 INTRODUCTION

Public transportation improves overall accessibility as it increases the capability to deliver goods and services and it is crucial to expand the reach to acquire education, employment, and services. More than 300 million passengers travel first- and last-mile connectivity through the public transport sector in India daily. It is the largest mover of passengers nationally and provides livelihood to about 40 million

DOI: 10.1201/9781003138020-4

71

people. Additionally, it enlarges the country's economy, overall productivity, and development by reducing transportation costs, road and parking facility costs, vehicle operating costs, accidents, and pollution. It contributes almost 8% to GDP (www. financialexpress, August 2020). During the phase of COVID-19 lockdown, the public transportation sector halted their services, affecting people's income significantly. Hence, efficient operations of public transportation are important for the overall growth of the people.

The road network of India is the second largest across the world, spread over 5.89 million KM. About 90% people commute through road transportation and the road network transports 64.5% of all goods in the country (www.ibef, July 2020). The road transportation system has grown rapidly over time, improving connectivity between cities, towns, and villages. The government will construct 65,000 km of highway roads at the cost of over Rs. 5,350,000 million by 2022 across the country (www. economictimes.indiatimes, February 2019). In India, after Maharashtra and Uttar Pradesh, Rajasthan has the third-highest national highways state till March 2019. The road density of Rajasthan state is 77.21 per $100 \, km^2$ (www.rajras.in, January 2018). Rajasthan is a popular tourist destination and has different topography as compared to the other state of India. The economy of Rajasthan depends on tourism; therefore, the state needs to have a good network of passenger transportation. In some places, road transport is the only choice to transport goods and people. Simultaneously, inefficient transportation enlarges these costs and efficient transportation reduces costs in many economic sectors. Concern for an adequate and efficient transport system in rural areas is essential as it will act as a catalyst for the economic and social progress of the state. Public road transport is one of the basic requirements for commuting in rural areas at low operating costs.

As we know, Rajasthan State Road Transport Corporation (RSRTC) is the sustainable provider of inter-city public bus transportation. The current fleet size has more than 4500 buses categorized into Ordinary, Express, Deluxe, Volvo/Mercedes, AC Sleeper, and Volvo LCD. RSRTC carries around 0.9 million passengers daily. Also, the Government of Rajasthan invested around 40,410 million in 2016–2017 (RSRTC, Annual Report, 2016–2017) to improve the performance of public transport services. Even with the increase in population, passenger traffic of RSRTC has decreased about 0.157 million every year for the period 2007–2017. Issues like slow-moving traffic, air and noise pollution, safety, unorganized and inconvenient bus services lead to increased private modes of transport and app-based taxi services. RSRTC is incurring high losses because of subsidized fares and inefficient depots. The overall financial performance of RSRTC appears to be gloomy, and it would not be wrong to say that it is heading toward a severe financial crisis. It is plagued by inefficiency and usually provides a poor service of quality. Therefore, there is a need to reform the existing and intensive steps should be taken to address the grievances of the public.

In this study, we have used a sustainable benchmarking analysis to improve the level of performance of RSRTC depots using the frontier technique. The frontier technique depends on the production level, which maximizes the output level with the same value of inputs or minimizes the level of the input with the same level of outputs. Two different methodologies, parametric and nonparametric are accessible for the production frontier (Zhou et al., 2014). The parametric methods need a specific functional form of the relationship between parameters, whereas no such assumptions are required on functional form for nonparametric methods. Nonparametric methods are useful in various applications for measuring the performance level of decision-making units (DMUs). Charnes et al. (1978) summarized nonparametric and linear programming techniques to measure productive efficiency of DMUs well-known method as data envelopment analysis (DEA). DEA approach estimates the efficiency of a homogeneous group of DMUs. It handles multiple inputs and outputs in a single linear equation. Several articles have been published in various fields using different DEA models and have provided insights to improve performance levels. Gandhi et al. (2018), Alatawi et al. (2020), and Barpanda et al. (2020) employed various DEA models for obtaining the performance scores for the hospital sectors. Vikas et al. (2019) assessed the technical efficiency (TE), pure technical efficiency (PTE), and scale efficiency (SE) of the Indian oil and gas sector. Tamatam et al. (2019) and Say et al. (2020) applied DEA models for the banking sector. Malhotra et al. (2020) analyzed the efficiency of public and private universities.

Due to the challenging geographic conditions of Rajasthan, the Thar Desert makes it difficult to create infrastructure for public transport, especially the construction of roadways. Further, it is necessary to conduct the study region-wise to identify and compare the performance level of the depots. We have classified the depots into low, medium, and most advantageous conditions based on the operating circumstances. There have been no studies in the literature review on the performance of the transport sector of Rajasthan. This study is explained in three sections. In Section 2, a concise literature review related to the transport sector analysis by DEA techniques is done. Section 3 contains results and discussions. The conclusion has been given in the last section.

4.1.1 BACKGROUND WORK

In the previous study, DEA has been used in 461 articles dealing with the public transport sector (1989–2016) (Cavaignac et al., 2017). Cowie et al. (1999) addressed the scale and technical efficiencies of British bus industry. They suggested that privately owned companies are more technically efficient due to significantly higher organizational efficiency than publicly owned companies. Husain et al. (2000) illustrated the efficiency for 46 service units of the Road Transport Department (RTD) of Malaysia. They identified the excess inputs of inefficient units by the Charnes–Cooper–Rhodes (CCR) DEA model. Barnum et al. (2011) expanded a Banker–Cooper–Charnes (BCC) DEA procedure for measuring public transportation overall efficiency of the USA. They estimated the TE to allocate resources effectively. Hilmola (2011) evaluated public transportation efficiency in big cities using multiple DEA models. He obtained support from the regression analysis that a high-efficiency score resulted in a lower share of private car use in big cities. Jarboui et al. (2012) reviewed the literature of 24 articles published between the period 2000–2011 on public transport efficiency using DEA and stochastic frontier analysis approaches. They found that the financial input and output variables are essential for efficiency studies. Carvalho et al. (2015) optimized the superefficiency DEA model and then analyzed the effective strategy for the competitiveness of urban space and their harmful effects on growth by taking the secondary data of the year 2005–2010 of 50 Brazilian cities. Wu et al. (2016) suggested that DEA is a more efficient technique than SFA in the transport sector of China's provincial-level region. Subsequently, various efficiencies have designed and optimized the transformation of CCR DEA model in the transportation field. Fanou et al. (2018) proposed three different types of comparative efficiency analysis for the six years (2008–2013) of some selected landlocked African countries (LLAC) corridor of transport systems with DEA models. They achieved that the corridor of Swaziland has the efficiency value 1, while the most inefficient corridor was the Central African Republic. Accordingly, Karim et al. (2019) assessed overall effectiveness by applying the constant returns to scale (CRS) model to six public bus companies in Morocco in 2013. Concurrently, they used Tobit regression to eliminate external variables affecting the efficiency score of the transport sector.

In India, active research has been done to evaluate the efficiency. Ramanathan (1999) evaluated 29 State Transport Undertakings (STUs) efficiency of India by 1993–1994 for applying the DEA model. Bhagavath (2006) determined TE using the BCC DEA model of 44 State Road Transport Undertakings (SRTUs). The findings of the study are that SRTUs operating as companies were relatively more efficient than others. Bishnoi et al. (2007) measured the TE of 20 Haryana depots from 2006 to 2007. Agarwal et al. (2010) analyzed the efficiencies of 35 STUs from 2004 to 2008 using an input-oriented fundamental DEA model. Calculations depicted that the average value of efficiency is 83.26% and they produced no change in output even with the reduction in inputs 16.74% lesser than their existing level. Further, Nagadevara et al. (2010) introduced the inter-temporal variations in the DEA efficiency of the sub-units of 25 road transport organizations of Karnataka SRTC over the period 2004–2009. Kumar (2011) used three popular DEA models namely CCR, BCC, and Andersen and Petersen's superefficiency. He computed various TE values for 31 individual SRTUs and conducted Tobit regression analysis to see the inter-SRTU variations in the efficiency and significantly explained the effect of input and output factors on efficiency. Also, Venkatesh et al. (2018) investigated the minimum cost efficiency using variable returns to scale (VRS) for short and longrun in Indian STUs. The key feature of this paper is fleet strength taken as the "quasi-fixed" input. Saxena (2019) discussed the efficiency of Delhi Transport Corporation (DTC), applying DEA technique and regression analysis. A data set of 46 depots has been considered for the year 2015–2016 to identify the technical inefficiency of 50.94%. Mahmoudi et al. (2020) pointed out that DEA is a useful approach for evaluating the transport sector's efficiency.

The review of literature has shown that there are limited studies about public transport sector efficiency in India. The remainder of the study is organized to frame a benchmarking analysis model in terms of the assessment for RSRTC depots' performance level. Benchmarking analysis involves analyzing the resources and determinants of operational efficiency. The main objective of this study to help policy-makers in the formulation of appropriate policies that enhance the overall health and competitiveness of the RSRTC depots. This study will also assist them with the perspective of how to develop satisfactory outcomes for passengers.

4.2 METHODOLOGY FRAMEWORK

4.2.1 DEA BACKGROUND

The DEA method is widely adopted to examine the performance levels for nonprofit-able organizations that cannot be measured by a single scale only in terms of profit. Although Farrell (1957) introduced the frontier analysis for the ratio of single output to input, after that, Charnes et al. (1978) invented a CCR model. He calculated the efficiency of DMUs with a maximum ratio between the aggregate weights of output and input. Banker et al. (1984) established a VRS DEA model based on the assumption after adjoining the convexity constraint in the CCR model. By adding an epsilon (non-Archimedes constant) in the objective function of the fundamental DEA model, the values of slacks have been recognized. In computation, the numerical value for ε is a very small value. Slack based measure (SBM) model proposed by Tone (2001) which deals with with input–output slacks and does not satisfy radial properties. The new slack model (NSM) follows all the radial properties and directly deals with the input–output slacks, which is given by Agarwal et al. (2011).

4.2.2 NEW SLACK MODEL

Each $DMUj\,(j=1,2,\dots o,\dots,J)$ consumes M inputs to produce N outputs defined by $x_{mj}\,(m=1,2,\dots,M)$ and $y_{nj}\,(n=1,2,\dots,N)$, respectively. Agarwal et al. (2011) measured overall technical efficiency (OTE) with the direct impact of input slacks on the value of efficiency at oth DMU.

The underlying mathematical formulation of the input-oriented NSM DEA model for DMU_o is given as follows:

$$\text{Min } \theta_o^* = \theta_o - \frac{1}{M+N}\left[\sum_{m=1}^{M}\frac{S_m^-}{x_{mo}} + \sum_{n=1}^{N}\frac{S_n^+}{y_{no}}\right] \tag{4.1}$$

subject to

$$\sum_{j=1}^{J}\lambda_{jo}y_{nj} - S_n^+ = y_{no} \quad \forall\,(n=1,2,\dots,N)$$

$$\sum_{j=1}^{J}\lambda_{jo}x_{mj} + S_m^- = \theta_o x_{mo} \quad \forall\,(m=1,2,\dots,M)$$

$$\lambda_j \geq 0 \quad \forall\,(m=1,2,\dots,M)$$

θ_o is unrestricted in sign

$$S_n^+,\; S_m^- \geq 0$$

where

θ_o^*: total input-oriented efficiency

θ_o: Reduction applied to all inputs of *DMU* to improve efficiency

S_n^+: The amount of deficiency for *n*th output

S_m^-: The amount of excess resources used for *m*th output

λ_j: Intensity variables for *o*th *DMU*.

Remarks:

1. When *o*th DMU satisfies both the conditions $\theta_o^* = 1$ and all the input slacks and output slacks $\left(S_n^+, \ S_m^-\right)$ equal to zero value if and only if *DMU* is called efficient. Otherwise, *o*th *DMU* is inefficient, we can also interpret that if $\theta_o^* \leq 1$ and/or nonzero value of slacks $\left(S_n^+, \ S_m^- \neq 0 \right)$ seek either excess or deficient, the resources exist inefficiency in the performance of *DMU*.

2. The collection of indices corresponding to positive λ_j is defined as the reference set of *o*th inefficient DMU_o the references set R_o explained as:

$$R_o = \left\{ j: \lambda_j > 0, \quad \forall \ (j = 1, 2, \dots, J) \right\} \tag{4.2}$$

Depots are known as radial efficient but not robust efficient because of the presence of some amount of input and output slacks. It signifies that the slack values are available in input parameters to influence the efficiency values. The fundamental DEA methods are not able to assess the exact impact of slacks on efficiency. The NSM model directly deals with the slacks of inputs and outputs. To estimate the total input saving of OTE by using an input-oriented NSM DEA model.

4.3 PERFORMANCE EVALUATION OF DEPOTS

This study is exposed to overcome the complex nature of the organizational operation of passenger transportation. In the present work, DEA is utilized to sweep the changes in performance to evaluate service efficiency. Various studies have used multi-criteria techniques to assess the effectiveness of public transport sectors. DEA is a nonparametric and linear programming-based technique that constructs a holistic perspective of organization performance. DEA evaluates the relative efficiency of comparable organizations and set the benchmarks to identify the resource of inefficient depots of RSRTC. Besides, many researchers have comprehensively applied DEA in the transport sector in various countries and identified inefficient depots.

4.3.1 DATA COLLECTION

The data set is given by the annual government report of RSRTC for the year 2016–2017. It is collected for the 52 bus depots that are located in 33 districts of Rajasthan. Appendix 1 shows a list of 52 depots, whereas D1 is Abu Road, and so on until D52, which represents Vidhyadhar Nagar.

4.3.2 REGION-WISE CLASSIFICATION OF DEPOTS

Extensively part of region varying the rocky terrains, rolling sand dunes, wetlands, barren tracts, and river-drained plains, the topography of Rajasthan can be partitioned into the following regions—the Aravalli or Hilly regions, Thar Desert, and other arid regions. It is necessary to study region-wise comparison to know the performance level of depots and thus is classified into three categories, low, medium, and most advantages condition based on the operating conditions. The plain region has all facilities such as well-connected roads, population density, schools, universities and hospitals. This category is known as the most advantageous condition. Further, medium advantageous conditions contained both plain and Thar Desert topography and good road connectivity. In contrast, the low advantageous conditions implanted the inhospitable Thar Desert topography; bad road connectivity of the villages; and lack of other facilities such as schools, hospitals, and businesses.

Low Advantageous Condition: There are 25 depots in this range, namely Anoopgarh, Banswara, Baran, Barmer, Beawar, Bundi, Churu, Dhaulpur, Didwana, Dungarpur, Falna, Hindaun, Jaisalmer, Jalore, Karauli, Khetri, Lohagarh, Pali, Phalaudi, Partapgarh, Sardarshahar, Shapur, Sirohi, Srimadhopur and Tijara.

Medium Advantageous Condition: There are 14 depots in this range namely Abu Road, Bharatpur, Bhilwara, Bikaner, Chittorgarh, Dausa, Hanumangarh, Jhalawar, Jhunjhunu, Kotputli, Nagore, Rajasamand, Sawai Madhopur, and Tonk.

Most Advantageous Condition: There are 13 depots in this range namely Ajaymeru, Ajmer, Alwar, Deluxe, Ganganagar, Jaipur, Jodhpur, Kota, Matsyanagar, Sikar, Udaipur, Vaishalinagar, and Vidhyadhar Nagar.

4.3.3 INPUT AND OUTPUT PARAMETERS

Once the model and DMUs were defined, the next step was to establish which parameters would be considered in the model. This step plays an important role in implementing the DEA approach. The list of commonly used parameters in literature review is shown in Table 4.1 (Agarwal et al., 2010; Markovits et al., 2011; Hanumappa et al., 2015). The aim to comprehensively review the performance of depots can improve results in choosing key features such as number of vehicles, employees, revenue and road connectivity. We have selected four inputs and single output parameters to evaluate efficiency and describing the significant factors of depots. It will help the policymakers in the legislative frameworks for improving the performance level.

Input Parameters Comprised:

(I*1) The number of buses that are indicative of capital input.

(I*2) The number of employees that are indicative of labor input.

(I*3) Fuel consumption that is indicative of energy input and is calculated as

$$\text{Total Fuel Consumption} \left(100 \text{ kl}\right) = \text{Description of km} / \text{Average dissel Consumptio}$$

(4.3)

(I*4) The number of routes that are described as network (connectivity) size.

TABLE 4.1
Defining the Relevant Parameters

Inputs	Outputs
Number of buses	Accident rate
Number of employees	Total no. of passengers
Number of routes	Vehicle utilization
Description of kilometers	Rate of breakdown
Diesel consumption	Load factor
	Operating income
	Total income per km
	Operating income per km
	Punctuality
	Fleet utilization

Output Parameter Comprised:
(O*1) Passenger km occupied is the cumulative distance travelled by each passenger which is defined as below:

$$\text{Passenger km Occupied}\left(\text{Lakh km}\right) = \text{Average no. of Buses}$$

$$\times \text{Description of km} \times \text{Load Factor} \quad (4.4).$$

Table 4.2 presents the statistics values for all parameters of 52 RSRTC depots for the period 2016–2017. Statistics summary suggests imperative variation in selected inputs and output for all the depots.

4.3.4 EMPIRICAL RESULTS

This study focuses on framing the benchmark model for RSRTC to improve organizational performance and deliver effective services. Besides, factors that are essential, to be controlled, and improved are analyzed to increase the efficiency of

TABLE 4.2
Statistics Summary of Parameters

	Inputs				Output
	Buses (Number)	Employees (Number)	Fuel Consumption (100 kl)	Routes (Number)	Passenger km Occupied (Lakh km)
Max	147	709	48.57	75	88.94
Min	16	64	4.455	12	8.19
Mean	19.346	315.654	22.182	43.731	39.097
Stdev	25.472	128.559	9.111	15.509	15.576

depots for RSRTC. A specific number was assigned for each depot, as explained in Appendix 1. The obtained main results are shown in Tables 4.3–4.5 for the year 2016–2017. Table 4.3 evidences the results for low advantageous condition depots, whereas Tables 4.4 and 4.5 show the results for medium and high advantageous condition depots, respectively. The tables depict the input slacks, OTE, and peer weights for the 52 depots of the RSRTC as follows.

After analysis, only four depots were efficient according to the NSM model, namely Barmer, Bhilwara, Deluxe, and Jaipur. It indicates that only four depots have an OTE value equal to 1.00. These depots are set as an instance of best performing

TABLE 4.3

Input Slacks, OTE, and Peer Weights of Low Advantageous Condition Depots

Depot Name	Inputs				OTE	Peer Weight[a]
						D8
	Buses (Number)	Employees (Number)	Fuel Consumption (100 kl)	Routes (Number)		
Anoopgarh	12.585	32.144	0	19.922	0.837	0.799
Banswara	31.505	99.550	0	14.686	0.905	1.007
Baran	23.660	94.107	0	12.991	0.860	0.876
Barmer	0.000	0.000	0	0.000	1.000	1
Beawar	26.800	39.890	0	13.392	0.908	1.081
Bundi	30.627	129.538	0	23.292	0.840	0.859
Churu	22.196	40.118	0	0.575	0.910	0.884
Dhaulpur	34.894	147.688	0	24.377	0.858	0.969
Didwana	26.024	84.232	0	20.401	0.836	0.708
Dungarpur	35.581	98.277	0	19.414	0.822	0.965
Falna	25.878	36.430	0	25.010	0.786	0.505
Hindaun	45.853	148.420	0	33.561	0.860	0.833
Jaisalmer	6.375	6.821	1.390	0.000	0.914	0.411
Jalore	28.638	47.389	0	5.841	0.855	0.841
Karauli	6.148	21.662	0	13.140	0.831	0.209
Khetri	30.200	52.867	0	17.820	0.816	0.78
Lohagarh	43.319	186.200	0	28.325	0.902	1.145
Pali	15.924	39.797	0	17.759	0.870	0.715
Phalaudi	12.799	15.042	0	8.042	0.900	0.772
Partapgarh	20.922	0.000	0.183	6.845	0.807	0.403
Sardarshahar	16.448	64.572	0	25.247	0.858	0.92
Shahpura	32.452	128.460	0	31.541	0.820	0.611
Sirohi	21.352	16.807	0	14.024	0.859	0.636
Srimadhopur	34.829	112.158	0	18.444	0.753	0.888
Tijara	32.458	73.777	0	14.618	0.951	0.794
Mean					0.862	

[a] D8 (Barmer) is a peer set for the inefficient depots

TABLE 4.4

Input Slacks, OTE, and Peer Weights of Medium Advantageous Condition Depots

	Inputs				OTE	Peer Weight[a]
						D11
	Buses	**Employees**	**Fuel Consumption**	**Routes**		
Depot Name	(Number)	(Number)	(100 kl)	(Number)		
Abu Road	6.969	1.123	0.000	11.426	0.813	0.452
Bharatpur	14.713	114.443	0.000	17.016	0.860	0.822
Bhilwara	0.000	0.000	0.000	0.000	1.000	1.000
Bikaner	1.743	56.531	3.315	0.000	0.934	1.133
Chittorgarh	7.721	19.235	0.000	13.749	0.824	0.891
Dausa	23.606	113.032	0.000	37.010	0.806	0.725
Hanumangarh	8.071	4.391	0.118	0.000	0.952	1.227
Jhalawar	22.374	30.078	0.000	26.793	0.754	0.680
Jhunjhunu	3.652	40.927	0.000	1.509	0.864	0.848
Kotputli	29.056	120.260	1.954	0.000	0.941	0.558
Nagaur	24.069	124.922	7.261	0.000	0.989	0.767
Rajasamand	12.658	13.122	0.000	25.324	0.826	0.442
Sawai Madhopur	14.152	35.800	0.000	4.380	0.818	0.280
Tonk	15.877	63.048	0.000	15.776	0.854	0.811
Mean					0.874	

[a] D11 (Bhilwara) is a peer set for the inefficient depots

unit for the remaining 48 inefficient depots and also called "reference set" for the other inefficient depots. Further, the OTE score has a value less than 1, meaning that there is an excessive reduction in input with the same production level of output.

The average OTE score of the depots working in the most advantageous condition is 0.853, with the lowest value of 0.701 of Vidhyadhar Nagar. The average of OTE scores is 0.874 and 0.862 in medium and low advantageous condition depots; the lowest value is 0.754 for Jhalawar and 0.753 for Srimadhopur.

It is noteworthy that, after having the most advantageous condition depot, the efficiency of depots is better than the other category depots. The average values of 33 depots were found to be below 0.863 efficiency scores. The perusal of Table 4 gives the higher number of peer count of depots differentiated the robust of efficiencies into other efficient depots. As we can conclude that Barmer (peer count = 24) has highly efficient robust. However, Bhilwara (peer count = 13) and Jaipur (peer count = 12) are in the middle level, and Deluxe (peer count = 4) is in the low level of efficient robust.

Under this perspective, the NSM model measures the efficiency score of depots since reducing the slack values in radial efficiency. In assessing the performance evaluation of a system, the slack of inputs and outputs was utilized to endeavor the reasons of an inefficient system (Cao et al., 2011). The analysis of slack confers a

TABLE 4.5
Input Slacks, OTE, and Peer Weights of Most Advantageous Condition Depots

	Inputs				OTE	Peer Weight[a]	Peer Weight[a]
	Buses	Employees	Fuel Consumption	Routes		D17	D25
Depot Name	(Number)	(Number)	(100 kl)	(Number)			
Ajaymeru	31.908	7.583	0	40.929	0.777	0	0.491
Ajmer	10.983	96.404	0	42.392	0.784	0	0.462
Alwar	41.356	54.007	0	54.864	0.829	0	0.630
Deluxe	0.000	0	0	0.000	1.000	1	0
Ganganagar	44.172	0	0	31.483	0.916	0.486	0.272
Jaipur	0	0	0	0.000	1.000	0	1
Jodhpur	43.418	0	0	43.308	0.927	0.286	0.460
Kota	23.647	66.205	0	22.209	0.804	0	0.526
Matsyanagar	23.218	20.051	0	13.418	0.929	0	0.546
Sikar	30.328	0	0	38.882	0.828	0.114	0.771
Udaipur	32.368	27.855	0	25.164	0.791	0	0.610
Vaishali Nagar	28.876	97.362	0	51.049	0.802	0	0.628
Vidhyadhar Nagar	55.211	228.209	0	75.158	0.701	0	0.292
Mean					0.853		

[a] D17 (Deluxe) and D25 (Jaipur) are peer set for the inefficient depots

new insight into inefficiency for the under-performed depots. The results described the average value of input slacks of the inefficient depots attained 22.45, 58.55, 0.27, and 18.67.

4.3.5 INPUT TARGETS FOR INEFFICIENT DEPOTS

The usages of an input-oriented DEA model yield different values of efficiency value and raise a new perception to enhance the service because of specific inherent characteristics in the productivity of inefficient depots. Thus, the input-oriented model structure reveals that the quantity of over-resources used would be controllable after implementation to reduce input in excess shown in Tables 4.6–4.8. Table 4.6 indicates the input target values and reduction for low advantageous condition depots, whereas in Tables 4.7 and 4.8, the input target values and reduction for medium and high advantageous condition depots, respectively, are appeared. Each nonzero value of slacks expostulates the underperforming depots. It gives a new insight to input parameters for the "development" stage of the suggesting performance scores. Input targets are calculated using as follows:

$$\text{Inputs Target} \left(t^{*-} \right) = \text{Actual Input} * \text{technical efficiency} - \text{Input Slacks} \left(s^{*-} \right) \quad (4.5)$$

TABLE 4.6

Input Target Values and Reduction (%) for Corresponding Inefficient Low Advantageous Condition Depots

Depot Name	Buses (Number)		Employees (Number)		Fuel Consumption (100 KL)		Routes (Number)	
	(t_1^{*-})	(%)	(t_2^{*-})	(%)	(t_3^{*-})	(%)	(t_4^{*-})	(%)
Anoopgarh	35.983	37.96	156.27	30.547	16.054	16.259	16.086	62.591
Banswara	39.973	49.401	176.411	42.16	19.468	9.518	20.6	47.179
Baran	35.686	48.281	152.738	46.781	16.987	13.99	17.972	50.078
Barmer	48.596	41.451	217.186	23.256	22.165	9.16	24.76	41.048
Beawar	29.047	59.089	123.447	58.988	15.49	15.948	12.848	70.121
Bundi	43.338	39.808	189.251	24.9	19.193	8.982	23.09	11.192
Churu	32.91	58.342	139.837	58.258	17.53	14.17	15.103	67.167
Dhaulpur	24.147	59.755	108.927	52.845	12.867	16.383	10.537	71.522
Didwana	35.084	59.205	160.555	49.03	18.101	17.831	17.561	60.976
Dungarpur	14.197	72.163	79.867	46.036	8.902	21.423	4.063	89.019
Falna	19.528	74.305	96.76	66.049	14.031	13.973	6.871	85.381
Hindaun	21.054	29.82	91.009	14.945	8.987	20.812	10.972	8.567
Jaisalmer	36.326	52.203	165.454	33.553	17.154	14.521	19.803	33.99
Jalore	7.148	55.325	31.521	50.748	3.702	16.902	0.986	94.2
Karauli	28.551	60.346	137.259	41.091	14.758	18.401	14.003	64.095
Khetri	37.878	57.913	160.238	58.271	20.59	9.78	17.686	65.322
Lohagarh	30.191	43.036	133.351	32.989	13.987	12.989	13.564	62.322
Pali	37.583	32.888	163.998	17.589	16.465	10.032	18.949	36.837
Phalaudi	15.388	65.804	80.69	19.31	7.987	21.116	8.486	55.337
Partapgarh	39.341	39.475	168.025	37.998	17.916	14.171	16.808	65.698
Sardarshahar	13.478	75.932	60.181	73.834	10.033	17.977	2.086	94.912
Shahpura	25.87	52.964	123.145	24.451	12.538	14.141	12.592	59.381
Sirohi	30.646	64.775	138.453	58.423	16.362	24.741	15.423	65.727
Srimadhopur	28.43	55.578	134.577	38.549	14.962	4.858	14.875	52.016
Tijara	35.983	37.96	156.27	30.547	16.054	16.259	16.086	62.591

Thus, regulating the suggested input target value for each inefficient depots can make them efficient. It can be noticed from Tables 4.6 to 4.8 that there is a significant scope to reduce the inputs, relative to the best performing depots.

In the most advantageous condition, Vidhyadhar Nagar has the highest buses and employee input reduction values 97.18% and 90.54% respectively, while Nagaur has the highest fuel consumption input reduction 31.48% in the medium advantageous condition and Shahpura has the highest route input reduction 94.91% in the low advantageous condition. Appropriate steps can be taken by policy-makers to increase overall performance of RSRTC. Based on the obtained results, inefficient old buses should be regularly maintained and replaced (if required) so as to avoid fuel wastage and reduce existing slacks in fuel consumption. Inefficiency of employees can also

TABLE 4.7

Input Target Values and Reduction (%) for Corresponding Inefficient Medium Advantageous Condition Depots

Depot Name	Buses (Number) (t_1^{*-})	(%)	Employees (Number) (t_2^{*-})	(%)	Fuel Consumption (100 KL) (t_3^{*-})	(%)	Routes (Number) (t_4^{*-})	(%)
Abu Road	33.661	32.68	138.644	19.39	11.189	18.74	17.015	51.39
Bharatpur	56.667	31.73	220.097	43.42	19.183	14	28.564	46.11
Bhilwara	91.607	8.39	367.278	19.1	29.674	16.03	48.542	6.65
Bikaner	68.958	25.85	279.234	22.86	22.642	17.55	35.721	40.47
Chittorgarh	42.503	48.17	167.525	51.86	15.552	19.38	17.005	74.62
Dausa	102.35	11.77	413.494	5.81	33.345	5.15	54.258	4.81
Hanumangarh	43.987	50.01	187.857	35	15.565	24.59	20.716	67.12
Jhalawar	68.94	17.93	277.098	24.7	22.481	13.58	36.516	17.01
Jhunjhunu	34.024	49.22	136.769	49.9	12.387	18.68	20.713	5.85
Kotputli	50.083	33.22	197.394	39.45	16.392	31.48	28.672	1.13
Nagaur	28.637	42.73	122.325	25.41	10.052	17.41	11.841	73.69
Rajasamand	16.936	55.43	72.189	45.31	6.342	18.19	9.528	43.95
Sawai Madhopur	57.541	33.09	232.333	32.85	19.418	14.63	29.47	44.4

TABLE 4.8

Input Target Values and Reduction (%) for Corresponding Inefficient Most Advantageous Condition Depots

Depot Name	Buses (Number) (t_1^{*-})	(%)	Employees (Number) (t_2^{*-})	(%)	Fuel Consumption (100 KL) (t_3^{*-})	(%)	Routes (Number) (t_4^{*-})	(%)
Ajaymeru	46.608	53.85	275.391	24.34	18.972	22.26	7.268	88.28
Ajmer	47.034	36.44	236.807	44.28	17.638	21.6	5.437	91.09
Alwar	57.256	51.89	334.649	28.65	23.753	17.13	7.285	90.29
Deluxe	49.307	51.66	344.604	8.35	29.048	8.35	14.339	71.32
Ganganagar	53.007	49.03	369.026	7.28	31.237	7.28	13.248	78.28
Jaipur	57.173	42.83	352.156	23.94	21.772	19.61	22.758	49.43
Jodhpur	58.142	33.93	273.032	25.2	22.054	7.06	7.534	76.46
Kota	91.459	37.78	511.185	17.15	36.354	17.15	23.253	69
Matsyanagar	63.39	47.61	354.389	26.63	24.601	20.86	15.988	69.25
Sikar	60.16	45.8	325.353	38.26	23.759	19.79	8.31	88.77
Udaipur	2.311	97.18	35.555	90.54	8.6	29.84	7.45	90.07
Vaishali Nagar	46.608	53.85	275.391	24.34	18.972	22.26	7.268	88.28
Vidhyadhar Nagar	47.034	36.44	236.807	44.28	17.638	21.6	5.437	91.09

be handled if employees are given proper technical training at regular time intervals. We also suggest that RSRTC can work on its existing bus route network so as to further increase the connectivity and decrease frequency of buses on most inefficient routes.

4.4 CONCLUSION

This study aimed to measure the total potential for enhancing the performance of RSRTC. The study finds a large gap between actual performance and achievable performance in the operating condition in most of the depots. We have identified only Barmer (D8), Bhilwara (D11), Deluxe (D17), and Jaipur (D25) having the efficiency score 1, which are in low, medium, and most advantageous categories, respectively. It shows that 48 depots have scope for improvement by reducing the input slacks. The number of routes has the highest amount of input slack (35.17%) while consuming fuel compared to other input slack behaving most efficiently (0.95%). Since the average slack is very high for the number of routes, therefore, this study concludes revamping the whole route map for depots of RSRTC.

ACKNOWLEDGEMENT

The authors thank to DST for providing financial support under the DST Rajasthan Grant No. 7 (3)(2)/DST/R/&D/2018/7631, FIST Grant No. SR/FST/MSI-090/2013(C) and Department of Mathematics, BITS Pilani.

REFERENCES

https://www.financialexpress.com/auto/industry/government-public-transport-indian-economy-delhi-transport-bus-delhi-metro-local-trains/2063737, (Accessed on 23 September 2020).

https://www.ibef.org, (Accessed on 24 September 2020).

https://economictimes.indiatimes.com/news/economy/infrastructure, (Accessed on 23 September 2020).

https://www.rajras.in/index.php/road-network-rajasthan/, (Accessed on 22 September 2020).

Annual Report Rajasthan State Road Transport Corporation (2016–2017), https://transport.rajasthan.gov.in/content/transportportal/en/RSRTC/public-relation/AnnualReport.html, (Accessed on 23 September 2020).

Agarwal, S., Yadav, S.P. and Singh, S.P. 2010. DEA based estimation of the technical efficiency of state transport undertakings in India. *OPSEARCH*. 47, 3, pp. 216–230.

Agarwal, S., Yadav, S.P. and Singh, S.P. 2011. A new slack DEA model to estimate the impact of slacks on the efficiencies. *International Journal of Operational Research*. 12, 3, pp. 241–256.

Alatawi, A.D., Niessen, L.W. and Khan, J.A. 2020. Efficiency evaluation of public hospitals in Saudi Arabia: an application of data envelopment analysis. *BMJ Open*. 10, p. 1.

Banker, R.D., Charner, A. and Cooper, W.W. 1984. Some models for estimating technical and scale inefficiencies in data envelopment analysis. *Management Science*. 30, 9, pp. 1078–1092.

Barnum, D.T., Karlaftis, M.G. and Tandon, S. 2011. Improving the efficiency of metropolitan area transit by joint analysis of its multiple providers. *Transportation Research Part E: Logistics and Transportation Review.* 47, 6, pp. 1160–1176.

Barpanda, S. and Sreekumar, N. 2020. Performance analysis of hospitals in Kerala using data envelopment analysis model. *Journal of Health Management.* 22, 1, pp. 25–40.

Bhagavath, V. 2006. Technical efficiency measurement by data envelopment analysis: an application in transportation. *Alliance Journal of Business Research.* 2, 1, pp. 60–72.

Bishnoi, N.K. and Sujata, U. 2007. Efficiency assessment of Haryana State Roadways: a data envelopment analysis. *Indian Journal of Transport Management.* 32, 1, pp. 9–23.

Cao, Q. and Hoffman, J.J. 2011. A case study approach for developing a project performance evaluation system. *International Journal of Project Management.* 29, 2, pp. 155–164.

Carvalho, M. and Syguiy, T. 2015. Efficiency and effectiveness analysis of public transport of Brazilian cities. *Journal of Transport Literature.* 9, 3, pp. 40–44.

Cavaignac, L. and Petiot, R. 2017. A quarter century of data envelopment analysis applied to the transport sector: a bibliometric analysis. *Socio-Economic Planning Sciences,* 57, pp. 84–96.

Charnes, A., Cooper, W.W. and Rhodes, E. 1978. Measuring the efficiency of decision-making unit. *European Journal of Operational Research.* 2, pp. 429–441.

Cooper, W.W., Seiford, L.M. and Zhu, J. eds., 2011. *Handbook on Data Envelopment Analysis,* Vol. 164. Springer Science & Business Media, New York.

Cowie, J. and Asenova, D. 1999. Organisation form, scale effects and efficiency in the British bus industry. *Transportation.* 26, 3, pp. 231–248.

Farrell, M.J. 1957. The measurement of productive efficiency. *Journal of the Royal Statistical Society: Series A (General).* 120, 3, pp. 253–281.

Fanou, E.H. and Wang, X. 2018. Assessment of transit transport corridor efficiency of land-locked African countries using data envelopment analysis. *South African Journal of Science.* 114, 1–2, pp. 1–7.

Gandhi, A.V. and Sharma, D. 2018. Technical efficiency of private sector hospitals in India using data envelopment analysis. *Benchmarking: An International Journal.* 25, 9, pp. 3570–3591.

Hanumappa, D., Ramachandran, P., Sitharam, T.G. and Lakshmana, S. 2015. Performance evaluation of Bangalore metropolitan transport corporation: an application of data envelopment analysis. *Journal of Public Transportation.* 18, 2, pp. 1–19.

Hilmola, O.P. 2011. Benchmarking efficiency of public passenger transport in larger cities. *Benchmarking: An International Journal.* 18, 1, pp. 23–41.

Husain, N., Abdullah, M. and Kuman, S. 2000. Evaluating public sector efficiency with data envelopment analysis (DEA): a case study in Road Transport Department, Selangor, Malaysia. *Total Quality Management.* 11, 4–6, pp. 830–836.

Jarboui, S., Forget, P. and Boujelbene, Y. 2012. Public road transport efficiency: a literature review via the classification scheme. *Public Transport.* 4, 2, pp. 101–128.

Karim, Z. and Fouad, J. 2019. An analysis of technical efficiency of public bus transport companies in Moroccan cities. *Management Research Practice.* 11, 1, pp. 56–73.

Kumar, S. 2011. State Road Transport undertakings in India: technical efficiency and its determinants. *Benchmarking: An International Journal.* 18, 5, pp. 616–643.

Mahmoudi, R., Emrouznejad, A., Shetab-Boushehri, S.N. and Hejazi, S.R. 2020. The origins, development and future directions of data envelopment analysis approach in transportation systems. *Socio-Economic Planning Sciences.* 69, p. 100672.

Malhotra, R., Malhotra, D.K. and Nydick, R. 2020. A comparative analysis of public and private universities in the United States using data envelopment analysis models. *Applications of Management Science.* 20, pp. 143–156.

Markovits-Somogyi, R. 2011. Measuring efficiency in transport: the state of the art of applying data envelopment analysis. *Transport.* 26, 1, pp. 11–19.

Nagadevara, V. and Ramanayya, T.V. 2010. Inter-temporal shifts in efficiency in a road transport organization. *Journal of the Academy of Business and Economics.* 10, 1, pp. 139–144.

Ramanathan, R. 1999. Using data envelopment analysis for assessing the productivity of the state transport undertakings. *Journal of Transport Management.* 23, 5, pp. 301–312.

Say, J., Zhao, H., Agbenyegah, F.S., Nusenu, A.A., Boadi, E.A. and Egbadewoe, S.M. 2020. Regional efficiency disparities in rural and community banks in Ghana: a data envelopment analysis. *Journal of Psychology in Africa.* 30, 3, pp. 249–256.

Saxena, P. 2019. A benchmarking strategy for Delhi Transport Corporation: an application of data envelopment analysis. *International Journal of Mathematical, Engineering and Management Sciences.* 4, 1, pp. 232–244.

Tamatam, R., Dutta, P., Dutta, G. and Lessmann, S. 2019. Efficiency analysis of Indian banking industry over the period 2008–2017 using data envelopment analysis. *Benchmarking: An International Journal.* 26, 8, pp. 2417–2442.

Tone, K. 2001. A slacks-based measure of efficiency in data envelopment analysis. *European Journal of Operational Research.* 130, 3, pp. 498–509.

Venkatesh, A. and Kushwaha, S. 2018. Short and long-run cost efficiency in Indian public bus companies using Data Envelopment Analysis. *Socio-Economic Planning Sciences*, 61, pp. 29–36.

Vikas, V. and Bansal, R. 2019. Efficiency evaluation of Indian oil and gas sector: data envelopment analysis. *International Journal of Emerging Markets.* 14, 2, pp. 362–378.

Wu, J., Zhu, Q., Chu, J., Liu, H. and Liang, L. 2016. Measuring energy and environmental efficiency of transportation systems in China based on a parallel DEA approach. *Transportation Research Part D: Transport and Environment.* 48, pp. 460–472.

Zhou, G., Chung, W. and Zhang, Y. 2014. Measuring energy efficiency performance of China's transport sector: a data envelopment analysis approach. *Expert Systems with Applications.* 41, 2, pp. 709–722.

APPENDIX (A)

S. No.	Depot Name	District	S. No.	Depot Name	District
D1	Abu Road	Sirohi	D27	Jalore	Jalore
D2	Ajaymeru	Ajmer	D28	Jhalawar	Jhalawar
D3	Ajmer	Ajmer	D29	Jhunjhunu	Jhunjhunu
D4	Alwar	Alwar	D30	Jodhpur	Jodhpur
D5	Anoopgarh	Sri Ganganagar	D31	Karauli	Karauli
D6	Banswara	Banswara	D32	Khetri	Jhunjhunu
D7	Baran	Baran	D33	Kota	Kota
D8	Barmer	Barmer	D34	Kotputli	Jaipur
D9	Beawar	Ajmer	D35	Lohagarh	Bharatpur
D10	Bharatpur	Bharatpur	D36	Matsyanagar	Alwar
D11	Bhilwara	Bhilwara	D37	Nagaur	Nagaur
D12	Bikaner	Bikaner	D38	Pali	Pali
D13	Bundi	Bundi	D39	Phalaudi	Jodhpur
D14	Chittorgarh	Chittorgarh	D40	Partapgarh	Partapgarh
D15	Churu	Churu	D41	Rajasamand	Rajasamand

(Continued)

S. No.	Depot Name	District	S. No.	Depot Name	District
D16	Dausa	Dausa	D42	Sardarshahar	Churu
D17	Deluxe	Jaipur	D43	Sawai Madhopur	Sawai Madhopur
D18	Dhaulpur	Dhaulpur	D44	Shahpura	Jaipur
D19	Didwana	Nagaur	D45	Sikar	Sikar
D20	Dungarpur	Dungarpur	D46	Sirohi	Sirohi
D21	Falna	Pali	D47	Srimadhopur	Sikar
D22	Ganganagar	Sri Ganganagar	D48	Tijara	Sirohi
D23	Hanumangarh	Hanumangarh	D49	Tonk	Sikar
D24	Hindaun	Karauli	D50	Udaipur	Udaipur
D25	Jaipur	Jaipur	D51	Vaishalinagar	Jaipur
D26	Jaisalmer	Jaisalmer	D52	Vidhyadhar Nagar	Jaipur

5 Weight-Based Codes—A Binary Error Control Coding Scheme—A Machine Learning Approach

Piratla Srihari
Geethanjali College of Engineering and Technology

CONTENTS

5.1 Introduction 89
5.2 Encoding..90
5.3 Decoding (Machine Learning Approach) 90
    5.3.1 Principle of Decoding...90
    5.3.2 Algorithm...91
5.4 Output Test Case..92
5.5 Conclusion 93
References..93

5.1 INTRODUCTION

An (n, k) binary linear block code is a channel coding scheme and transforms each k-tuple message word (constituted by the information from the source) $MW = \begin{bmatrix} mb_1 & mb_2 & - & - & mb_k \end{bmatrix}$ into an n-tuple code word $CW = \begin{bmatrix} cb_1 & cb_2 & - & - & cb_n \end{bmatrix}$ where all these $mb_i s$ (message bits) and $cb_i s$ (code bits) are selected from binary alphabets 0 and 1. In this transformation of 'k' bit word into 'n' bit word, 'r' numbers of parity bits collectively referred to as parity word $rw = \begin{bmatrix} pb_1 & pb_2 & - & - & pb_{n-k} \end{bmatrix}$ are appended to every message word of length 'k' (pb_i is the ith parity bit). The result is the code word CW of length $n(= r + k)$ bits. The parity bits are also referred to as redundant bits.[1,2]

In a systematic block code, the code word can be either $CW = \begin{bmatrix} cb_1 & cb_2 & - & - & cb_n \end{bmatrix} = \begin{bmatrix} mb_1 & mb_2 & - & - & mb_k pb_k & pb_2 & - & - & pb_{n-k} \end{bmatrix}$

or $\left[pb_1 \quad pb_2 \quad - \quad - \quad pb_{n-k} \quad mb_1 \quad mb_2 \quad - \quad - \quad mb_k \right]$, i.e., the code word contains the 'k' bit message block unaltered.[3,4]

The parity bits are computed using a prescribed combination (Modulo-2 addition) of message bits. The expressions for this linear combination are (n, k) code specific.[5]

The number of nonzero elements of a binary word is referred to its weight. Hamming distance (d_{min}) of the code (minimum of nonzero weights of all possible code words of the code) is a measure of the error detecting and correcting ability of the code.[6]

An (n, k) binary linear block code with Hamming distance d_{min}, can correct $\dfrac{d_{min} - 1}{2}$ number of errors [can detect $(d_{min} - 1)$ number of errors].[6]

The present chapter discusses a systematic single error correcting binary forward error correction scheme or channel coding scheme which is referred to as 'weight-based code,' also as 'data inverting code.'

An (n, k) weight-based code is in a code word format similar to a systematic (n, k) binary linear block code. In a weight-based code word, the structure of the parity word is not uniform across the code words of the code unlike block codes and is weight (of the message word) dependent.[5]

There is a unique mapping between the message and code words of an (n, k) weight-based code.[3]

5.2   ENCODING

The code word of a weight-based code is expressed as follows: $CW = \left[MW \; rw \right] = \left[mb_1 \quad mb_2 \quad - \quad - \quad mb_k pb_1 \quad pb_2 \quad - \quad - \quad pb_{n-k} \right]$ with $n = 2k$.[5,7]

If MW (Message word) is of 0 or even weight, $pb_i = mb_i$, $i = 1, 2, \dots k$ for the mapped CW (Code word).[8]

If MW is odd weight, $pb_i = \overline{mb_i}$, (complement) $i = 1, 2, \dots k$ for the mapped CW. Thus, the proposed coding scheme is also referred to as data inverting code.[5,7]

Similar to Hamming codes, all the proposed weight-based codes are single error correcting codes.[5,7]

Table 5.1 specifies all the possible message words and the corresponding code words of a (6, 3) weight-based code.

5.3   DECODING (MACHINE LEARNING APPROACH)

5.3.1   PRINCIPLE OF DECODING

The decoding of the received weight-based code word can be implemented using predictive modeling of machine learning, specifically classification algorithms.[9,10] Since each message word is mapped uniquely to a code word, the set of message words and the corresponding code words will be the data set used for training the decoder. A Bayesian classifier can be used to predict class membership, i.e., whether a given n-tuple belongs to a class.[11-13]

TABLE 5.1

Code Words of a (6,3) Weight-based Code

Message Word			Code Word					
			cb_1	cb_2	cb_3	cb_4	cb_5	cb_6
mb_1	mb_2	mb_3	mb_1	mb_2	mb_3	pb_1	pb_2	pb_3
0	0	0	0	0	0	0	0	0
0	0	1	0	0	1	1	1	0
0	1	0	0	1	0	1	0	1
0	1	1	0	1	1	0	1	1
1	0	0	1	0	0	0	1	1
1	0	1	1	0	1	1	0	1
1	1	0	1	1	0	1	1	0
1	1	1	1	1	1	0	0	0

The K-nearest neighbor or '*KNN*' algorithm, which is one of the simplest classification algorithms, is applied to the received code word to classify its class, i.e., to find its nearest neighbor among the members of the code word set (which is the one that is already transmitted) the training data.[13,14] The distance metric used in the process of grouping the neighbors is Hamming distance. This metric for the received code word will be computed with reference to all the code words in the training data set. The one among the code words of the training set, that is nearest to the received (with minimum distance), will be its neighbor. This distance will be decided by the value of 'K'. The maximum value of K will be the maximum number of errors that an (n, k) weight-based code can correct. For weight-based codes, maximum $K=1$, since they are capable of correcting single error. Reception can be error free also (without error/s), which can be decided using $K=0$. The nearest neighbor of the received CW among the members of the CW set (training data set) will be the transmitted code word, and from this, the uniquely mapped message word can be retrieved.

5.3.2 ALGORITHM

- Number of training data samples: 2^k
- $D = \left[2^k \ X \ k \right]$ binary array that consists of all the possible 'k' bit message words
- Code word length $= 2k$ bits
- $C = \left[2^k \ X \ 2k \right]$ binary array consisting of 2^k code words, with a one-to-one mapping with the 2^k message words
- Code word = [message word parity word]
  - Message word with zero/even weight—parity word is same as message word
  - Message word with odd weight—parity word is the inverted (bitwise) message word
- Training data will be D and C with a unique mapping between i^{th} row of D and i^{th} row of C

- R = Received code word
- For $i = 1$ to 2^k, compute the Hamming distance $dis[R, C(i)]$
- $K = 1$ (error correcting ability of the proposed *code*)
- Z = Number of the row vector of C, which is the nearest neighbor of R
  - Nearest neighbor satisfies the constraint $dis[R, C(i)] \leq K$
  - Nearest neighbor with zero distance implies error free reception
- The first 'k' bits of the Zth row vector of C is the message word.

5.4 OUTPUT TEST CASE

Let the code be $(8, 4)$ weight-based code.

- Number of training data samples: 16
- The message word array of the code $D =$
- 0. 0. 0. 0.
  0. 0. 0. 1.
  0. 0. 1. 0.
  0. 0. 1. 1.
  0. 1. 0. 0.
  0. 1. 0. 1.
  0. 1. 1. 0.
  0. 1. 1. 1.
  1. 0. 0. 0.
  1. 0. 0. 1.
  1. 0. 1. 0.
  1. 0. 1. 1.
  1. 1. 0. 0.
  1. 1. 0. 1.
  1. 1. 1. 0.
  1. 1. 1. 1.
- Code word length = 8 bits
  The code word array of the code $C =$
- 0. 0. 0. 0. 0. 0. 0. 0.
  0. 0. 0. 1. 1. 1. 1. 0.
  0. 0. 1. 0. 1. 1. 0. 1.
  0. 0. 1. 1. 0. 0. 1. 1.
  0. 1. 0. 0. 1. 0. 1. 1.
  0. 1. 0. 1. 0. 1. 0. 1.
  0. 1. 1. 0. 0. 1. 1. 0.
  0. 1. 1. 1. 1. 0. 0. 0.
  1. 0. 0. 0. 0. 1. 1. 1.
  1. 0. 0. 1. 1. 0. 0. 1.
  1. 0. 1. 0. 1. 0. 1. 0.
  1. 0. 1. 1. 0. 1. 0. 0.
  1. 1. 0. 0. 1. 1. 0. 0.

    1. 1. 0. 1. 0. 0. 1. 0.
    1. 1. 1. 0. 0. 0. 0. 1.
    1. 1. 1. 1. 1. 1. 1. 1.
- $R = [1\ 1\ 1\ 1\ 1\ 1\ 0\ 0]$
- $dis = [5\ 3\ 5\ 7\ 5\ 3\ 5\ 3\ 5\ 3\ 5\ 3\ 1\ 3\ 5\ 3]$
- $Z = 13$
- The nearest neighbor of R is 13th row of C, i.e., $[1\ 1\ 0\ 0\ 1\ 1\ 0\ 0]$
- The message word is $[1\ 1\ 0\ 0]$.

5.5 CONCLUSION

The training data set for the algorithm effectively reduces various stages such as:
   For the received code word

    i. identifying the parity word structure,
   ii. estimating the error location,
  iii. Locating the error and processing of error correction are essential in the conventional decoding of a weight-based code word.

This concept of classification algorithm can even applied for multiple error correction under the suitable channel encoding–decoding scheme.

REFERENCES

1. P. Sweeney, *Error Control Coding: From Theory to Practice*, John Wiley & Sons, Guildford, 2002.
2. T.K. Moon, *Error Correction Coding-Mathematical Methods and Algorithms*, John Wiley & Sons, NJ, 2005.
3. B. Sklar, *Digital Communications: Fundamentals and Applications*, Second Edition, Prentice Hall, Upper Saddle River, NJ, 2001.
4. J.G. Proakis, M. Salehi, *Digital Communications*, Fifth Edition, McGraw-Hill Higher Education, New York, 2008.
5. P. Srihari, B.C. Jinaga, "Data Inverting Codes". Ph.D. Thesis, Jawaharlal Nehru Technological University, Hyderabad, India, 2007.
6. D. Le Ruyet, M. Pischella, *Digital Communications 1: Source and Channel Coding*, Networks and Telecommunications Series, John Wiley & Sons, New York, 2015.
7. P. Srihari, B.L. Prakash, "Weight Based Codes – A Binary Channel Coding Scheme". *International Journal of Electronics, Communication and Instrumentation Engineering Research & Development*, Vol. 3, No. 4, pp. 23–36, ISSN No. 2249-7951, TRANS STELLER Journal Publications, 2013.
8. P. Srihari, "SUMULTI CODES – A Binary Forward Error Correcting Scheme". *International Journal of Future Generation Communication and Networking*, Vol. 13, No. 1, pp. 1170–1175, ISSN No. 2233-7857, 2020.
9. M. Paluszek, S. Thomas, *MATLAB Machine Learning*, Apress, New York, 2017.
10. E.S. Gopi, *Algorithm Collections for Digital Signal Processing Applications Using Matlab*, Springer, Dordrecht, 2007.
11. J.O. Berger, *Statistical Decision Theory and Bayesian Analysis*, Second Edition, Springer-Verlag, New York Inc., New York, 1993.

12. B.C. Levy, *Principles of Signal Detection and Parameter Estimation*, Springer Texts in Statistics, Springer, New York, 2008.

13. P.D. Hoff, *A First Course in Bayesian Statistical Methods*, Springer Texts in Statistics, Springer, New York, 2009.

14. K.P. Murphy, *Machine Learning – A Probabilistic Perspective*, The MIT Press, Cambridge, MA, 2012.

6 Massive Data Classification of Brain Tumors Using DNN

Opportunity in Medical Healthcare 4.0 through Sensors

Rohit Rastogi and Akshit Rajan Rastogi
ABES Engineering College

D.K. Chaturvedi
DEI

Sheelu Sagar
Amity International Business School

Neeti Tandon
Vikram University

CONTENTS

6.1 Introduction 96
    6.1.1 Brain Tumor..96
    6.1.2 Big Data Analytics in Health Informatics 96
    6.1.3 Machine Learning (ML) in Healthcare 97
    6.1.4 Sensors for Internet of Things 97
    6.1.5 Challenges and Critical Issues of IoT in Healthcare...........................97
    6.1.6 Machine Learning (ML) and Artificial Intelligence (AI) for
            Health Informatics 98
    6.1.7 Health Sensor Data Management 99
    6.1.8 Multimodal Data Fusion for Healthcare...99
    6.1.9 Heterogeneous Data Fusion and Context-Aware Systems—a
            Context-Aware Data Fusion Approach for Health-IoT 99

DOI: 10.1201/9781003138020-6

6.1.10 Role of Technology in Addressing the Problem of Integration of
        Healthcare System 100
6.2 Literature Survey 100
6.3 System Design and Methodology 102
     6.3.1 System Design 102
     6.3.2 CNN Architecture 102
     6.3.3 Block Diagram 103
     6.3.4 Algorithm(s) 104
     6.3.5 Our Experimental Results, Interpretation, and Discussion 105
     6.3.6 Implementation Details 105
     6.3.7 Snapshots of Interfaces 106
     6.3.8 Performance Evaluation 108
     6.3.9 Comparison with Other Algorithms 109
6.4 Novelty in Our Work 109
6.5 Future Scope, Possible Applications, and Limitations 109
6.6 Recommendations and Consideration 110
6.7 Conclusions 110
References 110

6.1 INTRODUCTION

6.1.1 BRAIN TUMOR

The human brain is naturally at the apex and a very complicated body part and can function in billions of cells. Brain tumors occur when ill and spurious cells replicate and grow in an uncontrolled manner. This dangerously increasing population of cells is capable of destroying healthy cells and affecting the normal functioning of the brain. Brain tumors are classified into categories, the first one is 'benign or low-grade' and the second one is 'malignant or high-grade.'

Because a benign tumor is non-progressive (noncancerous), it reflects vascular growth more slowly, and false exhibition in the brain. In addition, other parts of the body are not affected by this tumor. But, on the contrary, the malignant tumor is cancerous or destructive and grows rapidly at unknown boundaries.[1,2]

6.1.2 BIG DATA ANALYTICS IN HEALTH INFORMATICS

Live computer programming element is the main difference between big data health and traditional health analyses. When we look toward traditional systems, for analyzing big data, the healthcare industry is very dependent on other industries. Due to the significant impact of information technology, many healthcare professionals rely on it, because the operating systems are functional and capable of processing data in standard formats.[3,4]

Presently, the rapid development of large amounts of healthcare data is becoming a major challenge for the healthcare industry. With the expanding and ingesting field of big data analytics, useful insights into healthcare systems can be provided.

As mentioned above, most of the data generated by this system are stored and printed as there may be a requirement to be digitized.[4,5]

6.1.3 MACHINE LEARNING (ML) IN HEALTHCARE

Data mining uses the concept of scanning two pieces of data that helps in identifying scanning patterns. This concept of data mining is quite similar to machine learning (ML). ML does not use data extraction programs such as data mining programs that are based on human understanding to extract data, but it uses these data in order to enhance the understanding of the program. In addition, ML helps to recognize data patterns and modify the performance of programs accordingly.[6]

6.1.4 SENSORS FOR INTERNET OF THINGS

A brand-new model that facilitates our lives by connecting electronic devices and sensors through internal networks is termed as The Internet of Things (IoT). The IoT uses smart devices and the Internet to provide innovative solutions to a variety of challenges and problems related to various commercial, public, and public and private industries around the world. IoT has become an important aspect of our lives that we can feel around us. In general, the IoT is an innovation that integrates various intelligent systems, frameworks, and smart devices and sensors. In addition, it uses quantum and nanotechnology in terms of unimaginable memory, measurement, and processing speed. This can be seen as a prerequisite for creating an innovative business plan with security, reliability, and collaboration in mind.[7–9]

Here are the nine of the most popular IoT sensors

1. Temperature
2. Moisture
3. Pressure
4. Adjacent
5. Surface
6. Accelerometer
7. Gyroscope
8. Gas
9. Infrared.[7]

Let us look at Some Stats to See the Progress of IoT in Healthcare:
Business Insider forecasts more than 161 million healthcare IoT devices by 2020 (Figure 6.1).

6.1.5 CHALLENGES AND CRITICAL ISSUES OF IoT IN HEALTHCARE

Data security and privacy are the most important challenges of IoT. Smartphones or other smart devices like smart TVs, smart speakers, and toys, wearable are equipped with real-time IoT recording data, but most of them are not compliant with data protocols and standards. There are many ambiguities in data ownership and organization.

Estimated Healthcare IoT Device Installations
Global

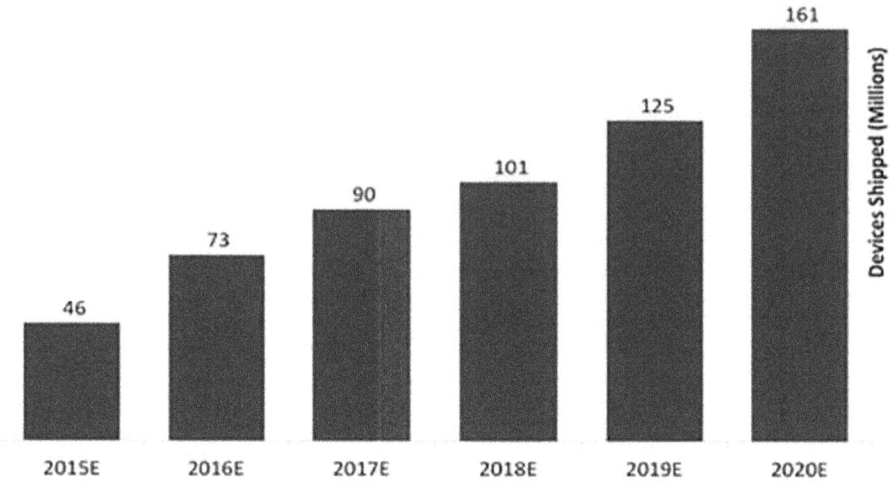

FIGURE 6.1   Estimated healthcare IoT device installation.[10]

As a result, the data stored on IoT devices are at risk of data theft, and this data can be exposed to cybercrime and penetrate the system, compromising personal health information. Fake IDs for health fraud and drug trafficking claims are some of the examples where misuse of IoT data can be seen.[10,11]

- **Integration**: Multiple Devices and Protocols
- Data Overload and Accuracy
- Cost.[10,12]

6.1.6   MACHINE LEARNING (ML) AND ARTIFICIAL INTELLIGENCE (AI) FOR HEALTH INFORMATICS

In the 1970s, artificial intelligence (AI) appeared in health care. The first AI system is essentially a knowledge-based decision support system, and the first ML method is used to approximate the classification rules of a label set. These first systems are working well. However, it is not commonly used in real patients. One of the reasons is that these systems are independent and have nothing to do with the patient's electronic medical records. Another reason is that the proficiency indicated in the knowledge field of these specialist systems expresses the non-acceptance of developed systems and that most systems are medically more academic practice.[13]

   After winning several championships, drug abuse has been transformed into a new mode of learning devices, by improving an intensive artificial neural network (NN), focusing on complex intensive learning. In May 2019, a team from the

New York University and Google reported that the accuracy can be improved by using deep learning models that are used in diagnosing lung cancer, and the study quickly covered magazines' headlines and many newspapers.[13,14]

6.1.7 HEALTH SENSOR DATA MANAGEMENT

Wide impact has been made on people's day-to-day lives by the latest technologies such as wearable sensor devices, cloud computing, and big data. These technologies have great future in the ecosystem based on the Internet provide personal consumption and sharing the information about the development of the health and welfare sector. Many new ways to collect information manually and automatically are provided by these tools. Numerous modern smartphones have multiple internal sensors, such as a microphone, camera, gyroscope, accelerometer, compass, proximity sensor, GPS, and ambient light.[15]

You can easily connect a new generation of new wearable medical sensors to smartphones and send measurement results directly. It will be a more effective and convenient set of personal health information such as blood oxygen saturation, blood pressure, pulse rate, blood sugar, electroencephalogram (EEG), electrocardiogram (ECG), and electrocardiogram (EKG) with all sensors and devices. In future, the process of collecting and interpreting health and activity data is not so far. The number of mobile phone sensor analysis and data collection is expanding rapidly. This graph of massive growth has created the ability to manage both data and challenge collaboration.[15]

6.1.8 MULTIMODAL DATA FUSION FOR HEALTHCARE

Healthcare is one of the areas of continuous improvement because of the proliferation of IoT technologies that are used to support the central functions of healthcare organizations. Thus, traditional hospitals have become the next generation of intelligent digital environments that use widely interconnected sensor systems and large-scale data acquisition and processing technologies. With the help of all of these scenarios, smart health can be supported as a complex ecosystem of smart spaces, such as ambulances, hospital wards, and pharmacies, and backed by strong framework stacks such as sensor networks and edge devices using innovative business models and rules.[16,17]

6.1.9 HETEROGENEOUS DATA FUSION AND CONTEXT-AWARE SYSTEMS—A CONTEXT-AWARE DATA FUSION APPROACH FOR HEALTH-IoT

Advances in low-cost sensor equipment and communications technologies are rapidly accelerating the evolution of smart homes and environments. The health industry is growing rapidly with the development of human body networks, big data technologies, and cloud computing. While using IoT, several challenges can be faced by an individual, such as text recognition, heterogeneous data mixing, reliability, complex query processing, and accuracy.

For more reliable, accurate, and complete results, personal data from sensor source is collected. In addition to the wearable sensor, an additional background sensor has been added to create the background. The IoT Health Program uses the potential benefits of combining knowledge data in this area. Background information can be used to tailor the behavior of the app to specific situations.[18]

6.1.10 ROLE OF TECHNOLOGY IN ADDRESSING THE PROBLEM OF INTEGRATION OF HEALTHCARE SYSTEM

Information and communication technology offers the opportunity to revolutionize health care. Technology-based healthcare coordination and care systems, including Web, mobile, measurement, computing, and bioinformatics technologies, offer great potential to enable a whole new model of health care both at home and abroad. They provide a formal care system and have the opportunity to have a major impact on public health. Increasingly, decision support tools are being built to help people better understand, access, and make decisions about treatment.[19]

The integration of behavioral health care into a care center that largely manages physical health is optimistic for improving care coordination, quality, and impact, but it also creates scenarios that physicians must now overcome. The limits may not feel the expertise, time, or resources that affect the client's behavioral health needs (e.g., substance use and mental health). In this way, technology reduces the quiet, dedicated care of illnesses, and provides countless opportunities for tailors to monitor the behavior and provide intervention to each individual in response to therapeutic behavior those changes over time.[19]

6.2 LITERATURE SURVEY

Various strategies for segmenting, locating, and identifying images of brain tumors have been proposed from the late request to group magnetic resonance imaging (MRI) images.

Parveen and Amritpal Singh proposed a method for data extraction. The grouping method is done in four stages: preprocessing, extraction, highlighting and grouping. The main tissues improve speed and accuracy by improving and eradicating the skull. Fluffy C Fluffy Pan (FCM) uses a portion of the grid in the dimension of the Dim dimension (GLRLM) to extract highlights in brain images. Sorting brain MRI images using the SVM (Support Vector Machine) method gives accurate accuracy and results.[1]

There are more than 120 types of brain tumors of different origin, location, size, and tissue characteristics, as per the reports of the World Health Organization (WHO). Articles on three types of malignancies are as follows:

Glioblastoma: An early malignant brain tumor consisting of astrocytes, called astrocytes, which support nerve cells and are classified as tetraploids. It usually starts in the brain.

Sarcoma: The degree varies from 1° to 4° and occurs in connective tissues such as blood vessels.

Metastatic Bronchial Cancer: A secondary malignant brain tumor that has spread from a lung tumor to a bronchial cancer.[1]

Maoguo Gong[20] and his team, in their article on Fuzzy Clustering Algorithm have provided improved Fuzzy image segmentation by introducing weight-lifting Fuzzy factors and core metrices. The fuzzy weight factor in the exchange depends on the distance in the space of all adjacent pixels and the difference between their gray surfaces simultaneously. The new algorithm uses fast bandwidth selection rules to determine core parameters based on the distribution of distances from all points in the dataset. In addition, both the business-based fuzzy weight factor and the core distance measurement are parameter free. Results of experiments on artificial and real images manifest that the new algorithm is successful, systematic and relatively independent of this type of noise.[20]

Rudie writes the paper; J. D. et al. titled 'emerging applications of AI in neuro-oncology.'[21] They showed that AI methods were developed to enhance the accuracy of medical and therapeutic diagnosis by increasing the computational algorithms. The field of radiology in neuroscience is now and perhaps at the forefront of this revolution. Advanced neuro-MRI data and the types of AI techniques are used in conventional tumors; permeability margins for glioma diffusion determine true progression assumptions and are used in routine clinics. Recurrence and survival are more predictable than what happens.

According to him, the radio genomics also facilitate understanding of cancer biology and enable high-resolution noninvasive sampling of the molecular environment, and providers understand the level of systems of heterogeneous cellular and molecular processes by making spatial and molecular heterogeneous markers in the body; these radiographs are based on AI and classify patients using more accurate diagnostic and treatment methods. It also can better monitor dynamic therapy. Although the basic challenges remain, radiology is changing dramatically given the ever-increasing development and validation of AI technologies for clinical use.

They revealed the purpose of this study that was to improve the outcomes in patients with central nervous system neoplasms through advances in diagnostics and therapeutics. Tools of AI that helps in combine clinical, radiological, and genomic information with predictive models promise critical guidance and guidance on personal care. However, there are many challenges and many things need to be done to keep promises.

Nevertheless, the use of AI technology will dramatically change the progress of radiology, improving the accuracy and efficiency of radiologists. As these powerful tools will be integrated into daily practice in the coming years, radiologists should need to know and use this powerful tools.[21]

According to Amin, Javeria and his team are diagnosing brain tumors, an active area for brain imaging research. This study proposes a methodology for presenting and classifying brain tumors using MRI. In order to segment the tumor, an architecture-based deep neural network (DNN) is used.[3,22]

He has given the proposed model; seven layers, including three concealers; three ReLUs; and a softmax layer was used for classification. The DNN determines the label based on the center pixel and performs the division. They described extensive experiments using eight large standard datasets including BRATS 2012 (image and artificial datasets), visual and artificial datasets, and ISLES (cerebral infarction ischemic stroke).

According to Amin, the results are proved with accuracy (ACC), attribute (SP), sensitivity (SE), accuracy, alopecia similarity coefficient (DSC), false-positive coefficient (FPR), true-positive rate (TPR), and Jaccuard similarity index (JSI).[3]

Gopal S. Tandel and His team member pointed out in a February 2018 WHO report that the death rate from the central nervous system (CNS) or brain cancer is the highest in Asia. Early diagnosis of cancer is critical, as it can save many lives. Cancer grading is a necessary feature of targeted therapy.[22,23]

According to him, the diagnosis of cancer is very aggressive, expensive, and time-consuming. There is an urgent need to describe and evaluate noninvasive, cost-effective, and effective brain cancer tools.

In this article, they tried to briefly describe the pathophysiology of brain cancer, a cancer treatment method, and an automatic computer method for describing brain cancer and deep learning of brain cancer patterns and equipment. It also aims to investigate issues with existing engineering methods and design future models. Besides, they emphasized the association of brain tumors with stroke, brain diseases (such as Alzheimer's, Parkinson's, and Wilson's), leukemia, and neurological diseases in other ML and deep learning paradigms. His main research areas are cancer pathway physiology, imaging technology, WHO's tumor classification guidelines, early detection methods, and existing computer algorithms that use deep equipment to classify brain cancer.

Finally, they contrasted brain tumors with other brain diseases. They made a conclusion that the ability to automatically extract DL (Deep Learning) based methods is once again more common than traditional medical image classification methods. If cancer is evaluated and treated quickly and cost-effectively, many lives can be saved. Therefore, there is a need for a rapid, noninvasive, and cost-effective diagnostic method. The DL method can play an important role here. According to him, the use of DL technology and its full potential to evaluate automatic tumors has been completed, but the work is still very little, but no research has been carried out.[23]

6.3 SYSTEM DESIGN AND METHODOLOGY

6.3.1 System Design

The analysis has been done on the benchmark dataset provided by the Apollo Cancer Hospital, NCR, India, and they have been tested and visualized in the study.

A block diagram that shows the working of our project in a step-by-step manner (as per Figure 6.2).

6.3.2 CNN Architecture

An evolutionary neural system (convolutional neural network (CNN) or ConvNet) is a profound neural system class, the most utilized in a visual examination. CNNs are

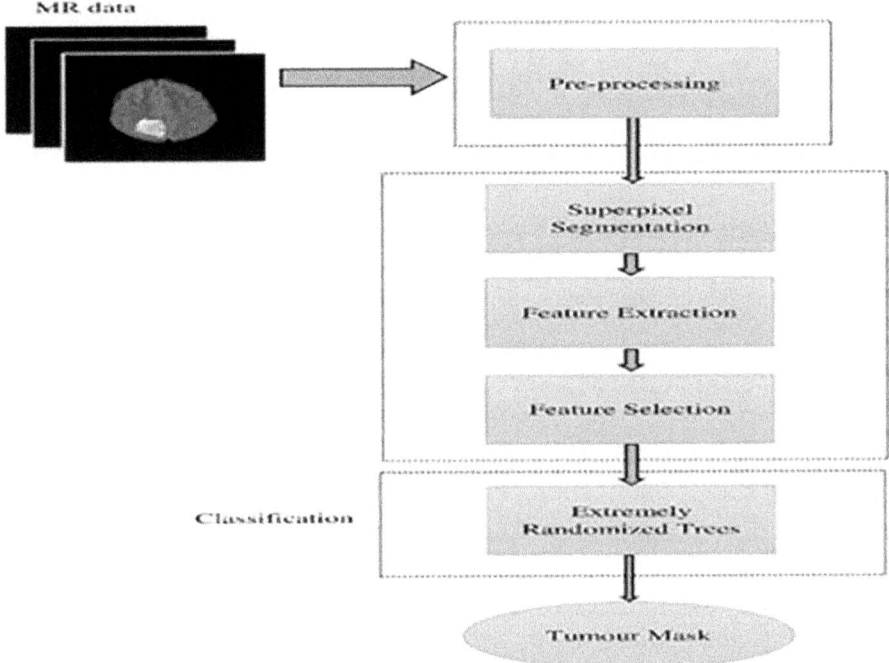

FIGURE 6.2   System design flowchart.

multilayer perceptron regularization renditions. Multilayer perceptron for the most part alludes to completely associated systems, which implies that each neuron in one layer is associated with the following layer everything being equal.

The CNNs, however, embrace another way to deal with regularization: they utilize the various leveled design in the information and utilize littler and increasingly basic examples to collect complex examples. Therefore, CNNs are on the lower extraordinary on the size of the association and multifaceted nature.

Natural procedures motivated progressive systems by the way that the example of the network between neurons resembles the creature's visual cortex association. The open fields of the different neurons mostly cover the whole field of vision.

Contrasted with other picture characterization calculations, CNNs utilize moderately little prehandling. This implies the system discovers that channels are produced by delivering customary calculations. This autonomy from past information and human exertion in the plan of highlights is a significant advantage.

Figure 6.3 is a representation of the CNN architecture and various layers.

6.3.3   BLOCK DIAGRAM

This picture represents the entire functioning of the project and how the CNN architecture is used to classify the type of tumor (as per Figure 6.4).

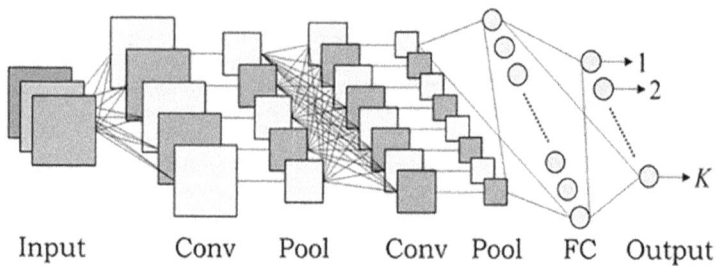

Input   Conv   Pool   Conv   Pool   FC   Output

FIGURE 6.3   CNN architecture.

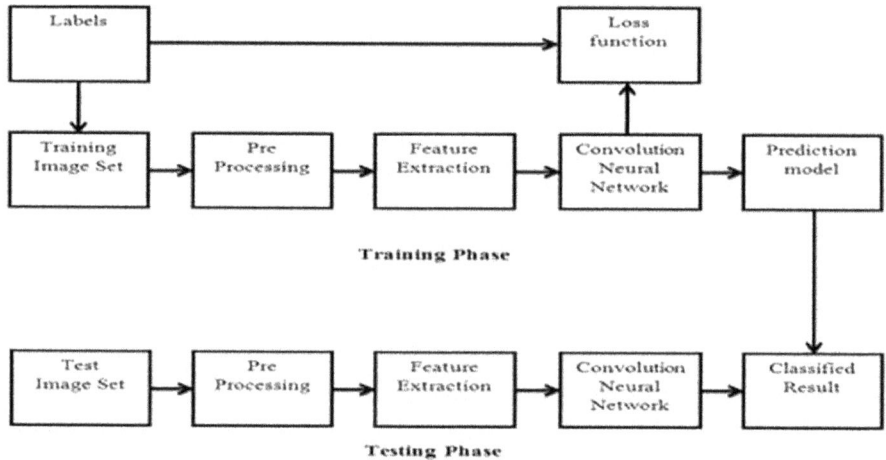

FIGURE 6.4   Block diagram.

6.3.4   ALGORITHM(S)

Gradient descent is a calculation of streamlining used to limit certain functionalities, moving to a lofty plummet toward the path characterized by the slope negative. We use the slope plunge in AI to refresh our model parameters. Parameters allude to straight relapse coefficients and neural system loads. According to our cost capacity, we can govern two parameters: m (weight) and b (inclination). We should consider fractional subsidiaries the impact everyone has on the last expectation. Concerning every parameter, the outcomes are stored in an inclination after we ascertain the halfway subordinates of the cost capacity.

The cost function is given as follows:

$$f(m, b) = \frac{1}{N} \sum_{i=1}^{n} (y_i - (mx_i + b))^2$$

The gradient can be calculated as follows:

$$f'(m, b) = \begin{bmatrix} \dfrac{df}{dm} \\ \dfrac{df}{db} \end{bmatrix} = \begin{cases} \dfrac{1}{N}\sum -2x_i\left(y_i - (mx_i + b)\right) \\ \dfrac{1}{N}\sum -2\left(y_i - (mx_i + b)\right) \end{cases}$$

To solve the gradient, we use our new m and b values to iterate through our data points and to calculate partial derivatives. This latest gradient expresses the slope of our cost function (current parameter values) to our current position and the direction to which we are to update our parameters. The learning rate controls the size of our update.

6.3.5 Our Experimental Results, Interpretation, and Discussion

Experimental Setup
Hardware Requirements
- 4 GB RAM
- Intel Core i3 or higher processor
- GPU (recommended)
- 2 GB available disk space

Software Requirements
- Python 3
- Spyder IDE
- Tensor Flow and Keras
- Scikit-Learn
- NumPy and Pandas
- Linux or Windows OS

Constraints
- Data limitation
- Hardware limitation
- Accuracy limitation

6.3.6 Implementation Details

Deep learning expanded the structure of traditional NNs between input and output layers to add hidden layers to the network architecture, creating a more complex and non-causal relationship. In recent years, the reflection neural network (CNN) is a popular architecture that allows you to carry out complex operations using concealer filters. A set of feedforward layers that implement loop filters and pool layers makes the typical CNN architecture. After the last layer, some fully connected layers are enhanced by CNN. Manipulate one-dimensional vectors for classification work with CNN. Get the results you need in four major steps. Data collection and preparation is the first step. We are going to collect labeled MRI images of tumor brains and prepare them so we can use them for training.

6.3.7 SNAPSHOTS OF INTERFACES

Below is the image showing the various classes of tumors our model can classify (as per Figure 6.5).

Code Snippet Below is the screenshot of the code that we have used to train our CNN model (as per Figure 6.6).

Welcome page Below is the snapshot of the welcome page. This is the first-page user will come across (as per Figure 6.7).

MRI selection Below is the snapshot of the page where the user can select and upload the MRI image (as per Figure 6.8).

```
-------------------------------------------------------------
Data source    | Tumor type          | No.of Images
-------------------------------------------------------------
REMBRANDT      | Astrocytoma         |          21307
               | Glioblastoma        |          17983
               | Oligodendroglioma   |          12460
               | Unidentified        |          13677
-------------------------------------------------------------
MIRIAD         | Healthy brain       |          30688
-------------------------------------------------------------
BRAINS         |                     |            556
-------------------------------------------------------------
Total          |                     |          96115
-------------------------------------------------------------
```

FIGURE 6.5 Various classes of tumors.

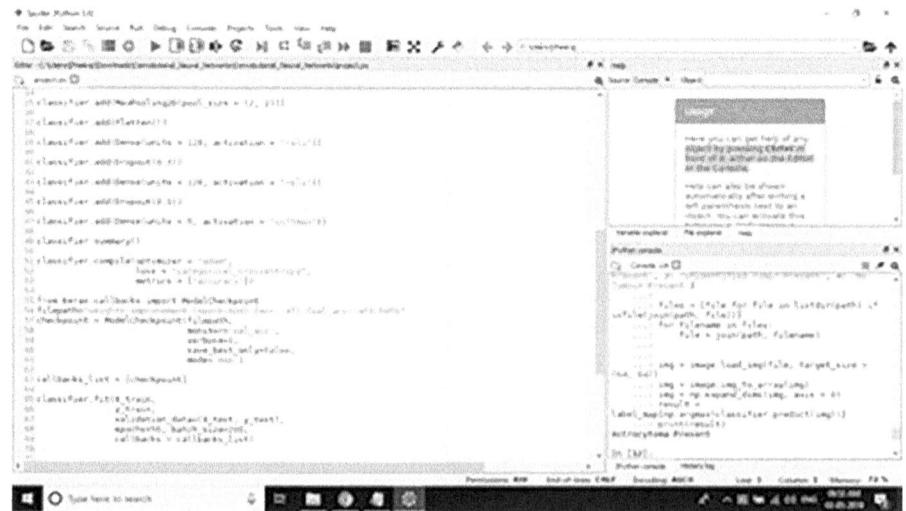

FIGURE 6.6 Trained our CNN model.

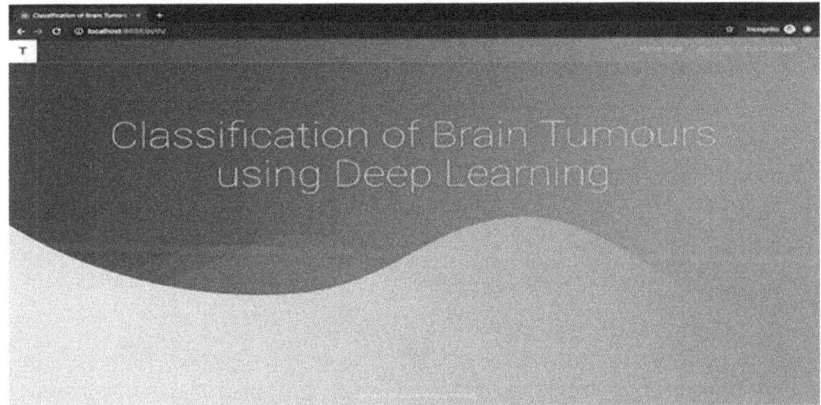

FIGURE 6.7   Loading page.

FIGURE 6.8   Result displayed.

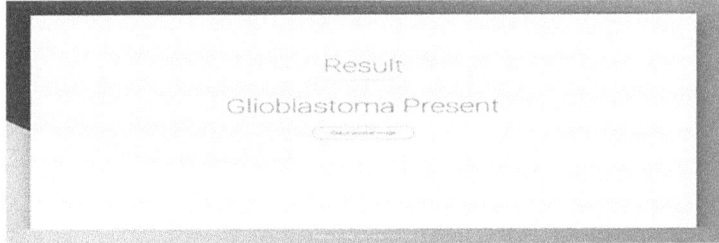

FIGURE 6.9   Result displayed.

The result displayed Below is the snapshot where the prediction result of the model is displayed (as per Figure 6.9).

About Section: The screenshot below is of the page of the about section where information about this project is displayed (as per Figure 6.10).

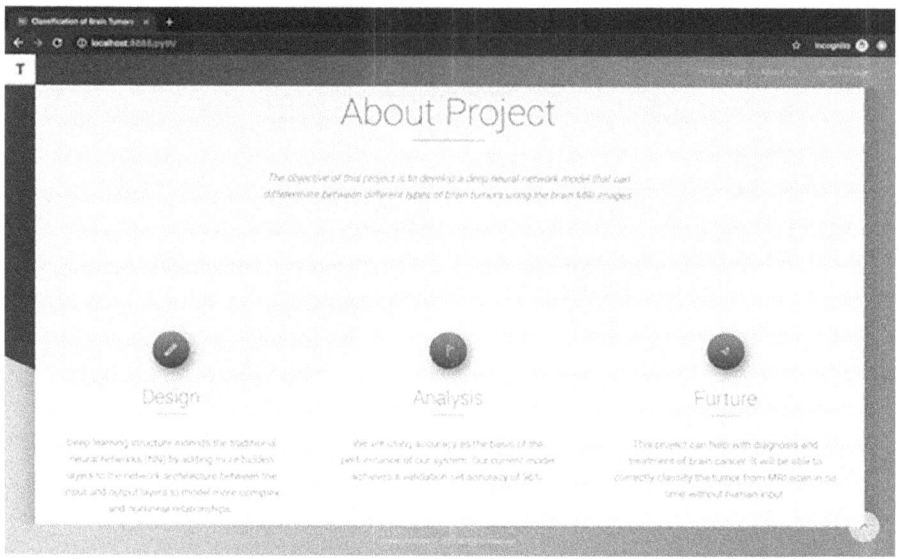

FIGURE 6.10   About section.

6.3.8   PERFORMANCE EVALUATION

We are using accuracy as the basis of the performance of our system. Our current model achieves a validation set accuracy of 96.95% in the training set and 96.03% in the validation set.

The accuracy score of the validation set is displayed (as per Figure 6.11).

```
IPython console                                                    ⊟  :
⌷  Console 1/A ⊠                                              ▣  ⬛  ⦃
train_test_split
    ...: X_train, X_test, y_train, y_test =
train_test_split(allX, allY, test_size = 0.2,
random_state = 0)
    ...:
    ...: from keras.utils import to_categorical
    ...: y_train = to_categorical(y_train)
    ...: y_test = to_categorical(y_test)
    ...:
    ...: classifier.evaluate(X_train, y_train)
    ...: classifier.evaluate(X_test, y_test)
Using TensorFlow backend.
76892/76892 [==============================] - 383s
5ms/step
19223/19223 [==============================] - 98s
5ms/step
Out[1]: [0.1271184872213701, 0.9603079644176247]
```

FIGURE 6.11   Accuracy score.

Algorithm	Classification rate
DNN	96.97%
KNN K = 1	95.45%
KNN K = 3	86.36%
LDA	95.45%
ᵃSMO	93.94%

FIGURE 6.12 Comparison chart.

We also have made a Webpage with options to upload the MRI and get results online. It has an upload button to choose an image from the system to check against the project.

6.3.9 COMPARISON WITH OTHER ALGORITHMS

Comparison Chart: This table shows the performance of our model in comparison with other conventional models (as per Figure 6.12).

6.4 NOVELTY IN OUR WORK

The DNN-based architecture and method will be the best method for most brain tumor classification contests. Many researchers are adding layers to CNNs to improve accuracy. This is a major aspect of scientist knowledge. Researchers and groups that work well with accurate learning methods and algorithms were able to do this using tools outside the grid, such as adding new data or implementing pre-processing techniques.

By adding a step before the normalization process, we improved the generalization capability of the network without changing the CNN architecture. We also tested the gradient descent algorithm for cost overoptimization, but these methods have not yet been implemented in the field of brain tumor classification.

6.5 FUTURE SCOPE, POSSIBLE APPLICATIONS, AND LIMITATIONS

I have too much fat now. This could be improved by expanding the dataset and updating it to identify new types of tumors. Implementing deep learning techniques and algorithms for classifying brain tumor data poses a myriad of challenges.

The lack of large training datasets is a challenging barrier to deep learning techniques. The Health Label Structure Report is expected to facilitate future analysis, especially in the analysis of brain tumors. Especially in the area of brain tumor analysis, the use of non-textual and structural reports for network training is expected to grow rapidly in the future.

Deep learning techniques, which effectively learn from a limited number of classified data, are another major limitation of deep learning algorithms. However, it is not useful if the input field for image data from the entire network is small.

6.6 RECOMMENDATIONS AND CONSIDERATION

Classification implementation in brain tumor analysis is a complex and rewarding task. This can be widely classified, processed, and postprocessed. The above methods have many challenges and complicate the problem. So far, there are no ideal computer tools for adaptation, tumor grade, or aggression.

Therefore, rapid, noninvasive, and cost-effective diagnostic techniques are essential. DL techniques can play a big part here. To understand this, much less work has been done to classify automatic tumors using Max DL techniques and the like. It is possible, but not studied.

6.7 CONCLUSIONS

The focus of data classification is brain cancer physiology, imaging techniques, tumor classification guidelines by WHO, early detection procedures, and existing computer algorithms for classifying brain cancer-using devices. The visualizations are clear to prove that the brain tumor can be more a less accurately detectable by the process defined above; however, accuracy can be increased with more refined dataset and improvised blend of algorithm.

REFERENCES

1. Parveen, Singh, A. (2016). 'Detection of brain tumor in MRI Images, Using Fuzzy C-means segmented images and artificial neural network', *Proceedings of the International Conference on Recent Cognizance in Wireless Communication & Image Processing: ICRCWIP-2014* (pp. 123–131). doi:10.1007/978-81-322-2638-3_14.
2. Mohsen, H., El-Dahshanbc, S.A., El-Horbaty, S.M., Salem, A.B. (2018). 'Classification using deep learning neural networks for brain tumors', *Future Computing and Informatics Journal*, vol. 3, no. 1, pp. 68–71, Received 26 October 2017, Accepted 5 December 2017, Available online 24 December 2017. doi:10.1016/j.fcij.2017.12.001.
3. Javeria, A., Sharif, M., Yasmin, M., Fernandes, S.L. (2018). 'Big data analysis for brain tumor detection: Deep convolutional neural networks', *Future Generation Computer Systems*, 87. doi:10.1016/j.future.2018.04.065.
4. Herland, M., Khoshgoftaar, T.M., Wald, R. (2014). 'A review of data mining using big data in health informatics', *Journal of Big Data*, vol. 1, no. 1, p. 2.
5. Sun, J., Reddy, C.K. (2013). 'Big data analytics for healthcare', in *Proceedings of the 19th ACM SIGKDD International Conference on Knowledge Discovery and Data Mining*, Chicago, IL, 2013, pp. 1525–1525.
6. Mike, C., Hoover, W., Strome, T. and Kanwal, S. (2013). 'Transforming health care through big data strategies for leveraging big data in the health care industry'.
7. Wang, H., Choi, H.-S., Agoulmine, N., Deen, M.J., Hong, J.W.-K. 'Information-based sensor tasking wireless body area networks in U-health systems', in *Proceedings of the 2010 International Conference on Network and Service Management*, Niagara Falls, ON, Canada, 25–29 October 2010; pp. 517–522.

8. Zanella, A., Bui, N., Castellani, A., Vangelista, L., Zorzi, M. (2014). "Internet of Things for smart cities', *IEEE Internet Things J*, vol. 1, no. 1, pp. 22–32.

9. Bauer, H., Patel, M., Veira, J. (2016). *The Internet of Things: Sizing Up the Opportunity.* McKinsey & Company, New York, NY. Available from: https://www.mckinsey.com/ industries/semiconductors/our-insights/the-internet-of-things-sizing-up-the-opportunity.

10. Amyra. (2019). 'Internet of Things in healthcare: Challenges and applications', Technology and Apps. https://www.valuecoders.com/blog/technology-and-apps/ iot-in-healthcare-benefits-challenges-and-applications/.

11. Kumar, S., Maninder, S. (2019). 'Big data analytics for the healthcare industry: impact, applications, and tools', *Big Data Mining, and Analytics*, vol. 2, pp. 48–57. doi:10.26599/ BDMA.2018.9020031.

12. Devendran, T., Archana, A.A., Suseela, S. (2018). 'Challenges and issues of healthcare in the Internet of Things', *International Journal of Latest Trends in Engineering and Technology*, pp. 86–91. Special Issue April 2018, e-ISSN: 2278-621X.

13. Singh, Y., Chauhan, A.S. (2005). 'Neural networks in data mining', *Journal of Theoretical and Applied Information Technology*, vol. 5, no. 6, pp. 37–42.

14. Dey, A. (2016). 'Machine learning algorithms: A review', *International Journal of Computer Science and Information Technologies*, vol. 7, no. 3, pp. 1174–1179.

15. Li, Y., Wu, C., Guo, L., Lee, C.-H., Guo, Y. (2014). 'Wiki-Health: A Big Data platform for health sensor data management'. doi:10.4018/978-1-4666-6118-9.ch004.

16. Haghighat, M., Abdel-Mottaleb, M., Alhalabi, W. (2016). 'Discriminant correlation analysis: real-time feature level fusion for multimodal biometric recognition', *IEEE Transactions on Information Forensics and Security*, vol. 11, no. 9, pp. 1984–1996.

17. Rastogi, R., Chaturvedi, D.K., Satya, S., Arora, N., Trivedi, P., Singh, A., Sharma, A., Singh, A. (2020). 'Intelligent personality analysis on indicators in IoT-MMBD enabled environment', S. Tanwar, S. Tyagi, N. Kumar (Eds), *Multimedia Big Data Computing for IoT Applications: Concepts, Paradigms, and Solutions*, Springer Nature, Singapore, pp. 185–215. doi:10.1007/978-981-13-8759-3_7.

18. Baloch, Z., Shaikh, F., Unar, M. (2018). 'A context-aware data fusion approach for health-IoT', *International Journal of Information Technology*, vol. 10. doi:10.1007/ s41870-018-0116-1.

19. Lisa, A., Gustafson, D.H. (2013). 'The role of technology in health care innovation: A commentary', *Journal of Dual Diagnosis*, vol. 9, no. 1, pp. 101–103. Published online 2012 November 27. doi:10.1080/15504263.20.

20. Gong, M., Liang, Y., Shi, J., Ma, W., Ma, J. (2013). 'Fuzzy C-means clustering with local information and kernel metric for image segmentation,' *IEEE Transactions on Image Processing*, vol. 22, no. 2, pp. 573–584.

21. Rudie, J.D., Rauschecker, A.M., Bryan, R.N., Davatzikos, C., Moha, S. (2019). 'Emerging applications of artificial intelligence in neuro-oncology', *Radiology*, vol. 290, no. 3, pp. 607–618. doi:10.1148/radiol.2018181928.

22. Rastogi, R., Chaturvedi, D.K., Satya, S., Arora, N., Trivedi, P., Gupta, M., Singhal, P., Gulati, M. (2020) *MM Big Data Applications: Statistical Resultant Analysis of Psychosomatic Survey on Various Human Personality Indicators ICCI 2018 Paper as Book Chapter*, Chapter 25, © Springer Nature Singapore Pte Ltd., Singapore, *Proceedings of Second International Conference on Computational Intelligence 2018*, ISBN: 978-981-13-8221-5, Online ISBN 978-981-13-8222-2, Will Publish in Early 2020. doi:10.1007/978-981-13-8222-2_25.

23. Tandel, G.S., Biswas, M., Kakde, O.G., Tiwari, A., Suri, H.S., Turk, M., Laird, J.R., Asare, C.K., Ankrah, A.A., Khanna, N.N., Madhusudhan, B.K., Saba, L., Suri, J.S. (2019). 'A review on a deep learning perspective in brain cancer classification', *Cancers*, vol. 11, no. 1, p. 111. doi:10.3390/cancers11010111.

7 Deep Learning Approach for Traffic Sign Recognition on Embedded Systems

A. Shivankit and Gurminder Kaur
BM Institute of Engineering & Technology

Sapna Juneja
IMS Engineering College

Abhinav Juneja
KIET Group of Institutions

CONTENTS

7.1 Introduction 114
7.2 Literature Review 114
7.3 General Challenges 114
7.4 Proposed Solution 115
    7.4.1 Hardware 116
7.5 Models 117
    7.5.1 YOLOV3 117
    7.5.2 Tiny-YOLOV3 119
    7.5.3 Darknet Reference Model 120
7.6 Flowcharts 123
7.7 Key Features of the System 125
7.8 Technology Stack 125
7.9 Dataset 125
    7.9.1 Labeling/Annotating the Dataset 130
7.10 Training the Model 131
7.11 Result 133
7.12 Future Scope 133
References 133

DOI: 10.1201/9781003138020-7

113

7.1   INTRODUCTION

India holds the worst road safety record in the world with 17 fatalities every hour happening on the Indian roads. The driver may fail to notice the signboard depending on the state of their mind. Ignorance of important signs like "STOP," "ONE WAY," and "ACCIDENT-PRONE AREA" can be the cause of major accidents. Due to the climatic conditions or angles viewed by the driver, it often happens that these signboards are missed or seen too late for the driver to make a decision and act on it. Some inexperienced drivers lack the knowledge about these signboards and the system provides an aid for them to maintain the correct speed. An extra eye in such cases that can detect such signs and alert the driver can be a lifesaver saving a lot of precious lives, thus improving India's road safety record. Such systems reduce drivers' burdens to make decisions based on their sight of the signboard and promote safe driving.

7.2   LITERATURE REVIEW

Earlier methods that have been proposed by Jian-He Shi[1] uses the concept of feature vectors made from histogram of oriented gradients, and second-stage detection is done by support vector machines (SVMs). Further, bilateral Chinese transform, vertex, and bisector transform are used to extract the exact area of traffic sign from images. Lastly, a neural network is adopted to identify the traffic sign information.

Huda Noor Dean[2] proposed a method that primarily focuses on segmenting the signboard from the background based on certain color and shape properties as color and shape can easily distinguish a sign from its background. The multilayer perceptron classifier module determines the type of detected road signs.

Other methods proposed include converting the image to grayscale and applying transformations on it, then using edge detection to get the region of interest (ROI) and then processing this ROI with SVM, neural networks, K Nearest Neighbors (KNN), etc.

The previously proposed methods include multiple steps for extracting the sign before passing the ROI (in these case signboards) through the neural network for final detection.[3] All these algorithms are able to achieve high accuracy, but they come at a great computational cost. Since the Raspberry Pi does not have much processing power, an algorithm is needed that can give acceptable accuracies but at the same time, it must be able to perform well on low-powered CPUs. For this task, the highest accuracy was not required as it does not matter if the signboard detected was recognized with 90% accuracy or 80% accuracy. Hence, a trade-off between accuracy and computational time had to be made while keeping the threshold accuracy of 60%.

7.3   GENERAL CHALLENGES

- In general, the same kind of traffic signs have similar consistency in color, in outdoor/real-life environments, the color of the traffic signs as perceived by a camera is greatly influenced by illumination and light direction. Therefore, the color information cannot be relied upon.[4]

- Vehicle-mounted cameras are not always perpendicular to the traffic signs; therefore, the shape of traffic sign is often distorted in road scenes, and the moving car can further destabilize the image as well; therefore, the shape information of traffic signs is no longer fully reliable.
- An issue identified during image collection was that traffic signs in some road scenes are often covered by buildings, trees, and other vehicles, hence, considering shape as a distinguishing factor can cause issues.
- Traffic sign damage, rain, snow, fog, traffic sign discoloration, etc., can also lead to unexpected results.

7.4 PROPOSED SOLUTION

Since the system is using low-powered Advanced RISC Machines Central Processing Units (ARM CPUs) for its calculations and object detection, an algorithm was needed that was capable of predicting on destabilized images (at times) as well as good-quality images with a decent accuracy but with a small inference time. To reduce this computational cost but not compromising on accuracy, more efficient deep learning neural network is required.

The language of choice is Python because of its easy to use syntax, compatibility across different platforms and since a lot of machine learning (ML) algorithms are well written in Python.

Therefore, the system uses YOLO (You Only Look Once). YOLO has been designed for detection of the objects in hard real-time environment and is based upon convolutional neural network (CNN). In this algorithm, a single full image has been taken, and then, a neural network has been applied to the entire image. The results from the neural network are in the form of bounding boxes and confidence scores. Then based on the confidence score of a bounding box, we either classify that the image fed to the camera has a road sign or not.

The main features of YOLO are that its accuracy is very high, it is quick to predict, and the most important is that it can work in real-time environment.[5,17,20] The algorithm "only looks once" at the image defines that this needs only single forward propagation pass through the neural network for making predictions. Following nonmax crackdown, it then provides the outputs in the form of identified objects bounded with the bounding boxes. With YOLO, a single CNN forecasts various bounding boxes and class probabilities for these bounding boxes in parallel. YOLO trains on complete images and instantly enhances the performance of detection[6,18,19]. This model has numerous features over other previously available methods used for the detection of objects (Figure 7.1).

These features are as follows:

1. **Speed**: The speed of YOLO is very good and it is fast as compared to other models.
2. This model sees the whole image at the time of training and testing so it completely encodes contextual data about the classes as well as their emergence.[7,23,24]

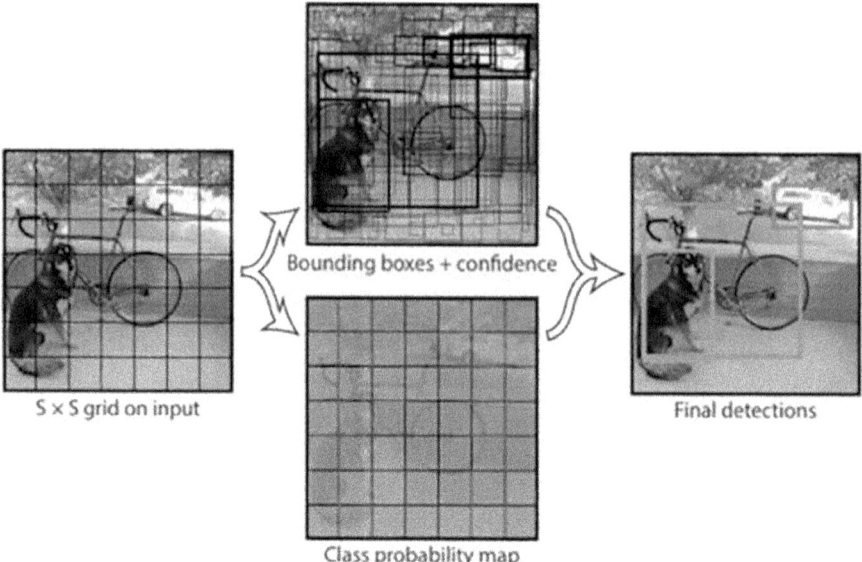

FIGURE 7.1 The complete process cycle of YOLO.

3. YOLO acquires knowledge about the observed representations of objects, as a result when trained on general images and then tested upon the sign-boards present on the roads, the algorithm surpasses other available detection models.[8]

YOLO model was originally introduced by Joseph Redmon and a team of three other people in 2016.[9] YOLO has been regularly improved to perform better than its previous versions and it has reached its fourth iteration.

The version of YOLO we will be using is tiny YOLOV3 (Figure 7.2).

YOLOV3-608 which is the full YOLO model got a mean average precision score of 57.9 with one of the lowest inference times which proves the effectiveness of YOLOV3.

7.4.1 Hardware

A Raspberry Pi 4b is a single-board computer, and it is the choice for this system for the below reasons:

1. **Great Specs**: The Raspberry Pi 4b has aquadcore Cortex-A72 (ARM v8) 64-bit SoC @ 1.5GHz CPU Broadcom BCM2711.
2. **Low Cost**: The Raspberry Pi 4b 2gb RAM variant costs only 3000 INR
3. **Easy Usage**: Using a Raspberry Pi is extremely simple. All you need to do is upload an operating system to a SDcard, attach a mouse and keyboard, and monitor or use Secure Shell (SSH) to access.

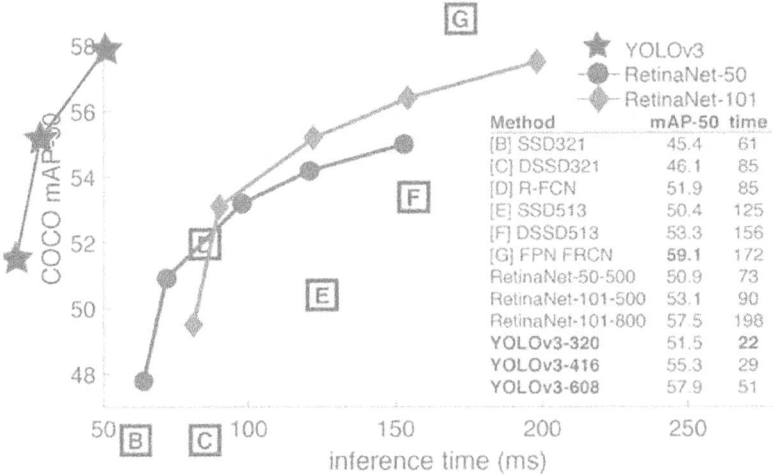

FIGURE 7.2 Comparison of similar computer vision architectures.

4. **Compatibility and Low Power Consumption**: Attaching a camera to the Raspberry Pi is very simple using the ribbon cable and the Raspberry Pi consumes very less power because of which it can be used as a standalone system with its own independent camera and power source.
5. **CSI Port for Camera Connection**: Raspberry Pi provides CSI connectors for easy connections with Raspberry Pi camera.

Other available options are the NVIDIA JETSON series but that cost 10,000 + INR, which would drive up the final cost by a huge margin.

Although the NVIDIA JETSON would give us better performance in terms of accuracy and inference times, the choice had to be made because of its high price.

7.5 MODELS

7.5.1 YOLOV3

YOLO itself is considered a very fast neural network, but to improve this even further the creators of YOLO introduced another architecture that is the tiny-YOLO.

As the name suggests, tiny YOLO is a smaller version of YOLO.

Darknet is a neural network framework written in CUDA and C. It is very easy to install, and supports CPU and GPU computation. Darknet is extremely fast since it is written in C and CUDA.[10,25,26]

Below is the architecture of YOLOV3 (Figure 7.3):

YOLO is a fully convolutional network and its eventual output is generated by applying a 1 × 1 kernel on a feature map.

The architecture of YOLO is basically that of darknet 53 but for detection purposes, another 53 layers have been stacked onto it to make a resulting 106-layer architecture for the complete YOLOV3 model which requires 140.69 bn FLOPS.[11]

```
78 conv    1024  3 x 3 / 1    13 x  13 x 512    ->    13 x  13 x1024
79 conv     512  1 x 1 / 1    13 x  13 x1024    ->    13 x  13 x 512
80 conv    1024  3 x 3 / 1    13 x  13 x 512    ->    13 x  13 x1024
81 conv      18  1 x 1 / 1    13 x  13 x1024    ->    13 x  13 x  18
82 detection
83 route   79
84 conv     256  1 x 1 / 1    13 x  13 x 512    ->    13 x  13 x 256
85 upsample            2x     13 x  13 x 256    ->    26 x  26 x 256
86 route   85 61
87 conv     256  1 x 1 / 1    26 x  26 x 768    ->    26 x  26 x 256
88 conv     512  3 x 3 / 1    26 x  26 x 256    ->    26 x  26 x 512
89 conv     256  1 x 1 / 1    26 x  26 x 512    ->    26 x  26 x 256
90 conv     512  3 x 3 / 1    26 x  26 x 256    ->    26 x  26 x 512
91 conv     256  1 x 1 / 1    26 x  26 x 512    ->    26 x  26 x 256
92 conv     512  3 x 3 / 1    26 x  26 x 256    ->    26 x  26 x 512
93 conv      18  1 x 1 / 1    26 x  26 x 512    ->    26 x  26 x  18
94 detection
95 route   91
96 conv     128  1 x 1 / 1    26 x  26 x 256    ->    26 x  26 x 128
97 upsample            2x     26 x  26 x 128    ->    52 x  52 x 128


 98 route   97 36
 99 conv     128  1 x 1 / 1    52 x  52 x 384    ->    52 x  52 x 128
100 conv     256  3 x 3 / 1    52 x  52 x 128    ->    52 x  52 x 256
101 conv     128  1 x 1 / 1    52 x  52 x 256    ->    52 x  52 x 128
102 conv     256  3 x 3 / 1    52 x  52 x 128    ->    52 x  52 x 256
103 conv     128  1 x 1 / 1    52 x  52 x 256    ->    52 x  52 x 128
104 conv     256  3 x 3 / 1    52 x  52 x 128    ->    52 x  52 x 256
105 conv      18  1 x 1 / 1    52 x  52 x 256    ->    52 x  52 x  18
106 detection
```

FIGURE 7.3   Complete YOLO model architecture.

The complete YOLOV3 model gives only 20–45 fps on a high-end Titan-X GPU. Detection using YOLOV3 on low-powered ARM CPUs is not really practical as the inference time would be huge if at all the detection is made without the system crashing completely.

Since YOLOV3 cannot be used on low-powered ARM CPUs, the tiny version of YOLO has been used for this system.

The most salient feature of YOLOV3 is that it makes detections at three different scales. YOLO being a fully convolutional network and its output is generated by applying a 1×1 kernel on a feature map. In YOLOV3, the detection finally is done by applying 1×1 detection kernels on feature maps of three different sizes at three different places in the network.[12]

YOLOV3 was a huge improvement over its previous version because of this architecture, it even improved detection of smaller objects. But it came at the cost of more computations because of a more complex architecture. Even though the base framework has been written in C and CUDA, the number of calculations was an issue and it cannot be implemented on a Raspberry Pi since predictions either take too long or just crash completely.

Although the complete YOLOV3 model is much more accurate than the tiny version, the trade-off had to be made to create a practical and viable solution for the problem at hand.[13]

7.5.2   Tɪɴʏ-YOLOV3

The tiny YOLOV3 architecture is given below (Figure 7.4):

Tiny-YOLOV3 has visible lower number of layers, i.e., 23 out of which only 7 are convolutional layers, which results in only 5.56 bn FLOPS that is a huge difference from the complete model.

The difference here is that YOLO used darknet53 architecture and then has multiple 1X1 and 3X3 convolutional layers stacked on top of it, whereas the tiny version has only seven convolutional layers.

This difference allows us to implement this model on an ARM CPU and achieve practical results that can be useful to us.

As mentioned earlier, it is critical to the system that the inference times are low for the predictions to be of any use.

8 conv	256	3 x 3/ 1	26 x 26 x 128 -> 26 x 26 x 256
9 max		2x 2/ 2	26 x 26 x 256 -> 13 x 13 x 256
10 conv	512	3 x 3/ 1	13 x 13 x 256 -> 13 x 13 x 512
11 max		2x 2/ 1	13 x 13 x 512 -> 13 x 13 x 512
12 conv	1024	3 x 3/ 1	13 x 13 x 512 -> 13 x 13 x1024
13 conv	256	1 x 1/ 1	13 x 13 x1024 -> 13 x 13 x 256
14 conv	512	3 x 3/ 1	13 x 13 x 256 -> 13 x 13 x 512
15 conv	21	1 x 1/ 1	13 x 13 x 512 -> 13 x 13 x 21
16 yolo			
[yolo] params: iou loss: mse (2), iou_norm: 0.75, cls_norm: 1.00, scale_x_y: 1.00			
17 route_13			-> 13 x 13 x 256
18 conv	128	1 x 1/ 1	13 x 13 x 256 -> 13 x 13 x 128
19 upsample		2x	13 x 13 x 128 -> 26 x 26 x 128
20 route_19 8			-> 26 x 26 x 384
21 conv	256	3 x 3/ 1	26 x 26 x 384 -> 26 x 26 x 256
22 conv	21	1 x 1/ 1	26 x 26 x 256 -> 26 x 26 x 21
23 yolo			

FIGURE 7.4   Tiny YOLO model architecture.

Even after we use tiny-YOLOV3, the Raspberry Pi 4b is just not powerful enough to give us the required inference times.

It must be noted that the creators of YOLO recommend the usage of a GPU for better results, but the Raspberry Pi does not have a very powerful GPU on board, but the Raspberry Pi 4b does have a quad core CPU.

Therefore, we took advantage of the quad core CPU on the Raspberry Pi to reduce the inference time to less than 2 seconds by using Neural Network Pack (NNPACK).

NNPACK is an acceleration package created by Marat Dhukan that can optimize existing neural networks to work better with multi core CPUs.

NNPACK is alone not intended to be used by ML researchers, as it only provides a low-level performance primitives leveraged from leading deep learning frameworks, such as darknet, Torch, Caffe, PyTorch, Caffe2, MXNet, tiny-dnn.

What this means is that NNPACK alone cannot perform tasks like classification, prediction, detection, but it is a tool that can work along with darknet to optimize the network for better inference times on low-powered CPUs.[14]

Without NNPACK, tiny-YOLOV3 on the Raspberry Pi4b takes 14 seconds for single image prediction, but with NNPACK, we get the prediction in under 2 seconds.

Although YOLO is heavily dependent on a GPU, it can perform the required calculations faster than a CPU, but along with NNPACK, we can shift these calculations to the 4 cores of the ARM CPU on the Raspberry Pi.

All this sounds great, but YOLO has its limitations. The biggest problem with YOLO is that it fails to accurately predict on smaller objects of interest.

The main reason behind this is that YOLO and even other single shot detectors (SSDs) use a feature map of very low resolution as this allows the model to perform faster. A model with a feature map that has a higher resolution takes more time. This also translates to a model that can learn to detect smaller objects in a larger background.[15] In case of YOLO, in the feature map the smaller object features become too small to be able to contribute to the prediction, because of which the model struggles to detect them accurately.

Consider a speed limit sign (Figure 7.5):

In this sign, the ROI occupies a very small percentage of the image. Although tiny-YOLO will be able to successfully detect and predict that there is a speed limit sign in this image, it will struggle in accurately detecting the value of speed limit on the board.[16]

As explained earlier, when the feature map is drawn across the image, the values will get completely lost because of the low resolution of the feature map.

Accuracy cannot be compromised in such situations as it could lead to give incorrect information to the driver and then further noting incorrect violations. For this purpose, we propose a two-stage detection.

The second stage of this detector also uses a neural network that is much smaller than tiny-YOLOV3. The model we decided to use was the darknet reference model.

7.5.3 DARKNET REFERENCE MODEL

This model is even smaller and faster when compared to tiny-YOLOV3, although it must be noted that darknet reference model is only a classifier, whereas tiny-YOLOV3 is a detector and a classifier.

FIGURE 7.5   Sample image in dataset.

The architecture for darknet reference model is given below (Figure 7.6):

The darknet reference model has a visibly smaller architecture in terms of number of layers, but at the same time, this model does not have the fully connected layers at the end of the network that reduces the inference times exponentially.

Although drawing a comparison to tiny-YOLOV3 is unfair since they perform different tasks, but still to put into perspective how fast this model, tiny-YOLOV3 had 5.56bn FLOPS, whereas darknet reference model has only 0.81bn FLOPS.

Since we already have a huge network at the beginning of our process, we needed something that would be able to perform very well but at the same time be very quick so that the system still effective.

Although we could have trained another tiny-YOLO model, but it would be a waste of computation since the extracted region in the second stage will be the speed limit sign itself and feeding this to a model that does detection as well as classification was completely unnecessary (Figure 7.7).

This classification task was relatively easier when compared to the first-stage detection, which is the reason that we chose to use only a classifier without any detector.

Other probable methods included using edge detection or histogram of oriented gradients along with an SVM classifier (or some other classifiers).

However, it must be noted that before such methods could be employed for this task, we would have had to apply multiple filters like Gaussian, thresholding, and edge detection and these methods involve applying kernel filters to the image which can be time-consuming, on real-time problems where an image quality can get degraded due to multiple factors like speed blur, lighting conditions such filters will

layer	filters	size/strd(dil)	input	output
0 conv	16	3 x 3/ 1	64 x 64 x 3 ->	64 x 64 x 16
1 max		2x 2/ 2	64 x 64 x 16 ->	32 x 32 x 16
2 conv	32	3 x 3/ 1	32 x 32 x 16 ->	32 x 32 x 32
3 max		2x 2/ 2	32 x 32 x 32 ->	16 x 16 x 32
4 conv	64	3 x 3/ 1	16 x 16 x 32 ->	16 x 16 x 64
5 max		2x 2/ 2	16 x 16 x 64 ->	8 x 8 x 64
6 conv	128	3 x 3/ 1	8 x 8 x 64 ->	8 x 8 x 128
7 max		2x 2/ 2	8 x 8 x 128 ->	4 x 4 x 128
8 conv	256	3 x 3/ 1	4 x 4 x 128 ->	4 x 4 x 256
9 max		2x 2/ 2	4 x 4 x 256 ->	2 x 2 x 256
10 conv	512	3 x 3/ 1	2 x 2 x 256 ->	2 x 2 x 512
11 max		2x 2/ 2	2 x 2 x 512 ->	1 x 1 x 512
12 conv	1024	3 x 3/ 1	1 x 1 x 512 ->	1 x 1 x1024
13 conv	3	1 x 1/ 1	1 x 1 x1024 ->	1 x 1 x 3

14 avg			1 x 1 x 3 ->	3
15 softmax				3
16 cost				3

FIGURE 7.6   Darknet reference model architecture.

FIGURE 7.7   Second-stage samples.

need to have kernel filters that can adapt and change accordingly which is not a very easy task for such problems, all this combined with the fact that such methods are never able to reach the accuracy of neural networks, we went ahead with the darknet reference model.

A key feature of this system was that it had to be efficient which meant that a lot of work had to be put into optimization. To improve the classification by this model further, we could have added another step before starting the second-stage detection which would have been detecting the circular region of the signboard which actually has the values.

Although as pointed out earlier before such detections can be made on real-time images, multiple filters need to be applied which would further increase the time that one image takes in the system, which is why the model directly takes the extracted ROI from the first-stage detector as its input.

Since the backend of second-stage detection is a neural network, it was able to generalize to different scenarios (lighting, etc.) and give very good accuracy.

Therefore, the final system uses tiny-YOLOV3 with NNPACK + darknet reference model (in case of speed limit signboards) on a Raspberry Pi 4b.

7.6 FLOWCHARTS

A flowchart explaining the internal working of the software is given below (Figure 7.8).

Tiny YOLOV3 is fed a natural image that predicts a class of signboards (in case any is present). If, however, the signboards predicted is a speed limit signboard, then it will be passed on to the second stage of detection which is a darknet reference model.

This entire process is happening in under 2.5 seconds of clicking the image.

The darknet reference model is only taking 100 milliseconds (maximum) to classify the signboard, whereas tiny YOLO is taking around 2–2.5 seconds.

Working of the complete system along with hardware (Figure 7.9):

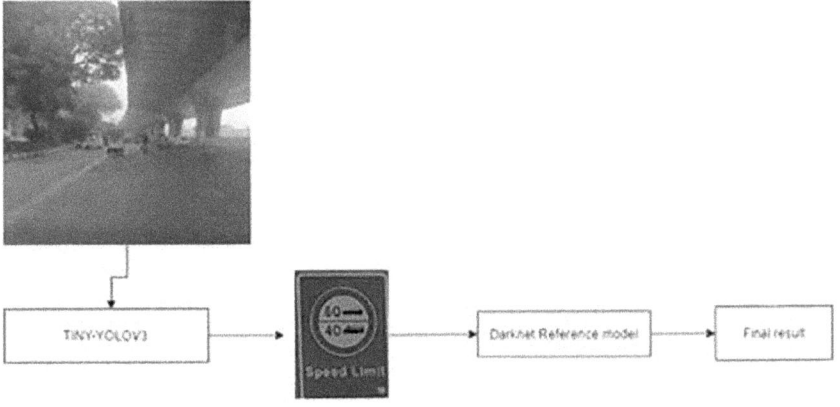

FIGURE 7.8   Flowchart for second-stage detection.

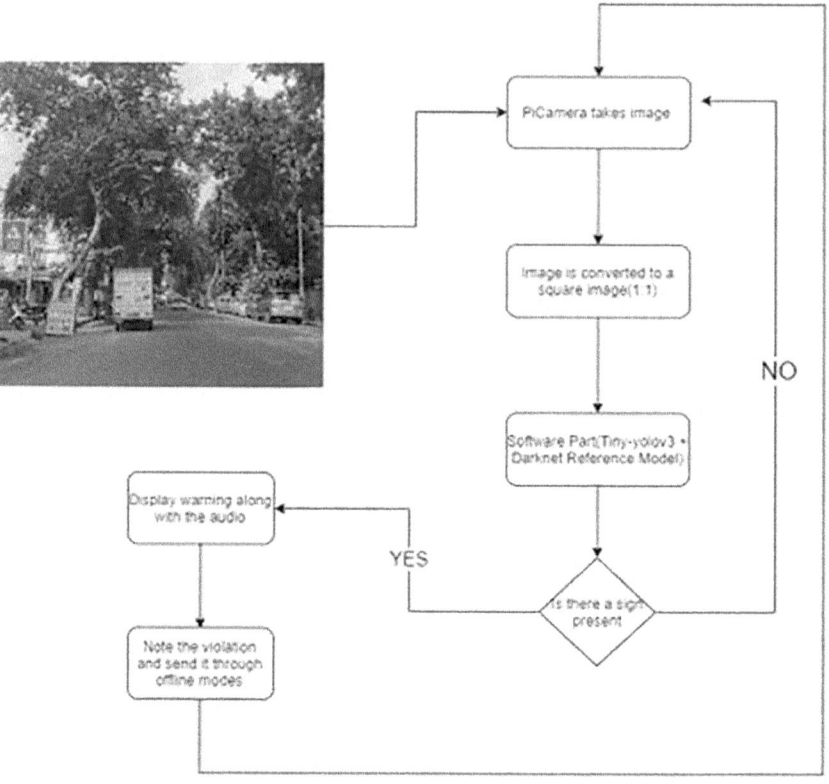

FIGURE 7.9   Flowchart for complete system.

1. Picamera, i.e., the chosen camera for the system taken an image of the road ahead.
2. Since YOLO requires the image to be square (equal height and width), we convert the image to 1:1 ratio by cropping the top and little bit of bottom from the image. Although YOLO can be given images of different sizes as it does resize the image on its own, this could cause a change in aspect ratio which will in turn mean a change in information given by the image. Hence, this is a crucial step.
3. The software part has been explained in the previous flowchart.
4. If the sign is not present in the sign, then the image will be taken again and this entire process will be repeated.
5. If the sign is present, the warning will be displayed along with an audio aid.
6. In case the speed is more than the prescribed speed by Indian Roads Congress (IRC), the violation will be noted and passed on the central servers through offline methods.

The images taken by the camera have a lot of information that is not really required, like the sky. Before training and even while testing, a portion of the image is cropped from the top and a little from the bottom as well as the signboard.

7.7 KEY FEATURES OF THE SYSTEM

- **The System Works Offline**: Since the backend software is a neural network, it does not require the use of any application programming interface (API) for its predictions which means that the system does not require an active Internet connection at any point of its functioning.
- **Budget Friendly**: The entire system costs under 7000 rupees.
- **Prediction Time**: The optimized version of YOLO is able to classify a sign from a distance and in under 2.5 seconds which gives the driver some crucial extra seconds to react to the warning.
- **Low Power Consumption**: Even on maximum load, the Raspberry Pi requires only 5.1 W of electricity that can be easily provided by the car batteries. The power supply needs to be only a 5V/2A power supply.
- **No Maintenance Cost/Subscription Fees**: Since the system is offline and does not rely on any APIs, it is just a one-time investment.
- **Effectiveness**: Since it is able to predict signs from a distance, even fast-moving vehicles will get the prompts in time which makes the system very effective in real life.

7.8 TECHNOLOGY STACK

The programming language used by the system will be Python because of its ease of use and compatibility with ML algorithms.

OpenCV: For all internal operations on the image like image cropping, grayscaling.

Numpy: For numerical calculations on arrays.

Pygame: For giving audio aids to the driver in case a sign is encountered.

Tkinter: Few graphical user interfaces (GUIs) had to be made for dataset cleaning.

Keras/Tensorflow: For implementing state-of-the-art neural network/SSDs to achieve quick and accurate results.

Darknet/Darkflow: Framework written in C++ that has YOLO implementation.

Hardware Components: Raspberry Pi camera and Raspberry Pi development board for capturing and processing of images.

SkLearn: Has the implementations for SVM and other basic algorithms like scaling.

7.9 DATASET

Training a model like YOLO requires a huge dataset for achieving good accuracy.

A total of six signs are predicted by the system:

1. School ahead
2. Pedestrian crossing
3. Merging traffic ahead
4. Accident-prone area
5. Speed breaker ahead
6. Speed Limit Signboards.

Signs like accident-prone area showed a lot of variability, for example, in some cases, the sign was of two vehicles in a head on collision; in some cases, it was an exclamation mark, or it was simple written "ACCIDENT-PRONE AREA."

Such variability can be handled by bigger datasets, i.e., adding lots and lots of images of each of these types under the same class.

Other classes showed very less variability which will easily by handled by our model.

While creating this dataset, we came across the following problems:

- **Conditions of the Signboards**: These signboards are often covered in dust, at some places people have stuck on advertisements to the signboard.
- **Maintenance**: A corollary to the previous point, the signboards are neglected after their installation, and hence, vegetation grows around them which hides them for a vehicle at a distance.
- **Placement**: In some cases, the warning signs like "Bump Ahead" was found right next to the bump or very close to it, which gives no time for the driver to react to the sign. A lot of time it was noticed that during driving these signboards were not really in the line of sight for a driver.
- **Size**: Warning signs must be visible from a decent distance so that the driver is able to act on it, but the signboards in NCR are of a smaller form factor. Also if we consider the speed limit signboards, it shows two speed limits for a car and for a truck, so the actual values are actually quite small. Maybe if these were divided into two separate signboards and the board itself was enlarged, it would be better as the driver would be able to clearly find the speed limit for that road.

Dataset collection was done from an ONEPLUS 6T camera which was attached to a tripod stand which was then taped on to the dashboard of the vehicle. The camera shot video of the road ahead of it in 4k resolution at 60fps.

These frames were then extracted from the video to create the dataset.

But since these images added a bit of motion blur and windshield glare, a part of the dataset was also created by taking images from outside the car. These were very clear images, and by applying data augmentation techniques, these images were then multiplied.

The creators of YOLO recommend that in case of custom object detection, the dataset must consist of negative samples (images with no signboard) equal to the number of images that have the signboard.

A few samples from the dataset (Figures 7.10 and 7.11):

As we can see in a few images, the signboard was covered by branches of trees, we added as many possible images of such cases in our dataset so that it could to an extent handle such scenarios.

The dataset contains all types of images for a certain class.

During dataset collection:

1. It was important to the quality of the dataset that all angles be covered for a signboard, so that the model can learn the data well and be able to generalize in real-time scenarios.

FIGURE 7.10   Sample images – 1.

2. Dataset must include images from a distance as well as high-resolution close up images, the close-up images help the model to learn the actual features of the required signboard.

3. Images must be square images (1:1): Although YOLO does not require us to feed the model square images, it allows us to feed the model images of different sizes as images to square. However, it must be noted that resizing an image of unequal dimensions YOLO itself resizes to equal dimensions (rectangular image to a square image) will lead to a change in aspect ratio which will then change the shapes of objects in the images. This could potentially lead to a problem; therefore, a Python GUI was created by using Tkinter that will crop the largest square from an image with a single click.

4. It was also important that for a particular class the images must be taken of signboards in different backgrounds, i.e., all images for a particular sign-board must not be taken off the same board; instead, multiple boards from multiple locations must be included in the dataset; this will improve the generalization even further.

FIGURE 7.11   Sample images – 2.

The dataset created had a total of over 8000 images and then a few more images were added from the available German and Belgium traffic sign datasets.

A little bit of bias was created since some signboards are seen more often than others, this bias is important since the model will then learn that some signs are more common than others which should improve the accuracy further.

The primary data augmentation technique used was of gamma correction.

Gamma correction in simple terms is changing the illumination of an image.

A gamma value of less than 1 when applied on an image will give a darker image, whereas a gamma value of more than one will yield an image brighter than the original image, a gamma value of 1 gives the exact same image.

To adjust the gamma for multiple images in our dataset, a simple Python script was written that would add various gamma values to the image and save the resulting image as a new image, this allowed us to create a minimum of 3–5 images from a single image.

To make labeling of our dataset, an easier task gamma correction was applied after the dataset had been completely annotated and the labels were just copied for the new images since the only change in them was that of lighting.

Labeling a dataset for YOLO is a very easy task as there are multiple labelers available on the Internet.

However, since there is a second-stage detection involved which required its own dataset, an open source labeling tool was found.

The dataset for the second-stage detection, i.e., darknet reference model was created from these images itself. The code for the labeling tool was modified in such a way that whenever the selected class is speed limit and a bounding box is made, the labeling tool will automatically extract that bounding box and save it as a jpeg file in a different folder.

This task was again a simple task as the entire tool was a single Python script and all we had to do was write a new function and call it at the right time.

A few samples from the second-stage detector dataset (Figure 7.12):

This dataset was created from the original images without gamma correction; again, this dataset was multiplied in number using gamma correction, and a

FIGURE 7.12   Sample images for second-stage detection.

minimum of 5–7 images were created from a single image. Samples were added to this dataset that had tree branch, stickers stuck to the signboard so the model is able to generalize well for real-time datasets.

Since we are using a neural network for this task, it becomes easier for the model to generalize.

7.9.1   LABELING/ANNOTATING THE DATASET

The original dataset had about 1500 images (without data augmentation), the only way to label this data was to create a GUI which would be able to display all these images one by one and bounding box be drawn over the ROIs which would then calculate the required anchor boxes based on the top left and bottom right corner of the bounding boxes.

Thanks to the popularity of YOLO as an object detection, several labeling tools are easily available on the Internet.

Although to reduce the time spent on labeling the images for the darknet classifier (i.e., the second-stage detector), the source code of the labeling tool was modified a little, so that it becomes capable of creating the labeled dataset for the second stage as well.

A Python implementation was found from GitHub which made it a trivial task to modify as per our requirements (Figure 7.13).

FIGURE 7.13   Snippet from labeling tool.

This tool made it a very easy task to label all of the images.

This is the format in which YOLO takes an annotation file:

object-id center_xcenter_y width height

Below is an example of how the annotation looks like.

0   0.3348623853211009   0.7194189602446484   0.1620795107033639
0.20948012232415902

The centre coordinates and the x and y coordinates are normalized to lie between the range of 0–1.

YOLO works on the concept of anchor boxes. Hence, knowing the center coordinates of the ROI along with the height and width allows YOLO to understand the bounding box.

These values are stored in a .txt file.

Each image gets its own .txt file with the same name of the image, during training when the network sees an image it opens the corresponding .txt file to extract the bounding box information.

For the second stage of detection, the dataset was created from this itself.

A few extra lines of code were added to the labeling tool source code that basically meant that whenever the class id corresponding to speed limit signboards is used, the tool will automatically save a copy of the cropped ROI.

The second-stage detector, i.e., darknet reference model does not require .txt files since it is just a classifier.

It just requires the name of the class in the image, so a labeled image for darknet reference model is the name of the image.

The name of the image is modified as below:

[class_name]_[base_name].jpg

This annotation was created by the same tool, which greatly reduced a lot of time for labeling and annotating data for the second stage of detection.

Once these images were labeled, we then applied gamma correction on our labeled dataset since the only thing we did was changing the lighting that would not change the bounding box information of the image, so the dataset was labeled first then multiplied in number.

7.10   TRAINING THE MODEL

Both the models are neural networks, hence require a good GPU for training. Training such neural networks on CPU is possible but not recommended. The CUDA cores of a GPU help speeding up the training process exponentially.

Training of our model cannot be done on a Raspberry Pi since it does not have the required resources to train a neural network, therefore the model is trained on another machine and then the trained weights file and other required files of the model are transferred to the Raspberry Pi for detection.

Training of our models was done on Google Colab which has a Tesla K10 GPU. The Tesla K10 GPU is a really capable and powerful GPU that is widely used for training and tuning of large neural networks.

A few more steps were involved before starting training.

1. The entire dataset had to be first transferred to Colab.
2. Additional files had to be created which had the absolute paths to each of the images. Python scripts were written for the same.
   Eg: data/obj/1000.jpg
   data/obj/1001.jpg
   and so on.
   This file contains the absolute paths to each of the image.
   Other files like names of each of the class and paths to where the training, testing, and validation images are along with the path to where the weights file should be stored.
3. The CFG file had to be modified as well, the filters and class values for the last layers had to be changed according to our need. This modification is well defined and easy to follow in the original GitHub repository.
4. For the reference model, the only change in cfg was equating the filters in the last layer to the number of classes; this is not specified in the repository, but in any neural network, the last layer must have filters equal to the number of classes on which the model is predicting; hence, this change was made.
5. Darknet framework has to be built for Google Colab, this is very simple since everything is specified in a makefile, so a single command does this although the first few lines of the makefile need to be tweaked by setting CUDA, GPU, OPENCV NNPACK equal to 1 which basically tells darknet that they are available for use.

Once these steps are completed actual training can start.

The creators recommend that for custom training the model must be trained for [2000*number of classes] iterations.

Since the model was originally trained for a completely different purpose we could not freeze some layers and train the others, complete retraining was required for our purpose.

Transfer learning on YOLO is also a very simple thing to do, just set the parameter "backward" equal to 1 from whichever layer you want to freeze.

Training the complete model on Colab takes about 15–16 hours for 12,000 iterations.

The model saves the trained weights after every 1000 iterations until the model is completely trained. The maximum number of iterations can be set in the cfg file as well, once the model is trained to the maximum number of iterations the final weight file gets stored in the specified folder and training stops.

Although the loss starts converging after 9000 iterations and from 9000 to 12,000, the MAP scores are very similar, but still to get the best results, the model must be allowed to completely train.

The darknet reference model is trained for around 9000 iterations on Google Colab.

Training this model takes only 2 hours since the architecture is smaller and less complex as compared to tiny YOLO.

7.11   RESULT

Once the training is complete, the final weight file was used in a Python script along with some logics to sequentially execute the required commands. The camera quality is the main deciding factor in the quality of detection. The cheapest Raspberry Pi camera is not able to capture the road ahead in a moving vehicle because of its low shutter speed although there are a few other Raspberry Pi compatible cameras that are able to do the job. For best quality detection, the new Raspberry Pi camera that has autofocus along with Sony IMX sensors is able to get the picture very well which further improves the detection.

The final script takes 2.6 seconds on average for a single image prediction, since the model was trained on images taken from a distance as well, which is why the model is able to detect the signboards from a distance.

The first-stage detector gives 96% accuracy in predicting the signboards while the second-stage detector is able to achieve 98% accuracy.

7.12   FUTURE SCOPE

This is merely a prototype for improved road safety. A few things that should be added to this prototype:

1. A larger dataset could further improve the accuracies achieved by the system. This dataset should cover a wider range of road conditions including rain and fog.
2. Night images are not currently a part of this dataset. These images could be slightly tricky to handle since there might be cases when the headlights of other vehicles could cause reflections from the signboard that could further be an issue for our model.
3. Better hardware should be used to make this prototype as a final product. This would greatly impact the accuracy and the prediction time. Getting a prediction time of under 1.5 seconds should be achieved which would make this system really effective.
4. This project requires extensive funding for research as well as implementation. This funding must be reduced as much as possible to make this a viable option.
5. This system should also add the detection of red lights.

REFERENCES

1. J. H. She and H. Y. Lin, "A vision system for traffic sign detection and recognition", *IEEE 26th International Symposium on Industrial Electronics (ISIE)*, Edinburgh, UK (2017).

2. H. N. Dean and K. V. T. Jabir, "Real time detection and recognition of Indian traffic signs using Matlab", *International Journal of Scientific & Engineering Research*, Vol. 4, No. 5, p. 684 (2013), ISSN: 2229-5518.

3. C.-H. Lai and C.-C. Yu, "An efficient real-time traffic sign recognition system for intelligent vehicles with smart phones", *International Conference on Technologies and Applications of Artificial Intelligence*, pp. 195–202, Hsinchu, Taiwan (2010).

4. J. N. Chourasia and P. Bajaj, "Centroid based detection algorithm for hybrid traffic sign recognition system", *Third International Conference on Emerging Trends in Engineering and Technology*, pp. 96–100, Goa, India (2010).

5. P. Priya, D. Modha and M. Agrawal, "Detection and recognition of Indian traffic signals", *International Journal of Engineering Research and Applications*, Vol. 2, No. 2, pp. 756–758 (2012), ISSN: 2248-9622, https://www.ijera.com.

6. M. Benallal and J. Meunier, "Real time color segmentation of road signs", *CCGEI*, Montreal, QC, Canada (2003).

7. X. W. Gao, L. Podladchikova, D. Shaposhnikov, K. Hong and N. Shevtsova, "Recognition of traffic signs based on their color and shape features extracted using human vision models", *Journal of Visual Communication and Image Representation*, Vol. 17, pp. 675–685 (2006).

8. V. Lihal, P. Paclik, r. B. Kova, P. Mosna, N. I. Vlcek and P. Zahradnik, "Road sign recognition system using TMS320C80", *2nd European DSP Education and Research Conference, ESIEE*, Paris, France (1998).

9. A. De La Escalera, L. E. Moreno, M. A. Salichs, and J. M. Armingol, "Road traffic sign detection and classification", *IEEE Transactions on Industrial Electronics*, Vol. 44, No. 6, p. 848859 (1997).

10. H. N. Dean and K. Jabir, "Real time detection and recognition of Indian traffic signs using Matlab", *International Journal of Scientific & Engineering Research*, Vol. 4, No. 5, pp. 684–690 (2013).

11. M. B. Blaschko and C. H. Lampert. "Learning to localize objects with structured output regression", *Computer Vision – ECCV 2008*, pp. 2–15, Springer, Berlin (2008).

12. L. Bourdev and J. Malik, "Poselets: body part detectors trained using 3D human pose annotations", *International Conference on Computer Vision (ICCV)*, Springer, Berlin (2009).

13. H. Cai, Q. Wu, T. Corradi and P. Hall. "The crossdepiction problem: computer vision algorithms for recognising objects in artwork and in photographs" (2015). arXiv preprint arXiv: 1505.00110.

14. N. Dalal and B. Triggs, "Histograms of oriented gradients for human detection", *IEEE Computer Society Conference on Computer Vision and Pattern Recognition (CVPR 2005)*, Vol. 1, pp. 886–893. IEEE, San Diego, CA (2005).

15. F. Moutarde, A. Bargeton, A. Herbin and L. Chanussot, "Robust on-vehicle real-time visual detection of American and European speed limit signs with a modular traffic signs recognition system", *IEEE Intelligent Vehicles Symposium*, pp. 1122–1126. IEEE, Istanbul, Turkey (2007).

16. A. De La Escalera, J. M. Armingol and M. Mata, "Traffic sign recognition and analysis for intelligent vehicles", *Journal of Image and Vision Computing*, Vol. 21, No. 3, pp. 247–258 (2003).

17. S. Waite and E. Oruklu, "FPGA-based traffic sign recognition for advanced driver assistance systems", *Journal of Transportation Technologies*, Vol. 3, pp. 1–16 (2013).

18. J. Greenhalgh and M. Mehdi, "Real-time detection and recognition of road traffic signs", *IEEE Transactions on Intelligent Transportation Systems*, Vol. 13, No. 4, pp. 1498–1506 (2012).

19. P. Karem and O. Tosun, "Real-time traffic sign recognition with map fusion on multicore/many core architectures", *Acta Polytechnica Hungarica*, Vol. 9, No. 2, pp. 231–250 (2012).

20. A. Karunalithika, R. P. Jayasundra, M. A. Rasamjan, D. N. Senayanke and V. N. Vithana, "Road sign identification application using image processing and augmented reality", *International Journal of Advanced Computer Technology*, Vol. 4, No. 11, pp. 79–93 (2015).

21. H. Li, F. Sun, L. Liu and L. Wang, "A novel traffic sign detection method via color segmentation and robust shape matching". *Neurocomputing*, Vol. 169, pp. 77–88 (2015). doi: 10.1016/j.neucom.2014.12.111.

22. C. Bahlmann, Y. Zhu, V. Ramesh, M. Pellkofer and T. Koehler, "A system for traffic sign detection, tracking, and recognition using color, shape, and motion information", *Proceedings of the 2005 IEEE Intelligent Vehicles Symposium*, Las Vegas, NV, 6–8 June 2005.

23. S. Ardianto, C. Chen and H. Hang, "Real-time traffic sign recognition using color segmentation and SVM", *Proceedings of the 2017 International Conference on Systems, Signals and Image Processing (IWSSIP)*, Poznan, Poland, 22–24 May 2017.

24. W. G. Shadeed, D. I. Abu-Al-Nadi and M. J. Mismar, "Road traffic sign detection in color images", *Proceedings of the 10th IEEE International Conference on Electronics, Circuits and Systems*, Sharjah, UAE, 14–17 December 2003.

8 Lung Cancer Risk Stratification Using ML and AI on Sensor-Based IoT
An Increasing Technological Trend for Health of Humanity

Rohit Rastogi and Mukund Rastogi
ABES Engineering College

D. K. Chaturvedi
DEI

Sheelu Sagar
Amity International Business School

Neeti Tandon
Vikram University

CONTENTS

8.1 Introduction 138
    8.1.1 Motivation to the Study 139
    8.1.2 Problem Statements 139
    8.1.3 Authors' Contributions 139
    8.1.4 Research Manuscript Organization 139
    8.1.5 Definitions 139
    8.1.6 Computer-aided Diagnosis System (CADe or CADx) 140
    8.1.7 Sensors for the Internet of Things 140
    8.1.8 Wireless and Wearable Sensors for Health Informatics 140
    8.1.9 Remote Human's Health and Activity Monitoring 141

DOI: 10.1201/9781003138020-8

        8.1.10 Decision-Making Systems for Sensor Data..................................... 141
        8.1.11 Artificial Intelligence (AI) and Machine Learning (ML) for
                Health Informatics... 141
        8.1.12 Health Sensor Data Management 142
        8.1.13 Multimodal Data Fusion for Healthcare... 142
        8.1.14 Heterogeneous Data Fusion and Context-Aware Systems—a
                Context-Aware Data Fusion Approach for Health-IoT 142
8.2   Literature Review 143
8.3   Proposed Systems.. 144
        8.3.1   Framework or Architecture of the Work 145
        8.3.2   Model Steps and Parameters 145
        8.3.3   Discussions 145
8.4   Experimental Results and Analysis.. 146
        8.4.1   Tissue Characterization and Risk Stratification 146
        8.4.2   Samples of Cancer Data and Analysis... 147
8.5   Novelties 148
8.6   Future Scope, Limitations, and Possible Applications 149
8.7   Recommendations and Considerations.. 149
8.8   Conclusions.. 149
References.. 150

8.1   INTRODUCTION

Non-small cell lung cancer (NSCLC) includes three types of cancers: squamous cell carcinoma, adenocarcinoma, and large cell carcinoma derived from lung tissue. Adenocarcinoma is a slow-growing cancer that first appears in the outer region of the lung. Lung cancer is more common in smokers, but the most well-known sort of lung cancer in nonsmokers. Squamous cell carcinoma is more normal in the focal point of the lung and all the more generally in smokers, but large cell carcinoma can be found anywhere in the lung tissue and grows faster than adenomas and lung cancer [9,20].

According to Choi, H. and his team members, lung cancer risk classification models with gene expression function are very interesting. Change are required in the previous models based on individual symptomatic genes.

They have revealed that the aim to develop a risk classification model was based on a novel level of gene expression network that was performed using multiple microarrays of lung adenocarcinoma (LUAD), and gene convergence network investigation was carried out to recognize endurance networks. Genes representing these networks have been used to develop depth-based risk classification models. This model has been approved in two test sets. The efficiency of the model was strongly related to patient survival in the two sets of experiments and training. In multivariate analysis, this model was related to persistent anticipation, autonomous of other clinical and neurotic highlights.

The researchers have shown that how the gene structures and expressions can be useful in early detection of the cancer and suitable steps can be taken to cure the patients with higher probability of saving the lives [4].

8.1.1 Motivation to the Study

The medical service industry is confronted with the test of the quick improvement of many medical service data. The field of big data investigation is extending—you can leverage your healthcare system to provide valuable insights. As mentioned above, most of the data produced by this system is digitally printed and stored.

The principle distinction between customary well-being analysis and big data well-being is the live programming component. In customary frameworks, the medical service industry depends on different ventures to examine big data. Many healthcare professionals rely on IT industry due to its huge impact. Their operating system is functional and capable of processing data in standard formats.

8.1.2 Problem Statements

Malignant lung tumor portrayed by sporadic development of lung tissue is known as lung cancer. Metastases can spread past the lungs to encompassing tissues and different pieces of the human body. Most cancers of the lung are called primary lung cancer, carcinoma. Small cell lung cancer (SCLC) and NSCLC are the important types of lung cancer. The most common symptoms of pesticides (including coughing blood) are fatigue, emphysema, and angina (coronary thrombosis). NSCLC accounts for approximately 81%–86% of lung cancers. By this study, we are classifying the lung cancer cases as per their medical parameters.

8.1.3 Authors' Contributions

Mr. Rohit Rastogi was the team lead and executed the experiment. Dr. D. K. Chaturvedi created the design of the experiment, Ms. Sheelu and Ms. Neeti conducted the experiments, Mr. Mukund analyzed the experiment, and all contributed in manuscript formation.

8.1.4 Research Manuscript Organization

This chapter has been started with Abstract and followed by Introduction that contains a short Literature Review and then the motivation of the study. After this problem, statement and definition have been introduced and then followed by Authors' Contribution and chapter organization.

The literature survey contains latest relevant papers and followed by proposed systems and experimental setup and analysis. After that, Results and Discussions have been presented which is succeeded by Recommendations and Considerations and then future research directions, limitations of our study, and Conclusions have been established.

It is followed by Acknowledgements and References. At last, in Annex, experimental dataset images and experimental snapshots have been given for readers.

8.1.5 Definitions

Some important terminologies and key components are being explained here in light of our experimental work.

8.1.6 Computer-aided Diagnosis System (CADe or CADx)

Computer-aided detection (CADe) or computer-aided diagnosis (CADx) is a type of system software that has been shown to be very helpful to physicians in the recent microscopic interpretation of medical images. X-ray diagnostics, magnetic resonance imaging (MRI), and ultrasound imaging technologies provide a wealth of information to help medical professionals to make comprehensive analyses and assessments in the short term. The computer-aided design (CAD) system processes the digital image to highlight the normal display or obvious areas such as possible illnesses and provide input to support a particular expert decision [14].

With the help of computers, all-slide imaging algorithms and machine learning (ML) have potential plans for digital pathology. So far, the program has been limited to physical safety, but is now being studied for standard spots. CAD is an interdisciplinary technology with artificial intelligence (AI) computer elements with radiation and pathological imaging. A common program is tumor diagnosis.

8.1.7 Sensors for the Internet of Things

The Internet of Things (IoT) encourages our lives by associating electronic gadgets and sensors through interior networks. IoT utilizes smart gadgets and the Internet to give inventive answers for different difficulties and issues identified with different business, public, and public and private enterprises around the world. IoT has become a significant part of our daily life that we can look about us. When all is said and done, IoT is an advancement that coordinates different perceptive frameworks, systems, shrewd gadgets, and sensors. We also use quantum and nanotechnology in terms of memory, measurement, and unimaginable speeds. This can be seen as a prerequisite for creating an innovative business plan with security, reliability, and collaboration [2, 21].

Here are the nine of the most popular IoT sensors

1. Temperature
2. Moisture
3. Pressure
4. Adjacent
5. Surface
6. Accelerometer
7. Gyroscope
8. Gas
9. Infrared [2].

8.1.8 Wireless and Wearable Sensors for Health Informatics

The Internet of Things (IoT) is a new concept that enables wearable devices to control healthcare. The IoT supports embedded technologies and is supported as a network of physical objects that connect data and sensors to communicate with the internal and external states of the object and its environment. Over the last decade, wearable have attracted the attention of many researchers and industries and have become very popular recently [7, 19].

8.1.9 REMOTE HUMAN'S HEALTH AND ACTIVITY MONITORING

Remote monitoring of healthcare allows you to stay at home instead of expensive medical centers like hospitals and nursing homes. Accordingly, it gives a proficient and practical option in contrast to clinical checking here. With a noninvasive, invisible, visible wearable sensors, such a system is an excellent diagnostic tool for healthcare professionals to diagnose physiologic critical conditions and real-time patient activity from remote centers. In this way, it is intelligible that handheld sensors assume a significant part in such observation frameworks. These reconnaissance frameworks have pulled in the consideration of numerous specialists, business visionaries, and goliath engineers [11].

Handheld sensor-based health monitoring systems include textile fibers, fabrics, elastic bands, or several kinds of adaptable sensors that can be straightforwardly associated to the human body. These sensors measure physiology such as electromyography, body temperature, electromyography activity, arterial oxygen saturation, heart rate, blood pressure, electrocardiogram, and respiratory rate and can measure physical symptoms [5].

8.1.10 DECISION-MAKING SYSTEMS FOR SENSOR DATA

Management decisions are very basic and are widely used in economics. It relies upon the information and experience of the administrator, however increasingly more on target data. There are advance tools available for demographical data measurement like wetland detection and real-time monitoring of mountains, rivers, and forests [15, 18].

Until now, management has focused only on intuitive facts from checking data, for example, the overall status of water quality pointers, cases without accurate secondary analysis. For effective management and decision-making [10, 22].

8.1.11 ARTIFICIAL INTELLIGENCE (AI) AND MACHINE
LEARNING (ML) FOR HEALTH INFORMATICS

Artificial intelligence (AI) showed up in medical services during the 1970s. The main AI frameworks are information-based decision support systems, and the principal AI methods are utilized to foresee the classification standards of label sets. These first frameworks function admirably. Nevertheless, it is not commonly used in real patients. One of the reasons is that these systems are independent and have nothing to do with the patient's electronic medical records. Another reason is that the skill communicated in the information on these master frameworks shows that the created framework is not worthy here [13].

After winning several championships in focusing on artificial neural networks and improving complex learning, substance abuse became a new learning method.

In May 2019, a team from Google and the New York University announced that deep learning models used to analyze lung cancer could improve precision, and the investigation immediately covered numerous newspaper and magazine title texts.

8.1.12   HEALTH SENSOR DATA MANAGEMENT

Trendsetting innovations, for example, cloud computing, wearable sensor gadgets, and big data, will affect individuals' day-to-day life, have extraordinary potential in Internet-based biological systems and provide personal and shared consumption information on the development of the health and welfare sector. These apparatuses give numerous better approaches to gather data physically and consequently. Many modern smartphones have some internal sensors such as a microphone, camera, gyroscope, accelerometer, compass, proximity sensor, GPS, and ambient light [12].

You can easily connect the new generation of wearable medical sensors to your smartphone and send the measurement results directly. This set is more effective and convenient than individual health measurement like BP, oxygen content in blood, and heart rate variability. Different sensors can be used for the analyses and visualizations of the patient details with accurate and fast speed. This dramatic development enables both data management and collaboration [12].

8.1.13   MULTIMODAL DATA FUSION FOR HEALTHCARE

Given the proliferation of IoT techniques, they can be used to help the critical functions of healthcare management. In this way, traditional hospitals with large-scale interconnected sensor systems and extensive data collection and collection technology have become the next generation of smart digital environments. From this point of view, Intelligent Health supports a complex ecosystem of intelligent spaces such as hospitals, ambulances, and pharmacies supported by powerful infrastructure stacks such as edge devices and sensor networks, using new business models and rules [3].

8.1.14   HETEROGENEOUS DATA FUSION AND CONTEXT-AWARE SYSTEMS—A CONTEXT-AWARE DATA FUSION APPROACH FOR HEALTH-IoT

The improvement of inexpensive sensor gadgets and correspondence advancements is quickening the improvement of elegant homes and conditions. With the development of human body networks, wireless sensor networks, big data technologies, and cloud computing, the healthcare industry is growing rapidly and uses the IoT. There are numerous difficulties, for example, heterogeneous data blending, text recognition, complex question preparing, unwavering quality, and exactness.

From this point of view, Intelligent Health supports a complex ecosystem of intelligent spaces such as hospitals, ambulances, and pharmacies supported by powerful infrastructure stacks such as edge devices and sensor networks, using new business models and rules [8].

8.2   LITERATURE REVIEW

According to Timor Kadir and his team members, that machine-based lung cancer prediction model was developed to help undiagnosed lung nodules and assist physicians on-screen. Such systems can reduce the number of node classifications, improves decision-making, and ultimately reduces the number of benign nodules that are tracked or manipulated unnecessarily [9].

This article outlines the main approaches to lung cancer prediction to date and highlights some of the relative strengths and weaknesses. The authors have discussed some of the challenges of developing and validating such technologies, as well as clinical acceptance strategies. They review the main approaches used to classify lymph nodes and predict lung cancer from computer tomography (CT) imaging data. In our experience, one should apply ML to get good accuracy by using the right training data and using a comprehensive convolutional neural network (CNN), achieving classification performance in regions with low 90s AUC (Area Under the Curve) points and sufficient training data [9].

According to Choi H., Na KJ in this study, a gene correlation network, we created a risk classification model for LUAD. An extension of future research is the use of this method in concurrent networks of cancer progression. Advances in technology change the DL (Deep Learning) design and the way toward choosing delegate qualities to improve expectation exactness. They found that NetScore was related with sex, status of smoking, phase, and subatomic subtype. In summary, a high NetScore trend in men, smokers, and KRAS mutants was delayed and observed to be positive [6].

Finally, they expected future clinical trials designed with all around controlled clinical and obsessive factors to help find clinical applications for their new danger grouping models.

Yin Li and his team members have predicted the risk of LUAD is important in determining subsequent treatment strategies. Molecular biomarkers may improve risk classification for LUAD [11].

Yin Li et al. analyzed the gene expression profile of LUAD patients by atlas cancer genome (TCGA) and omnibus gene expression (GEO) analyses. They first evaluated the prognostic relationship for each gene using three separate algorithms: notable function, random forest, and variable coke regression. Next, survival-related genes were included in the LASSO minimum and selection function models to create a LUAD risk prediction model [17].

They initially identified large dataset significant survival-related genes. A hybrid strategy was used to identify key genes associated with survival in large datasets. Enhancement analysis showed an association of these genes with tumor development and progression. A risk prediction model was created using the LASSO method. The risk model was approved with two outside sets and one free set. Patients in the high-hazard bunch had a lower danger of repeat (RFS) and in general endurance (OS) than patients at low risk. We also created a registry that predicts LUAD patient operating systems, including models and risk stages.

Hence, they conclude that risk models may serve as a pragmatic and reliable predictor of LUAD and may provide new experiences into the atomic instruments of infection [11, 16].

The paper written by Francisco Azuaje titled as "artificial intelligence for precision oncology: beyond patient stratification" described axial data from medical conditions and treatment options as a key challenge for accurate oncology [1].

Artificial intelligence (AI) offers an unparalleled opportunity to enhance such predictive capabilities in laboratories and clinics. AI, including ML, which is the most well-known area of research, has been able to accurately identify tumors beyond relatively well-known detection patterns such as single-source omics and supervised classification of imaging datasets.

According to his perspective, major developments and challenges in this regard argue that the scope and depth of AI research should be expanded in order to achieve geological advances in accurate oncology [1].

According to Xu, J. [20] in a large era of data on cancer genetics, wide availability of genetic information provided by the next-generation sequencing techniques and rapid development of medical journals integrates AI approaches such as ML, detailed learning, and natural language processing to challenge big data and high-dimensional scalability; uses this method to process clinical data to handle big data; brings the knowledge you have it is bent, using the base; and opens and lies down it is really medicine.

In this paper, they reviewed that the status and future guideline for using AI in cancer genomics in the field of workflow is genomic analysis for accurate cancer treatment. Existing AI solutions and their limitations in genetic testing for cancer and its diagnosis, including various contacts and interpretations, are being considered.

The tools or common algorithms available for the leading NLP technologies in literature extraction are reviewed and compared to evidence-based clinical recommendations.

According to him, this paper deals with data needs, algorithm transparency, the importance of preparing patients and physicians for real-time reproduction and assessment and modern digital healthcare. They believe that AI is the main factor in the evolution of healthcare into a precise drug, but of the precedent that needs to be created to ensure safety and beneficial effects on healthcare [20].

8.3 PROPOSED SYSTEMS

Based on histopathological image input, there are two steps:

- **Model Building**: The shape of this model is based on the extraction function.
- **Model Evaluation**: Form biological communication.

Modeling attempts to preserve the extracted shape by extracting shape-based features with full focus on the shape of the model. The fixed size is considered a constraint, and it focuses on all other faces found to achieve the form of the constraint, making it easy to extract the entire model.

8.3.1 FRAMEWORK OR ARCHITECTURE OF THE WORK

However, model evaluation includes some of the biological significance of the form, which requires the physician to have the precise and accurate information needed to evaluate the form.

8.3.2 MODEL STEPS AND PARAMETERS

Here we will focus more on size-based clinical models. There are several steps like segmentation, dimension reduction, sampling, enhancement, and active contour mapping and classification of the objects which are involved that help you to design your model in a very efficient way (As per Figure 8.1).

8.3.3 DISCUSSIONS

Cancer has been the reason for death for a very long time. A little less than half of the cancer happens in individuals matured ≤ 70–85 years. The maximum death-prone cancer is lung cancer where it threatens the all types of aged persons from

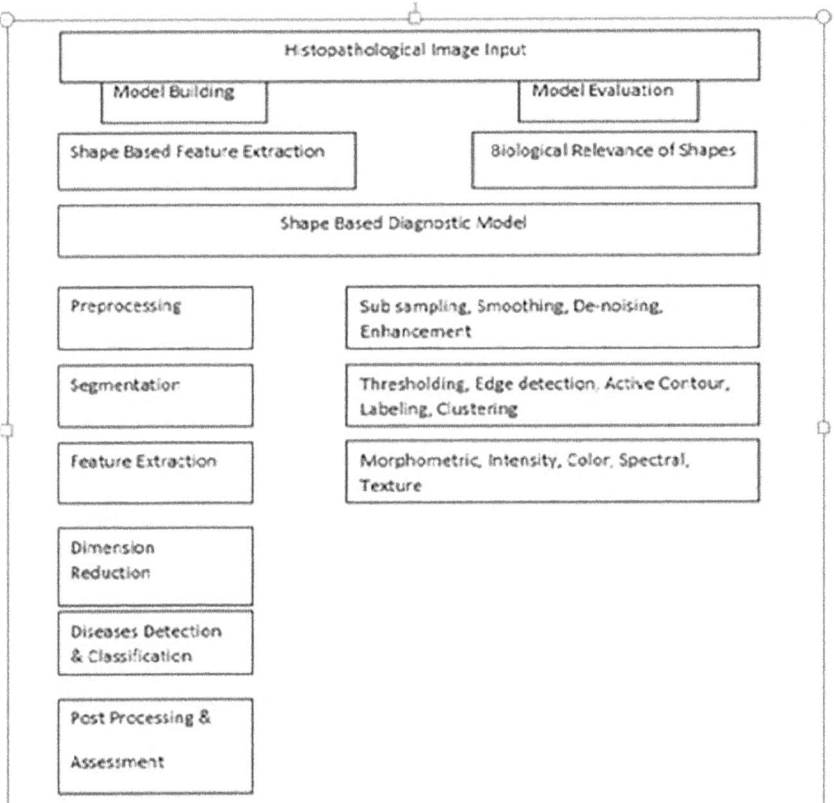

FIGURE 8.1 Framework of the experimental study of lung cancer stratification.

adolescents to middle age to adults. Those who are addicted with drinking alcohol or smoking or females who were nonsmoking, all types of subjects were included in the list of casualties unfortunately from this illness.

Thus, it is needed to separate lung cancer based on risk included (high-risk, okay, and rising-risk). This investigation depends on building up an arrangement conspired for lung cancers through examining minuscule images and ordering them utilizing deep convolutional neural network (DCNN), which is an AI (ML) structure.

Microscopic images of tissues will be characterized on the basis of risk involved using DCNN. CNNs are deep artificial neural networks that are utilized fundamentally for ordering images, group them by likenesses (photograph search), and perform object recognition inside the scenes. They are the calculations that can recognize faces, tumors, people, road signs, and numerous different parts of the graphic information.

The effectiveness of convolution networks in image recognition is one of the main reasons why the world has woken up for the productivity of deep learning. They are fueling significant progress in computer vision (CV), which has clear applications for self-driving vehicles, mechanical technology, security, clinical judgments, and medicines for the outwardly impaired.

8.4   EXPERIMENTAL RESULTS AND ANALYSIS

The term "tissue characteristics" covers a wide range of meanings, from qualitative assessment to various scientific measurements. In the extensive literature available, we expand the definition of terms, show the relationship between tissue features and images, identify some relevant physical parameters, and briefly describe the subject's medical history. Excerpts are provided for how to explain.

8.4.1   Tissue Characterization and Risk Stratification

According to an article titled "Automatic Classification of lung Cancer from Cellular Images Using Deep Symmetric Neural Networks," we evaluated three types of cancers and trained for classification. However, pictures from the same sample belonged to the same group.

While executing crossvalidation algorithm in different sets of images, we found that in Set1, there are 28 items and respective crossvalidation score is 5280 for adenocarcinoma. There are 42 items and crossvalidation score is 5478 for squamous cell carcinoma. For small cell carcinoma, there are 26 images and crossvalidation score is 5070.

In Set2, there are 28 items and respective crossvalidation score is 5184 for adenocarcinoma. There are 37 items and crossvalidation score is 5220 for squamous cell carcinoma. For small cell carcinoma, there are 33 images and crossvalidation score is 5280.

In Set2, there are 26 items and respective crossvalidation score is 5040 for adenocarcinoma. There are 46 items and crossvalidation score is 5310 for squamous cell carcinoma. For small cell carcinoma, there are 32 images and crossvalidation score is 5214.

8.4.2 SAMPLES OF CANCER DATA AND ANALYSIS

Figures 8.2 and 8.3 show a sample of cancer types correctly classified using the data amplification method, the classification confusion matrix. Squamous cell carcinoma is often mistaken for adenocarcinoma.

After successfully applying classification model in this dataset, we have to test the accuracy, precision, and recall of the model, so we observed the confusion matrix whose results are as follows: actual and predicted value of adenocarcinoma is 89% matched and only 11% items are not predicted correctly. Squamous cell carcinoma is predicted correctly 60% which is less as compared to adenocarcinoma. Actual and predicted value of small cell carcinoma is 70% matched and only 30% items are not predicted correctly.

It shows the original image classification and the wide range of image accuracy, respectively. Results obtained using magnification images show that the classification accuracy for adenocarcinoma, squamous cell carcinoma, and small cell cancer is 89.0%, 60.0%, and 70.3%, respectively, and the overall corrected rate is 71.1%. In addition, plug-in is applied to improve the classification.

Finally, we improved our classification model and accuracy score of trained set and augmented sets given below:

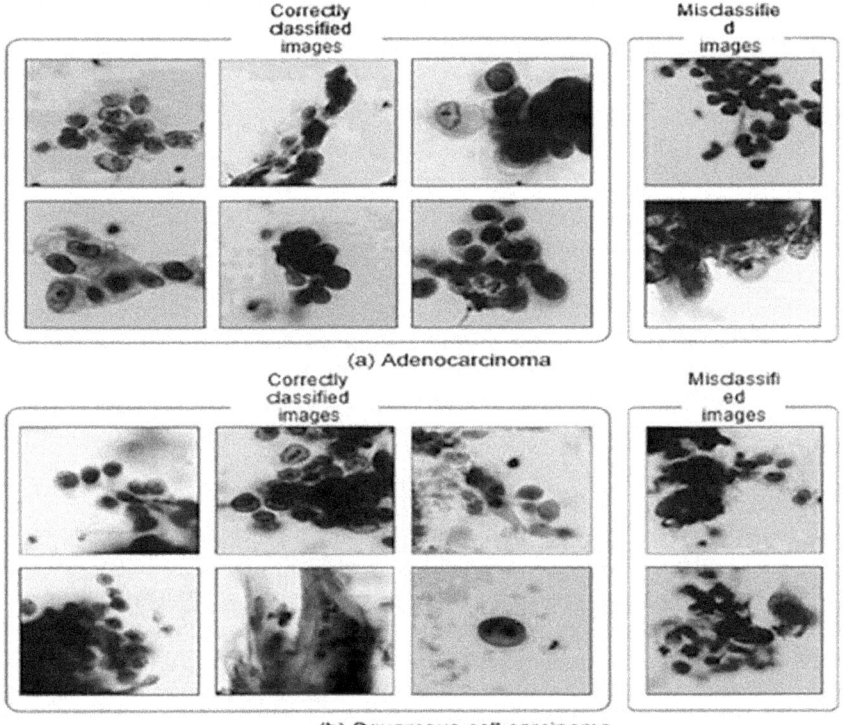

FIGURE 8.2   Sample images of correctly classified and misclassified carcinoma.

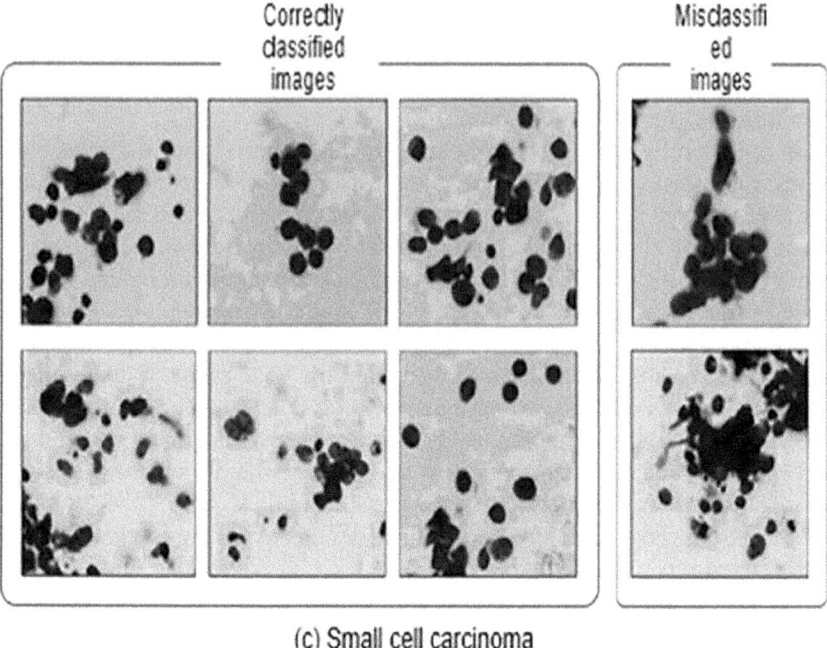

(c) Small cell carcinoma

FIGURE 8.3   More sample images of correctly classified and misclassified carcinoma.

Adenocarcinoma predicted 73% original images of lung cancer, whereas 89% for augmented images. We found that 45% original images are correctly classified and 60% augmented images are classified for squamous cell carcinoma. Small cell carcinoma is classified correctly 75% for original images and 70% for augmented images.

In the three types of lung cancer, the efficiency of the unsupervised clustering was maximum in adenocarcinoma and minimum in squamous cell carcinoma. Squamous cell adenocarcinoma requires more images. DCNN helped to correctly classify 70% of lung cancer cells.

8.5   NOVELTIES

Artificial intelligence (AI) will not only identify and predict at-risk patients, but will also be a large dataset available to hospitals and health care providers to identify changes in patient health and medical outcomes, as well as accurate diagnoses.

Treatment was planned especially for patients with chronic diseases. AI's information on infection states depends on learning calculations and doctor experienced in the treatment of patients with indications, signs, conclusions, medicines, and comparative results.

Most clinical data fall into a wide range of limited categories, although they are distant and may be limited by potential sampling, but future studies on the adoption of approved standards are not required. This allows AI algorithms to learn information and enhance the creation of reinforcement learning loops.

8.6 FUTURE SCOPE, LIMITATIONS, AND POSSIBLE APPLICATIONS

The forthcoming examination is required to investigate how AI can customize treatment choices for singular patients to a clinician. The nature of data that AI gains from is likewise significant and an expected boundary to the far-reaching selection of exactness medication. The size of data needed for deep learning and the variety of strategies utilized make it hard to obtain away from of how precisely AI frameworks may function in genuine practice or how reproducible they might be in various clinical contexts.

Forthcoming exploration openings are necessary to address "social inclination" in AI calculations, and sufficient advances should be taken to abstain from compounding medical care differences when utilizing AI apparatuses to save patients are famous. Tolerant security should be ensured and more noteworthy straightforwardness into algorithmic. Fairness is expected to guarantee acknowledgment of AI by suppliers and patients.

The IoT solutions for healthcare that collect, transmit, and visualize data in complex intelligent systems via wearable and field sensor networks can facilitate analytics, activity detection, and decision making. AI and ML technologies play a significant role in this transition, but their implementation requires computational power. It is often only available using cloud services.

In fact, with the increasing amount of data generated by sensors, the performance of ML-based cloud processing has several weaknesses for various reasons.

8.7 RECOMMENDATIONS AND CONSIDERATIONS

Artificial intelligence (AI) in healthcare is ready to bring change and disrupt medical care. While not giving up marketing and profitability of drug addiction is the wisest guide, it balances AI, the need for comprehensive healthcare to plan, manage, and reduce potential unexpected consequences.

It is wise to take AI, as the best solution, to start with a real healthcare issue, involving the relevant stakeholders, first-line users, patients and their families (including artificial and non-AI options). You need to find a solution and work on it, which should be implemented and extended by our five goals i.e. better health, better care experience, doctor's health, lower cost, and common rights.

8.8 CONCLUSIONS

The nature of administration is significantly influenced by the nature of your Internet association, making it difficult to use. Healthcare providers require shorter response times to address potential health risks, especially when the performance such as early detection, risk prevention, activity diagnosis, is guaranteed in real time.

Because of the huge measure of individual data that should be overseen, data stockpiling and security are additionally vital when managing medical care. For all of this, choosing purely local administration, especially for mobility, is not yet practical due to limited processing and storage capabilities.

REFERENCES

1. Azuaje, F. (2019). Artificial intelligence for precision oncology: beyond patient stratification. *NPJ Precision Oncology*, 3:6. doi:10.1038/s41698-019-0078-1.
2. Bauer, H., Patel, M., Veira, J. (2016). *The Internet of Things: Sizing Up the Opportunity.* New York (NY): McKinsey & Company. Available from: http://www.mckinsey.com/industries/high-tech/our-insights/the-internet-of-things-sizing-up-the-opportunity.
3. Baloch, Z., Shaikh, F., Unar, M. (2018). A context-aware data fusion approach for health-IoT. *International Journal of Information Technology*, 10. doi:10.1007/s41870-018-0116-1.
4. Choi, H. (2018). A risk stratification model for lung cancer based on gene coexpression network and deep learning. *Applications of Bioinformatics and Systems Biology in Precision Medicine and Immuno Oncology*, 2018:11, Article ID 2914280. doi:10.1155/2018/2914280, Received 13 October 2017|Revised 07 December 2017|Accepted 11 December 2017 |Published 16 January.
5. Deen, M.J. (2015). Information and communications technologies for elderly ubiquitous healthcare in a smart home. *Personal and Ubiquitous Computing*, 19:573–599. doi:10.1007/s00779-015-0856-x.
6. Gao, W., Emaminejad, S., Nyein, H.Y.Y., Challa, S., Chen, K., Peck, A., Fahad, H.M., Ota, H., Shiraki, H., Kiriya, D., Lien, D.H., Brooks, G.A., Davis, R.W., Javey, A. (2016). Fully integrated wearable sensor arrays for multiplexed in situ perspiration analysis. *Nature*, 529(7587):509–514.
7. Gyllensten, I.C., Gastelurrutia, P., Riistama, J., Aarts, R., Nunez, J., Lupon, J., Genis, A.B. (2014). A novel wearable vest for tracking pulmonary congestion in acutely decompensated heart failure. *International Journal of Cardiology*, 177(1):199–201.
8. Haghighat, M., Abdel-Mottaleb, M., Alhalabi, W. (2016). Discriminant correlation analysis: real-time feature level fusion for multimodal biometric recognition. *IEEE Transactions on Information Forensics and Security*, 11(9):1984–1996.
9. Kadir, T., Gleeson, F. (2018). Lung cancer prediction using machine learning and advanced imaging techniques. *Translational Lung Cancer Research*, 7(3):304–312. Retrieved from http://tlcr.amegroups.com/article/view/21998.
10. Kumar, S., Maninder, S. (2019). Big data analytics for healthcare industry: impact, applications, and tools. *Big Data Mining and Analytics*, 2:48–57. doi:10.26599/BDMA.2018.9020031.
11. Li, Y., Ge, D., Gu, J., Xu, F., Jhu, Q., Lu, C. (2019). A large cohort study identifying a novel prognosis prediction model for lung adenocarcinoma through machine learning strategies. *BMC Cancer*, 19:886. doi:10.1186/s12885-019-6101-7.
12. Li, Y., Wu, C., Guo, Li., Lee, C.H., Guo, Y. (2014). *A Big Data Platform for Health Sensor Data Management. Wiki-Health*. doi:10.4018/978-1-4666-6118-9.ch004.
13. Lisa, A., Gustafson, D.H. (2013). The role of technology in health care innovation: a commentary, *Journal of Dual Diagnosis*, 9(1):101–103. Published online 2012 November 27. doi:10.1080/15504263.20.
14. Macedo, F., Giovani, A., Higor, M., Marcel, R. D., Ricardo, V. (2016). Computer-aided detection (CADe) and diagnosis (CADx) system for lung cancer with likelihood of malignancy, *BioMedical Engineering Online*, 15. doi:10.1186/s12938-015-0120-7.
15. Mike, C., Hoover, W., Strome, T., Kanwal, S. (2013). Transforming healthcare through big data strategies for leveraging big data in the healthcare industry.

16. Rastogi R., Chaturvedi, D. K., Satya, S., Arora N., Trivedi P., M., Gupta, M., Singhal, P., Gulati, M. (2020). *MM Big Data Applications: Statistical Resultant Analysis of Psychosomatic Survey on Various Human Personality Indicators ICCI 2018 Paper as Book Chapter,* Chapter 25, © Springer Nature Singapore Pte Ltd., Singapore. doi:10.1007/978-981-13-8222-2_25.

17. Singh, Y., Chauhan, A.S. (2005). Neural networks in data mining. *Journal of Theoretical and Applied Information Technology,* 5(6):37–42.

18. Sun, J., Reddy, C.K. (2013). 'Big data analytics for healthcare', in *Proceedings of the 19th ACM SIGKDD International Conference on Knowledge Discovery and Data Mining,* Chicago, IL, 2013, pp. 1525–1525.

19. Wu, M., Luo, J. (Fall, 2019). Wearable technology applications in healthcare: a literature review. *Online Journal of Nursing Informatics,* 23(3). Available at http://www.himss.org/ojn.

20. Xu, J., Yang, P., Xue, S., Sharma, B., Martin, M.S., Wang, F., Beaty, K.A., Dehan, E., Parikh, B. (2019). Translating cancer genomics into precision medicine with artificial intelligence: applications, challenges and future perspectives. *Human Genetics,* 138:109–124. doi:10.1007/s00439-019-01970-5.

21. Zanella, A., Bui, N., Castellani, A., Vangelista, L., Zorzi, M. (2014). Internet of Things for smart cities. *IEEE Internet of Things Journal,* 1(1):22–32.

22. Zhao, W., Wang, H. (2016). Strategic decision-making learning from label distributions: an approach for facial age estimation. *Sensors,* 16:994–1013. doi:10.3390/s16070994.

9 Statistical Feedback Evaluation System

Alok Kumar and Renu Jain
University Institute of Engineering and Technology

CONTENTS

9.1   Introduction 153
9.2   Related Work 154
9.3   Types of Feedback Evaluation Systems 156
     9.3.1   Questionnaire-Based Feedback Evaluation System (QBFES) 156
     9.3.2   Star-Point-based Feedback Evaluation System (SBFES) 158
     9.3.3   Text-Based Feedback Evaluation System (TBFES) 159
9.4   Statistical Feedback Evaluation System 161
     9.4.1   Aspect Extraction 162
          9.4.1.1   Feedback Collector 163
          9.4.1.2   Feedback Preprocessor 163
          9.4.1.3   Aspect Validator 163
     9.4.2   Aspect Weight Estimation 169
     9.4.3   Sentiment Evaluation 171
          9.4.3.1   Sentiment Estimator 172
          9.4.3.2   Sentiment Aggregator 174
     9.4.4   Customized Evaluation 174
     9.4.5   Aspect-Based Questionnaire Design 176
9.5   Result Analysis and Discussion 177
9.6   Conclusion 179
9.7   Future Work 179
References 180

9.1   INTRODUCTION

Any individual's feelings, knowledge, and experiences about an issue or an item or a human being are termed as his or her opinion or feedback or views. Every person, irrespective of his age and gender, strives for others' opinions directly or indirectly. Opinions play a very important role in influencing people, provided they are properly expressed and rightly understood. It has been observed that others' opinions and perceptions may play an important role in making the final decision by anyone. For example, suppose someone is interested in purchasing any expensive item or taking admission in an institute or joining an organization etc. In that case, he or she will take opinion/feedback from their friends, family members, colleagues, and

DOI: 10.1201/9781003138020-9

from other persons having knowledge of that domain before taking the final call. A large amount of feedback data about an item is readily available to the users in the present time. Still, it becomes humanly impossible to go through all the feedback and then quantify the consolidated information. The correct opinion about each feature of the item can be gathered to make the right decision. In this competitive era, users' feedbacks are very important for all institutions, product manufacturers, and service providers to measure the performance/effectiveness of their policies, products, or services. It helps them to spot the weak and strong features of the institute, product, or service. On the basis of users' feedbacks, new policies are formed by the institutions, new features are added/removed by the product manufacturers, and the service providers do enhancements/variations. Hence, there is a need to automate the process of analyzing a huge collection of textual opinions/feedback. The digging of feedback and retrieving a concrete utility measure representing the goodness/badness is crucial before making any decision. Therefore, researchers are keeping this issue on the front burner. This field is known as opinion mining or feedback analysis, a subarea of natural language processing and text mining which has become an emerging field of research. However, it is still far away from its ultimate goal.

In this chapter, existing feedback evaluation systems are discussed with their merits and demerits. Existing feedback evaluation systems can be divided into three categories, i.e., questionnaire-based, text-based, and star-point-based feedback evaluation systems (SBFESs). Questionnaire-based feedback evaluation system (QBFES) deals with structured feedback, while text-based feedback evaluation systems (TBFESs) handle unstructured feedback. SBFES is for overall performance evaluation. Examining the existing feedback evaluation systems, it is found that a more effective feedback evaluation system can be designed by adding linguistic knowledge of the text and the statistical knowledge available in the data. In the proposed system, the first most prominent aspects are identified using statistical information, and then for each aspect, its sentiments are evaluated using linguistic knowledge. With the help of aspects and their sentiments, a customized opinion index is evaluated, a more relevant and effective questionnaire is designed, and an overall comprehensive rating is also evaluated.

9.2 RELATED WORK

This study has gone through the different feedback evaluation systems to understand its working and implementation details. Researchers have used different methodologies in different feedback evaluation systems. Some pioneer researches are done in this domain and mentioned below.

Hu and Liu [1] proposed a rule-based method to perform aspect-level sentiment analysis. In this research, tentative aspect terms are identified on the basis of the syntactic category of the words. Meaningless aspects are removed using association-based filters. Further, the sentiment of aspect terms is estimated on the basis of sentiment polarities of contextual words. Abdullatif Ghallab et al. [2] performed a systematic study of Arabic sentiment analyzers. In the study of one hundred eight research papers, authors found that the Arabic sentiment analyzer's performance can be enhanced by using the language-specific features and lexicons.

The authors also highlighted that unavailability of a standard dataset is the main hurdle in developing a good Arabic sentiment analyzer. A. Kumar and R. Jain [3] proposed a method to analyze textual feedbacks using supervised and semi supervised machine learning techniques. In this research, authors have experimented with teachers' textual feedback and explained the utility of the results, which can enhance the quality of academic and administrative activities of the institute/organization.

A. Kumar and R. Jain [4] proposed a method to extract important attributes from textual feedback. In this research, authors have used topic modeling and linguistic knowledge in identifying significant attributes. These attributes are further used in performing effective sentiment analysis. M. S. Neethu and R. Rajasree [5] proposed a machine learning-based technique for sentiment analysis of tweets related to electronics goods like mobile and laptop. In this research, the authors have used an additional subsystem to handle misspelled and slang words. Mohammad Al-Smadi et al. [6] implemented two different models using a deep neural network and support vector machine for aspect-based sentiment analysis of Arabic hotel reviews. Both models are trained using lexical, syntactic, morphological, and semantic features. In the result analysis, the support vector-based model's performance is better than deep neural network base model. Soujanya Poria et al. [7] proposed a rule-based approach for aspect extraction from the product reviews. In this approach, the author used commonsense knowledge and sentence dependency to identify implicit and explicit aspects. The proposed methodology is tested on two standard datasets, and high accuracy is achieved on both datasets.

Warih Maharani et al. [8] proposed a pattern-based method to extract the aspects in customer reviews. In this research, authors have experimented with many syntactic and semantic structure base patterns. Experimental results show that the pattern-based method's performance is not very good and not able to identify implicit aspects. A. Ilmania et al. [9] proposed a deep neural network-based method for aspect detection and sentiment classification. In this research, authors have experimented with several models with different input vectors and topologies. Jin et al. [10] proposed a lexicalized hidden Markov model (HMM)-based method to extract aspect and sentiment terms. In this research, the identification of aspect and sentiment terms is considered as a sequence labeling problem. To enhance performance of aspect-based sentiment analysis, a conditional random field (CRF) is suggested by Shariaty and Moghaddam [11]. Lia et al. [12] proposed a skip-tree CRF model to identify aspect and sentiment terms to incorporate more language-dependent features. In this research, a model is trained by using the word, dictionary, and sentence-related features.

Nandal et al. [13] proposed a machine learning-based method to perform aspect-level sentiment analysis of customer reviews available on electronic platforms. In this research, the authors give more attention to the bipolar words to improve the sentiment analyzer's performance. Sentiment polarity of bipolar words is decided on the contextual words. The proposed method is tested on Amazon's reviews. Chuhan et al. [14] proposed a hybrid unsupervised approach for fine-grained sentiment analysis. The proposed approach work is classified into two phases. First phase is aspect term extraction (ATE). It is responsible for extracting important aspects. The second phase of the proposed system is opinion target extraction (OTE), to extract sentiment of different aspects. In this research, rule base and supervised methodologies are combined to take the advantage of both methodologies. Abdullah Alsaeedi and Mohammd Zubair Khan [15] conducted a detailed study on sentiment analysis on Twitter data. The authors explained different techniques used for document-level sentiment analysis and sentence-level sentiment analysis Twitter data in this research. In this research, authors found that ensemble and hybrid approaches are more suitable for Twitter data sentiment analysis.

A. Shah et al. [16] suggest a method for review classification and star rating prediction. In this research, the authors have predicted star rating on the basis of sentiment analysis. Authors have used naïve Bayes and Maxent classifiers for the sentiment classification. Tun Thura Thet et al. [17] presented a method to perform aspect-level sentiment analysis of movie reviews. In this paper, the authors have used grammar-dependency structure for identifying aspects and sentiment words. Sentiments of different aspects are estimated with the help of 32,000 sentiment-bearing words extracted from the SentiWordNet [18]. In this research sentiment, negation words are handled with care. E. Cambria et al. [19] have briefly discussed different methodologies used for sentiment analysis and highlighted that concept-level methodologies are more efficient than others. Concepts are the clue realization of the human mind. This research also suggested that performance can be enhanced by using broader and deeper commonsense knowledge.

9.3   TYPES OF FEEDBACK EVALUATION SYSTEMS

Individuals, institutions, manufacturers, and service providers are always eager to know their users' opinions, and most of the big companies hire analysts to analyze the feedback of users keeping the various possible perspectives in mind and then submit a report to higher administration. Some commonly used methods for feedback evaluation are discussed below with their merits and demerits.

    a. Questionnaire-based feedback evaluation system (QBFES)
    b. Star-point-based feedback evaluation system (SBFES)
    c. Text-based feedback evaluation system (TBFES)

9.3.1   QUESTIONNAIRE-BASED FEEDBACK EVALUATION SYSTEM (QBFES)

Traditionally, QBFES [20] is used for feedback collection and evaluation. In this system, a list of questions (questionnaire) is given to the users, which directly relates to different features of the item. Users answer each question by giving some numeric score on the Likert scale (1–5 or 1–10) or relative score on a relative scale (good, average, below average, unsatisfactory) as per the instructions. Most product manufacturers, service providers, and employers, use QBFES to evaluate products, services, and employees' effectiveness, respectively. Aggregating responses of all users estimate the overall effectiveness of a product or service. It is also helpful in the identification of the strong and weak features of the item. Let us understand it with the help of an example:

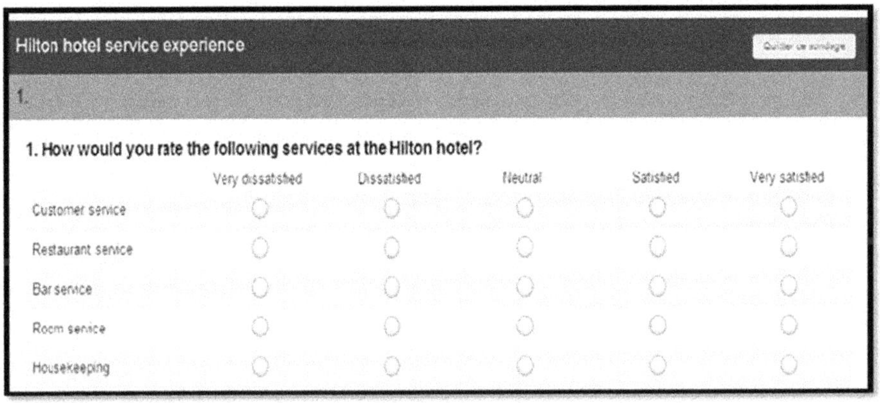

FIGURE 9.1 Sample questionnaire for feedback of a hotel.

The chain of Hilton Hotels uses QBFES for the evaluation of their hotels. Higher authorities design questionnaires to collect feedbacks/opinions about different services from the customers, and then, the performance of a hotel is measured on some scale. A sample of a questionnaire used in Hilton Hotels is given in Figure 9.1. Rows of the questionnaire contained a set of questions related to important features of the hotel, and columns explain the scale of satisfaction. There are six questions, and the customers evaluate each question on a Likert scale (1–5). Merits of mentioned aspects/features are estimated by taking the arithmetic mean of the customers' evaluation scores. The overall performance is evaluated by aggregating merit scores of all aspects/features included in the questionnaire. A QBFES is a user friendly, straightforward, and it can spot the weak and strong features of the product or service.

In the study of QBFES, the following observations were made [21]:

a. **No Standard Method for Questionnaire Design**: There is no standard method to choose the questions for the questionnaire. Most questions are decided by the senior members of the organization on the basis of their knowledge and experience, but it is possible that they may miss some key questions which may be very important from users' point of view.

b. **Equal Weight of all Features**: In the estimation of the overall effectiveness of an item (product/service/professional), all the questions mentioned in the questionnaire are generally considered of equal weight. Considering the equal weight of each question may not be appropriate because some features may be more important to users in comparison with other features. For example, in teacher's feedback questionnaire, *'knowledge'* and *'way of explanation'* are more important features than *'handwriting'* and *'teacher's dressing sense'*. Similarly, in a product questionnaire, *'quality of product'* and its *'usefulness'* are more important than *'packaging'* and *'delivery time'* of it.

c. **No Flexibility to Decide the Scale of Evaluation**: In questionnaire-based systems, each question is evaluated by the evaluator on the basis of a pre decided scale. The scale is fixed by the questionnaire, which cannot be changed by evaluators, for example, if there are only three categories like '*satisfactory*', '*good*', and '*excellent*' to evaluate a questionnaire, then if a user does not want to choose from that set, he or she cannot add new categories like '*average*', '*below average*', and '*unsatisfactory*'.

d. **Two Items are not Comparable**: Mostly, the questionnaires used by different organizations are different, and due to that, it is not possible to compare two items of different brands. If feedbacks are on the basis of features, then items of different brands can be compared easily aspect wise.

9.3.2 STAR-POINT-BASED FEEDBACK EVALUATION SYSTEM (SBFES)

In modern times, SBFES is a simple and popular method for capturing the users' opinion. After using some product or service, user is asked to rate the product or service through handheld device like a mobile phone by assigning 1–5 stars. This method is very simple and fast. Screenshots of star ratings used by WhatsApp and Uber are shown in Figure 9.2. But, with a SBFES, we can capture the only overall quality of a product or a service.

Recently, a modification has been done by popping up subsequent windows if a user does not give a good rating to a product or a service; the system gives a questionnaire to know which aspects are not liked by the user and why. After completing a Google Meet session, a pop-up appears on the screen to rate the current meeting by giving 1–5 stars; if a user does not give 4 or 5 stars, more questions are asked to the user related to meeting audio, video, and presentation, etc. Questions asked related to the audio of a meeting are shown in Figure 9.3.

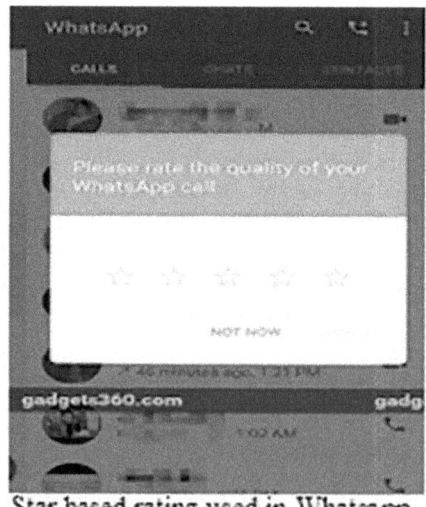
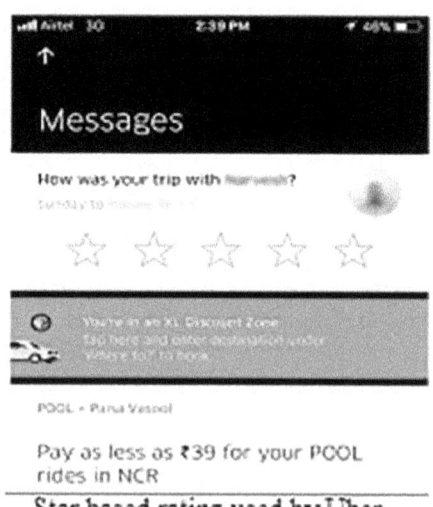

Star based rating used in Whatsapp   Star based rating used by Uber

FIGURE 9.2   Examples of star-based ratings.

FIGURE 9.3   Google Meet pop-up on giving negative feedback.

9.3.3   Text-Based Feedback Evaluation System (TBFES)

Text-based feedback evaluation system (TBFES) permits users to communicate their feelings and emotions in the form of a running text without any kind of restrictions [22]. Analyzing such a text to measure the authors' opinion concretely on an issue/item is known as sentiment analysis or opinion mining. Natural language processing and text mining-based techniques are used to estimate the merits and demerits of items from collecting texts received as feedback. Sentiment-based feedback evaluation system is flexible; users are free to express their opinions on any aspect or feature of the item/service with their degrees of the scale. A sample of textual feedbacks of 'Redmi 8A Dual (Sky White, 32 GB)' mobile phone collected from e-commerce platform 'Flipkart' is shown in Table 9.1.

At the granularity level, analysis of textual feedback can be divided into three categories, i.e., document-level sentiment analysis, sentence-level sentiment analysis, and aspect-based sentiment analysis. In document-level sentiment analysis, textual feedback of an item is treated as a single document, and whole document/feedback is processed to estimate the author's overall opinion. Sentence-level sentiment analysis first marks each sentence of feedback as subjective or objective. Objective sentences are ignored in sentiment analysis because they are factual statements and do not carry author's opinion. Each subjective sentence's sentiment score is evaluated based on the sentiment orientation of words presented in the sentence. Aspect-based sentiment analysis works in two phases, and the first phase identifies the aspects mentioned by the users in the feedback, and sentiment evaluation of those aspects is accomplished in the second phase. In aspect-based sentiment analysis, subjective sentences are processed one by one to estimate user opinion on each identified aspect, and then, a final sentiment score for each aspect is evaluated. However, textual feedback evaluation systems face the following difficulties [22]:

TABLE 9.1

Reviews of Redmi 8A Dual (Sky White, 32 GB)

Feedback 1	Feedback 2	Feedback 3
Gifted this to my father. This is his second smart phone so I have simple expectations with this but trust me it's beyond expectations. A cool buget oriented device with very good camera and battery backup for normal daily usage	It is best value for money mobile and it's camera is awesome. Battery back up is very good and it's work very smooth	Camera: 4/5 Display: 5/5 Battery: 5/5 Value for money: 5/5 Specs: 3.5/5
Feedback 4	**Feedback 5**	**Feedback 6**
Vary good quality	Very good phone in this budget	Fabulous! Thanks to mi, I love this phone
Feedback 7	**Feedback 8**	**Feedback 9**
Great product of Redmi thankyou for the fastest delivery as i didn't expected that i will get my product this soon again thankyou	Very disappointed Camera and flash low quality and Gorila 5 is only the name doesn't work as gorilla 2	Must buy! Good mobile with better performance and high battery power. But RAM and ROM capacity is less
Feedback 10	**Feedback 11**	**Feedback 12**
Very bad	Value for money	Mind-blowing purchase It's a very good mobile for soft work

a. **Feedback in One Word or Only in One Sentence**: Many times users' textual feedbacks are very short. Users write their views in one sentence or sometimes in one word also instead of expressing their views in few lines. More than 30% of feedbacks have only one or two words. With short feedbacks, we can capture overall user sentiment about the product or service like a SBFES but unable to know the users' opinion on different aspects of the item. For example, if one user writes 'value for money', while another user writes 'worst item' in the feedback of a mobile phone, then the only conclusion that can be made is that the first user likes the mobile and the second does not like it at all.

b. **Non Uniformity in Feedback**: TBFES **provides the users the freedom to express** an opinion in their own ways. There is no standard format or method for writing the feedbacks. Due to this liberty, feedbacks do not have a uniform structure. In one feedback, user may express his or her opinion on some aspects, while the other user may give his or her opinion on entirely new aspects. It may also be possible that one aspect is mentioned in very few feedbacks, while another aspect is mentioned in many feedbacks. As shown in feedback 9 and feedback 11 of Table 9.1, feedback 9 has opinions on 'battery' and 'memory', while feedback 11 has opinions only on 'cost'.

c. **Multilingual Feedback**: A few years back, Internet was limited to only some specific people, but now it is used worldwide. In online reviews, it is observed that users are using the English language mixed with their regional language common words to express their opinions. For example, a user might give an opinion about a mobile charger like this 'Charger travel ke liye perfect hai'. In this sentence, a user has used both Hindi and English words. In a country like India, it becomes even more complicated because there are more than 22 official regional languages. It has also come to notice that more than three languages are used in some feedback. Handling such multilingual feedbacks becomes even more challenging. An additional module can be added to the feedback evaluation system and by developing a small multilingual dictionary of common words used by users to tackle this issue.

d. **Use of Slang Words**: It is also observed that people frequently use slang words or abbreviations in their feedback. Slang words are informal and commonly used words that are not part of the vocabulary of any language. Some commonly used words are Dope, GOAT, Gucci, Lit, OMG, etc. Dope is used for cool or awesome, and the word GOAT is used for Greatest of All Time, Gucci for good/cool/going well, Lit for Amazing/exciting, and word OMG is used as an abbreviation for 'Oh My God'. To put weight on their expression, sometimes users use words like 'Toooooooooo… Good' or 'Gooooood' or use short form like 'nce' instead of nice. Handling such feedbacks is another challenge.

e. **Misspelled Words**: Misspelled words are also a big problem in the textual feedback system. Occurrences of such words are very high in feedbacks. Misspelled words may sometimes be due to keyboard interruption or sometimes due to lack of language expertise or simple carelessness. Due to keyboard interruption, the word 'good' may get spelled as 'bood'; similarly word 'computer' may get spelled as 'computre'. Artificial intelligence (AI)-based techniques can be used to predict the correct spelling of the misspelled word.

f. **No Proper Grammar is Used**: Most of the users do not follow the grammar rules of the language while writing the feedback. While doing the language processing, the meaning of the sentences may get changed due to that and might affect the sentiment analysis.

9.4 STATISTICAL FEEDBACK EVALUATION SYSTEM

In the study of different feedback evaluation systems, it is found that each feedback system has its merits and demerits. Questionnaire-based evaluations are quite straightforward and informative but very restrictive in nature. The SBFES is fast, users can share their opinion in just one click, but in this system, it is not possible to capture users' aspect-specific opinions. A TBFES allows users to express feature-based opinions without any restrictions. In a textual feedback evaluation system, we can estimate the sentiments for all those specific aspects that users explicitly state. It is not possible to provide users' opinions about other important aspects of the product or service.

On the basis of the study, it can be concluded that the questionnaire-based method is better than other systems if the formation process of questions can be done such that it is able to capture users' requirements. To improve the power of the existing QBFES, a new statistical feedback evaluation system is designed to remove most of the traditional questionnaire system's shortcomings, as discussed in Section 3.1. In this system, important aspects of an item are automatically extracted using natural language processing and text mining techniques, and then the questions related to the stated aspects are included in the questionnaire. Significant weights are assigned to all extracted aspects, and their importance is statistically estimated using the statistical information retrieved from textual feedbacks. An item's overall effectiveness is estimated by considering weights and sentiment scores for all the important aspects. With a statistical feedback evaluation system, we can also perform customized feedback evaluations to measure an item's effectiveness on some selected aspects. Statistical feedback evaluation system is implemented using a modular approach. There are five modules: aspect extraction, aspect weight estimation, sentiment evaluation, customized evaluation, and aspect-based questionnaire design. The functions of these modules are discussed in the next sections.

9.4.1 Aspect Extraction

Aspect extraction is one of the important modules of the statistical feedback evaluation system. This module's major responsibility is the identification of essential (important) aspects from text after the collection of textual feedback and preprocessing of feedback. A simple block diagram of this module is shown in Figure 9.4. It is implemented using three sub modules, i.e., 'feedback collector', 'feedback preprocessor', and 'aspect validator. Essential aspects identified by this module are passed to other modules for further processing.

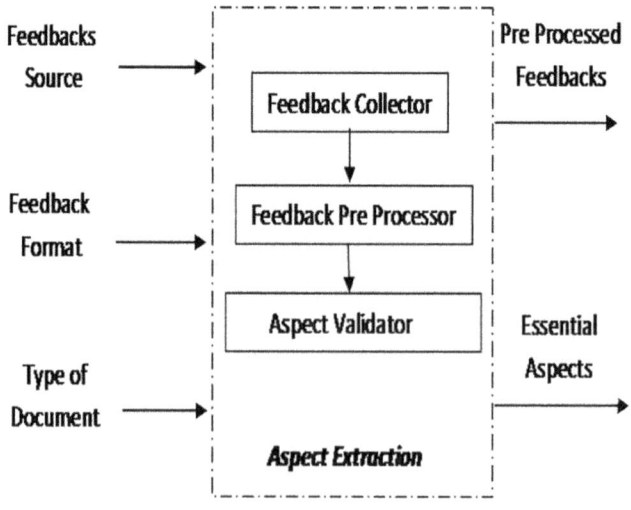

FIGURE 9.4   Block diagram of aspect extraction module.

9.4.1.1 Feedback Collector

This module has three inputs, i.e., source of feedback, feedback template, and type of feedback document. Source of feedbacks is the address of the remote location where feedbacks of an item is stored; feedback template is the template used to store the feedback at the source location, and type of document is the type of the sources document, i.e., PDF, doc, docx, HTML, or XML which are used at the source to store the feedbacks of an item. In the Internet world, a large number of feedbacks on an item (product/service/professional) is available on different online platforms, i.e., e-commerce Web sites, networking Websites, blogs, and forums. With the help of this module, feedbacks of an item are collected from different sources stored in different layouts and formats.

To implement the suggested methodology, feedbacks of teachers and laptops have been used. Feedbacks of laptops are collected from the public platform that was released by SemEval [23,24] and feedbacks of faculty members were collected from three different sources, i.e., public platform for sharing feedback of faculty from United States of America [25], public platform for sharing feedbacks of faculty from India [26], and textual feedback of faculty members collected from our institute [27]. This module's output is a raw text, and it is passed to 'feedback preprocessor' module to remove unwanted characters, symbols, etc. We have implemented this module using standard libraries of Python [28] and Natural Language Toolkit (NLTK) [29].

9.4.1.2 Feedback Preprocessor

Opinions/feedbacks collected by '*feedback collector*' module may contain many unwanted characters, symbols, tags, links, etc. Therefore, this module's feedback collected in raw form is processed to get the clean text and make it suitable for processing for other modules. This module's important functions are removing unwanted information, text normalization, standardization of decoding, spelling and grammar correction, etc. Removal of unwanted information includes removal of HTML tags, HTML entities, hyperlinks, images, tables, URLs, etc.

Feedbacks stored at different sources are mostly given in a variety of formats using different encoding schemes also. With the help of text normalization and standardization of decoding, feedbacks are converted into standard form. The grammar of feedback sentences is checked, and many misspelled words are also corrected by this module. We have executed this module with standard libraries of Python [28] and NLTK [29].

9.4.1.3 Aspect Validator

The entities, attributes, and properties of an item are known as aspects/features. For example, aspect terms of faculty members are '*qualification*', '*knowledge*', '*presentation*', '*discipline*', '*paper pattern*', etc., while aspect terms of laptops are '*price*', '*memory*', '*processing speed*', '*color*', '*camera resolution*', etc.

In this module, we have used statistical and linguistic properties to identify valid aspects of an item. The proper syntactic category of the word in a sentence extracted using NLP tools helps us in identifying the probable aspects, and mostly these are either a noun or adjective or adverb, which define valid aspect. It is also observed that the valid aspects have similar context (similar surrounding words) and are often associated with sentiment displaying words. Focusing on these points, this module works in two phases: in the first phase, a list of all possible aspect terms is generated from textual feedbacks, and in the second phase, a list of syntactically and statistically adequate terms called essential aspects is selected from the list generated in the first phase.

The aspect validator module's complete flowchart is given in Figure 9.5, and it is based on the method suggested by Hu and Liu [1].

In the first phase's implementation, all those terms are added in a list L1 that has syntactic category as noun or noun phrase or compound noun. Part of speech tagger of natural language toolkit (NLTK POS tagger) [29] has been used to identify the words' proper syntactic categories in preprocessed feedback.

Few adjectives and rare adverbs (excluding general sentiment-bearing words) present in the sentence as noun/adjective modifiers may be the aspects of the item/ service. In the feedbacks of faculty members, modifier adjectives and adverbs of words 'teacher', 'department', 'institute', etc., are valid aspects. For example, the terms '*regular*', '*knowledgeable*', '*punctual*', '*sincere*', etc., are adjectives but the modifiers of the word 'teacher' shall be considered as aspects for faculty. Therefore, in the next step, such adjectives and adverbs are added to the list L1. Adjectives and adverb modifiers of domain-specific terms are identified with the help of the Stanford Dependency Parser [30].

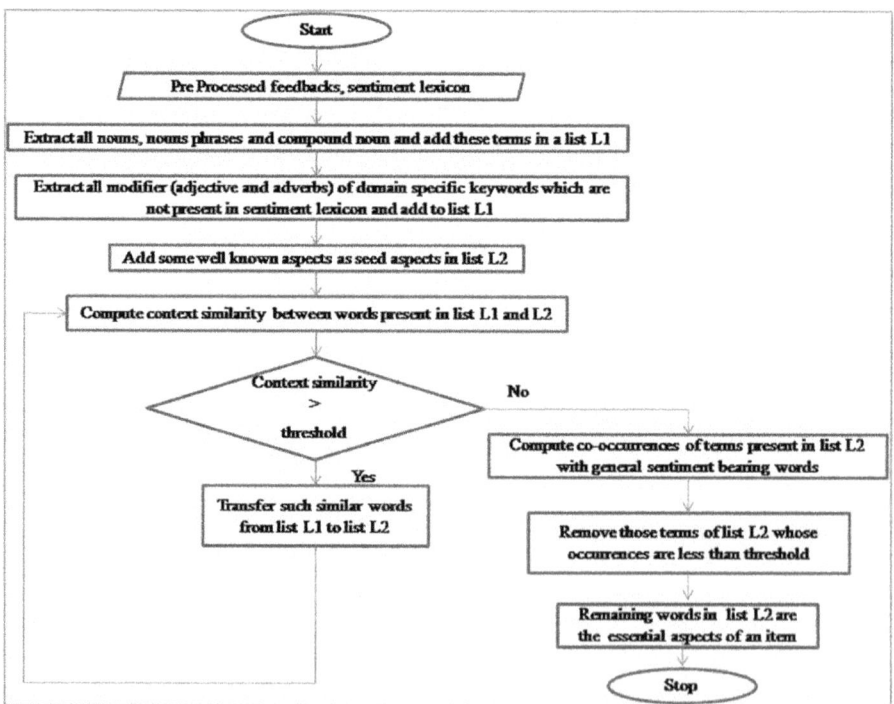

FIGURE 9.5 Flowchart for essential aspect extraction using linguistic and statistical knowledge.

In previous steps, the most probable aspects of an item are added into a tentative list of aspects, but they may contain many irrelevant aspects. A bootstrapping-based technique is used to remove irrelevant words from the tentative list of aspects. The technique is based on simple intuition that valid aspects are paradigmatic words and frequently co-occur with sentiment-bearing words. Words that have similar contexts and can be used in place of other contextually similar words without violating the syntactic and semantic structure of sentences are known as paradigmatic words.

To implement bootstrapping for removing the irrelevant words from list L1, a new list L2 is formed containing few seed aspects representing necessary aspects of the item. Then, contexts of all terms present in the list L1 and L2 are computed and compared. An N-dimensional vector represents a word's context, and the computational details are explained in the next paragraph.

After the context matrix computation, those terms of list L1 are added to the list L2, which have a high contextual similarity value for the terms present in the list L2. The process of adding highly contextual similar words from the list L1 to the list L2 is continued till more new words are added to list L2. After this step, L2 is considered as the list of valid aspects.

9.4.1.3.1   Evaluation of Context Similarity Measurement between Two Words:

The context of a word is addressed by an N-dimensional vector, where N express the number of total words in the vocabulary, and for each word 'w', the value of the ith entry in the vector shows the number of times that word has occurred in the context of word w. To form an N-dimensional context vector of a word 'w', all the words present to its left and right sides in the context window of four words are extracted. Extracted words are converted into N-dimensional vectors by using the Bag of Words (BOW) scheme. To compare contexts of two terms, cosine similarity is computed between two N-dimensional context vectors as shown in Figure 9.6. Let N-dimensional context vectors of two words 'w1' and 'w2' are V1 and V2.

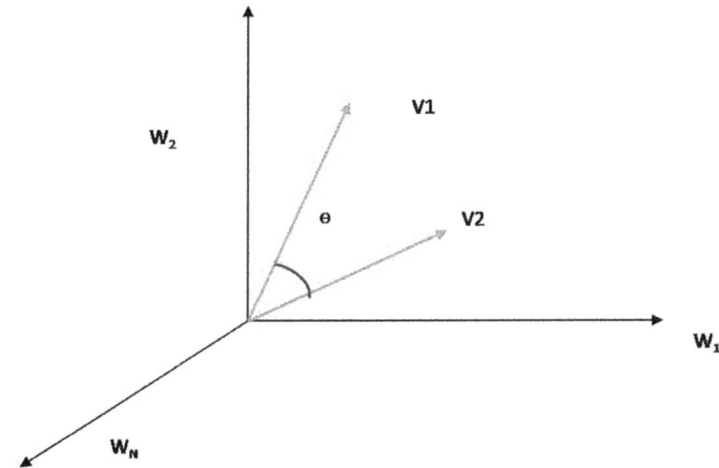

FIGURE 9.6   Vector representation of context.

$$V1 = [0,1,0,2,1,...1] \quad \text{and} \quad V2 = [1,0,0,2,1,...1].$$

Cosine similarity between two vectors '$V1$' and '$V2$' is computed by measuring cosine angle between them where cosine angle is derived from Euclidean dot product formula as shown in equations (9.1) and (9.2).

$$V1 \cdot V2 = |V1||V2|\cos\theta \tag{9.1}$$

$$\cos\theta = \frac{V1 \cdot V2}{|V1||V2|} = \frac{\sum_{i=1}^{N} V1_i V2_i}{\sqrt{\sum_{i=1}^{N}(V1_i)^2}\sqrt{\sum_{i=1}^{N}(V2_i)^2}} \tag{9.2}$$

$V1_i$ and $V2_i$ are the magnitudes of vectors $V1$ and $V2$ in ith dimension.

Three different methodologies have been experimented to compute the context vectors of words. In the first case, every dimension represents the number of occurrences of every word (the number of times a specific word has appeared in the context). In the second case, a binary contextual vector is created where each dimension of the contextual vector represents only the presence or absence of a specific word by a binary number. In the third case, each contextual vector dimension represents the term frequency and inverse document frequency (TF-IDF) score of individual words.

In TF-IDF scheme, each word's importance (term) is estimated, and high weights are assigned to those terms, which appear more frequently in some sentences and rarely in other sentences. Term frequency (TF) of a term 'w' in a sentence 'S' is the function of the total number of appearances of the term 'w' in the sentence 'S' normalized by the length of the sentence 'S' as shown in equation (9.3). The inverse document frequency (IDF) score of a term 'w' is computed by the logarithmic ratio of the total number of sentences in the feedback and the number of sentences in which term 'w' appears, as shown in equation (9.4). TF-IDF weight of term 'w' is the multiplication of TF and IDF scores as shown in equation (9.5). TF score gives weight to frequently occurring terms, while IDF score penalizes if the terms commonly occur in other sentences, as shown in Figures 9.7 and 9.8.

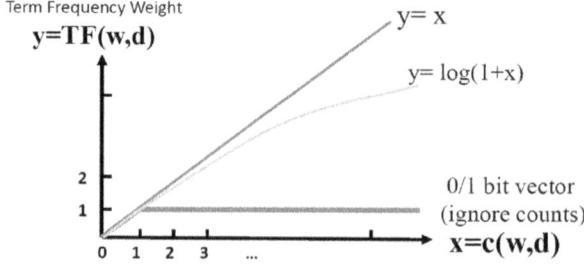

FIGURE 9.7 Variation in TF scores.

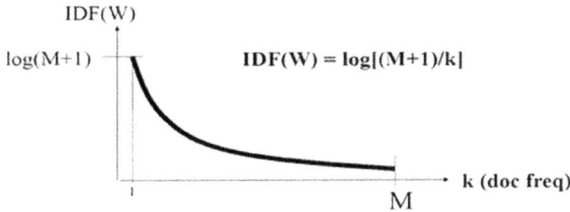

FIGURE 9.8 Variation in IDF scores.

$$\mathrm{TF}\big(w, S\big) = \log\left(1 + \frac{\text{Number of ``}w\text{'' appear in sentence ``}s\text{''}}{\text{Number of words in the sentence ``}s\text{''}}\right) \qquad (9.3)$$

$$\mathrm{IDF}(w) = \log\left(\frac{M+1}{k}\right) \qquad (9.4)$$

where
  M represents the total number of sentences in feedback collection,
  k represents the total number of sentences containing the term 'w'.

$$\mathrm{TF} - \mathrm{IDF}(w) = \mathrm{TF}\big(w, S\big) * \mathrm{IDF}(w) \qquad (9.5)$$

Results show that TF-IDF-based context representation gives better results than other methods. In teachers' feedback, contextually similar terms for the seed word 'punctual' are shown in Figure 9.9. In laptops' feedback, contextually similar terms for the seed word 'keyboard' are shown in Figure 9.10. The threshold value for each collection of feedbacks is decided through experiment.

9.4.1.3.2 Association Measurement between the Aspect and the Sentiment-Bearing Terms

In the next step, the list L2 is scanned to examine the frequency of co-occurrence of each word with the sentiment words, and the words not occurring with sentiment words within a predefined threshold are removed from the list L2. To measure

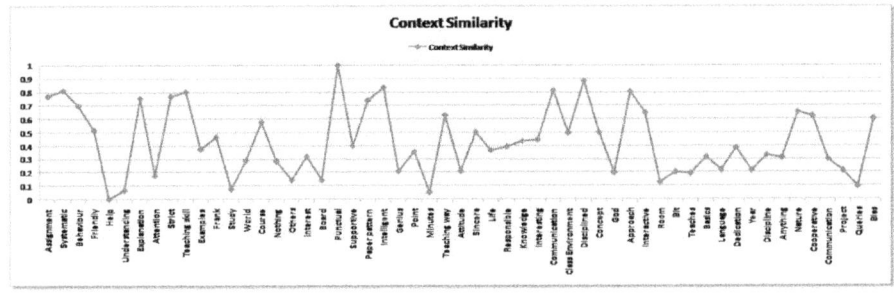

FIGURE 9.9 Contextually similar words with similarity for the word 'Punctual' in teachers' feedbacks.

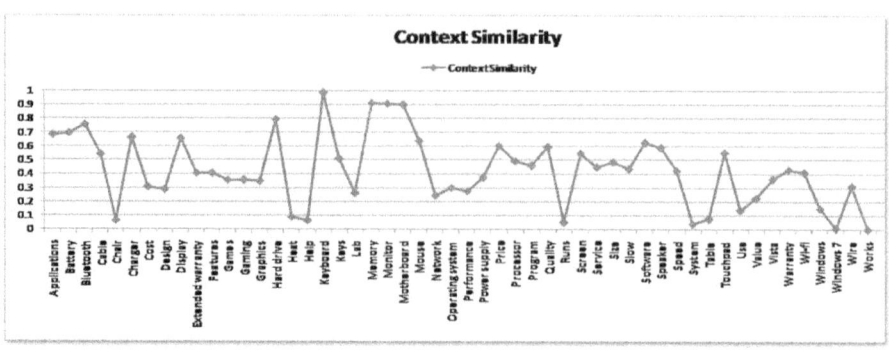

FIGURE 9.10 Contextually similar words with similarity for the word 'Keyboard' in laptops' feedbacks.

the co-occurrences of possible aspects with the sentiment words, point wise mutual information (PMI) is calculated between the words in L2 and the sentiment words taken from the sentiment lexicon. PMI between two words' 'w1' and 'w2' is computed by the formula given in equation (9.6).

$$\text{PMI}(w1, w2) = \log \frac{P(w_1, w_2)}{P(w_1) X P(w_2)} = \log \frac{P(w_1 \mid w_2)}{P(w_1)} = \log \frac{P(w_2 \mid w_1)}{P(w_2)} \quad (9.6)$$

In teachers' feedbacks, the degree of association of sentiment words and terms in list L2 is shown in Figure 9.11. Similarly, in laptops' feedbacks, the degree of association of sentiment words and terms in list L2 is shown in Figure 9.12.

In the final step of aspect identification, only those aspect terms are retained in list L2 that appear at least in minimum M feedbacks where M is a user-defined threshold but decided experimentally. The final list of essential aspects identified for faculty members and laptops is shown in Tables 9.2 and 9.3. Aspect terms of teacher and laptop that qualify all necessary conditions are shown in Figures 9.13 and 9.14.

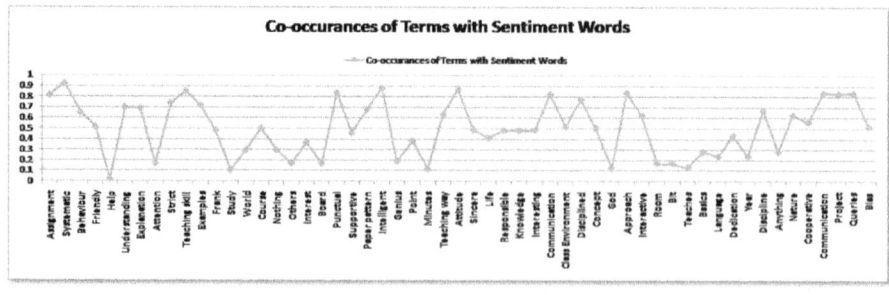

FIGURE 9.11 Degree of association of possible aspects with sentiment words in teachers' feedbacks.

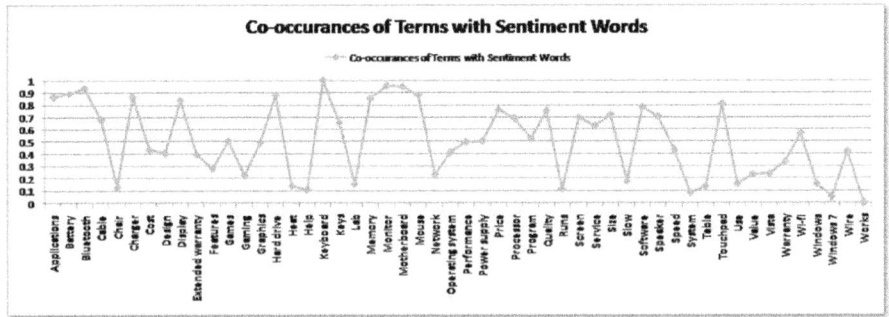

FIGURE 9.12 Degree of association of possible aspects with sentiment words in laptop's feedbacks.

TABLE 9.2
List of Essential Aspects for Teachers Identified from Teachers' Feedbacks

Punctual	Assignment	Teaching Way	Class Environment
Disciplined	Strict	Cooperative	Frank
Intelligent	Explanation	Bias	Interesting
Systematic	Paper pattern	Course	Knowledge
Communication	Behavior	Friendly	Supportive
Teaching skill	Nature	Concept	Responsible
Approach	Interactive	Sincere	Dedication

TABLE 9.3
List of Essential Aspects for Laptops Identified from Laptops' Feedbacks

Keyboard	Applications	Speaker	Program
Memory	Charger	Touchpad	Service
Monitor	Display	Screen	Power supply
Motherboard	Mouse	Cable	Warranty
Hard drive	Software	Keys	Speed
Bluetooth	Price	Processor	Wi-Fi
Battery	Quality	Size	Extended warranty

9.4.2 Aspect Weight Estimation

Some features of each item (product/service/professional) carry more dominance than other aspects. For example, for the item laptop, *'battery life'* and *'processor'* are more valuable in comparison with other aspects like *'laptop color'* and *'warranty'*. This module estimates the importance (weight) of every aspect of the item. In

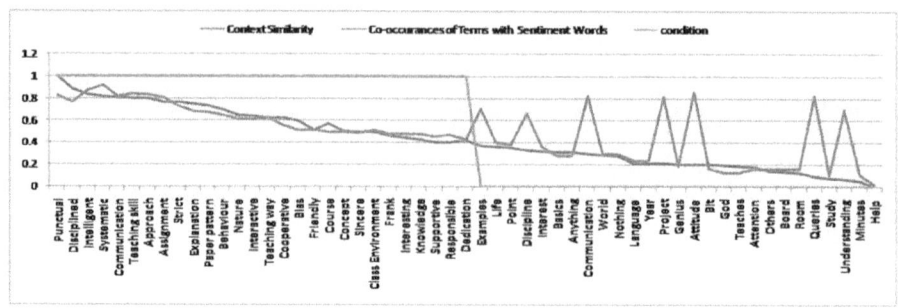

FIGURE 9.13   Selection of essential aspects of teachers.

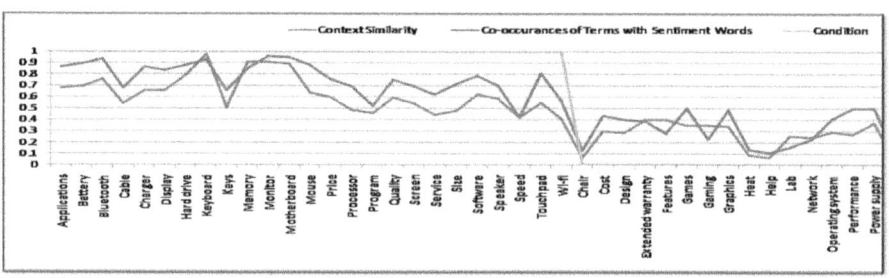

FIGURE 9.14   Selection of essential aspects of laptops.

the estimation of weight or importance of features, it is assumed that the users like to write about those features in their feedback which were the key features for them at the time of purchase or at the time of hiring. However, users tend not to explicitly mention those features that are not so important to them. Therefore, the degree of inclusion of aspect terms in the feedbacks represents the weight of aspects. The weight or importance of essential aspects is computed by Algorithm 1 given below. Weights of some aspects of laptops and teachers are shown in Figures 9.15 and 9.16.

Algorithm 1: Aspect Weight Estimation

Input:	VF : List of essential aspects, LF: List of preprocessed feedbacks
Output:	FW : Aspect weight vector

Step 1:	Set pointer i at the beginning of the list VF
Step 2:	f=aspect pointed by pointer i in list VF
Step 3:	count=0
Step 4:	Set pointer j at the beginning of the list LF
Step 5:	If aspect f is present in the feedback-pointed pointer j in list LF, then count=count+1
Step 6:	If j is not at the end of the list LF, then forward pointer j and goto Step 5
Step 7:	FW[i]= count/total number of feedback in the list LF (length of list LF)

Step 8:   If pointer i is not at the end of the list VF, then increment pointer i and goto Step 3

Step 9:   Return to vector FW, values in the list FW are the weight of essential aspects.

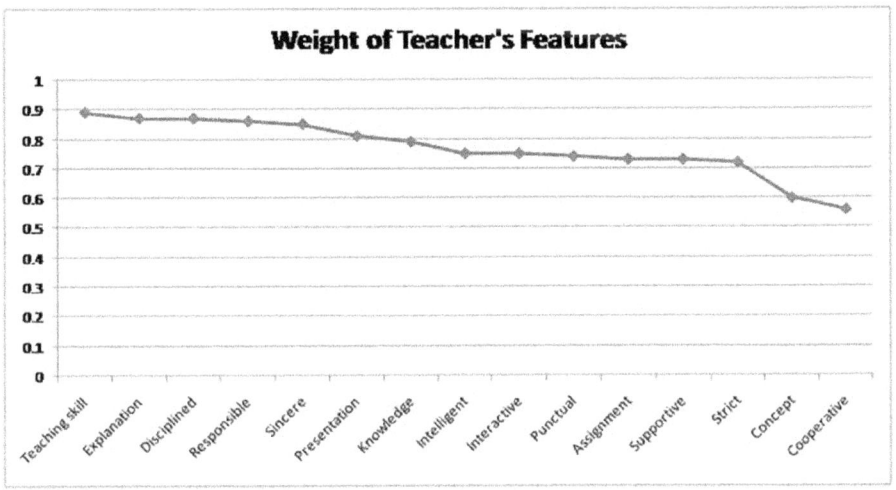

FIGURE 9.16   Weights of important features of teachers.

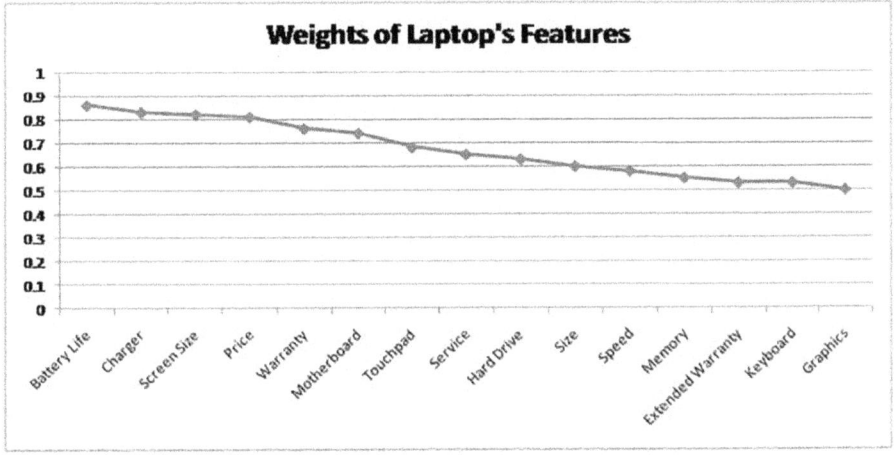

FIGURE 9.15   Weights of important features of laptops.

9.4.3 SENTIMENT EVALUATION

Sentiment evaluation module calculates a sentiment score for all the aspects identified in the previous module. The final score is computed in two phases. In the first phase, all aspects' sentiment score is evaluated for each feedback separately by taking into account all the occurrences related to that aspect in that feedback. In the second phase, each aspect's total sentiment score is calculated by aggregating sentiment scores of all feedback for that aspect. Sentiment evaluation is implemented with the help of two modules, i.e., sentiment estimator and sentiment aggregator. Implementation details of sentiment estimator and sentiment aggregator modules are given in the next section.

9.4.3.1 Sentiment Estimator

Sentiment estimator module processes all feedbacks one by one to estimate the sentiment score of all mentioned features and output sentiment score of each aspect for every user feedback. Block diagram of the sentiment estimator module is shown in Figure 9.17. Input to this module is the list of essential aspects identified by aspect extraction, preprocessed feedbacks generated by aspect extraction module, and sentiment lexicon (list of positive and negative sentiment-bearing words with sentiment scores). In this module, preprocessed feedback is processed user by user to estimate all essential aspects' sentiment score. Results generated by this module are kept in a table named User Aspect Sentiment Record.

In the sentiment estimator, only subjective sentences are processed, and the system ignores objective sentences. The subjectivity analyzer checks the subjectivity of sentences. The subjectivity analyzer module's basic function determines whether the sentence contains information describing aspects of it is just a factual statement that does not carry user opinion. Subjectivity analyzer module has used Naïve-based approach to tag the sentence as objective or subjective. Subjective sentences are processed to compute the sentiment score of aspects with the help of sentiment-bearing words and sentiment shifter words present in the sentence. This module assigns '+1' or '−1' or '0' as a sentiment score to each of the aspects having 'positive', 'negative', and 'neutral' sentiment, respectively. The assigned sentiment score sign is inverted if sentiment shifter words like 'no', 'not', 'never', 'do not', 'does not'

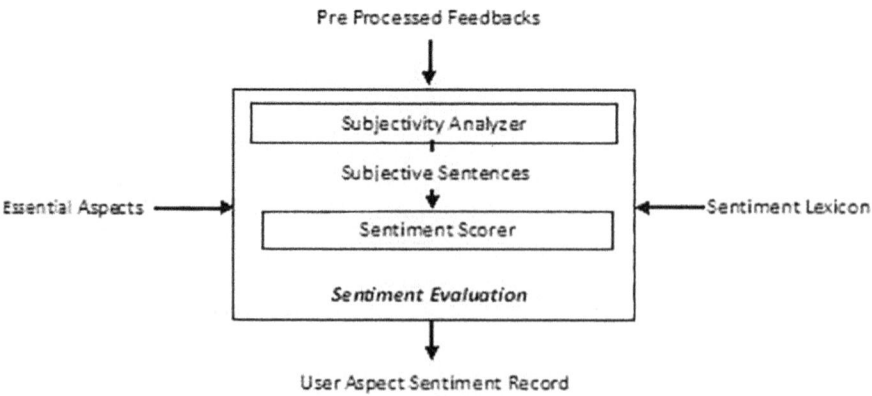

FIGURE 9.17 Block diagram of sentiment estimator model.

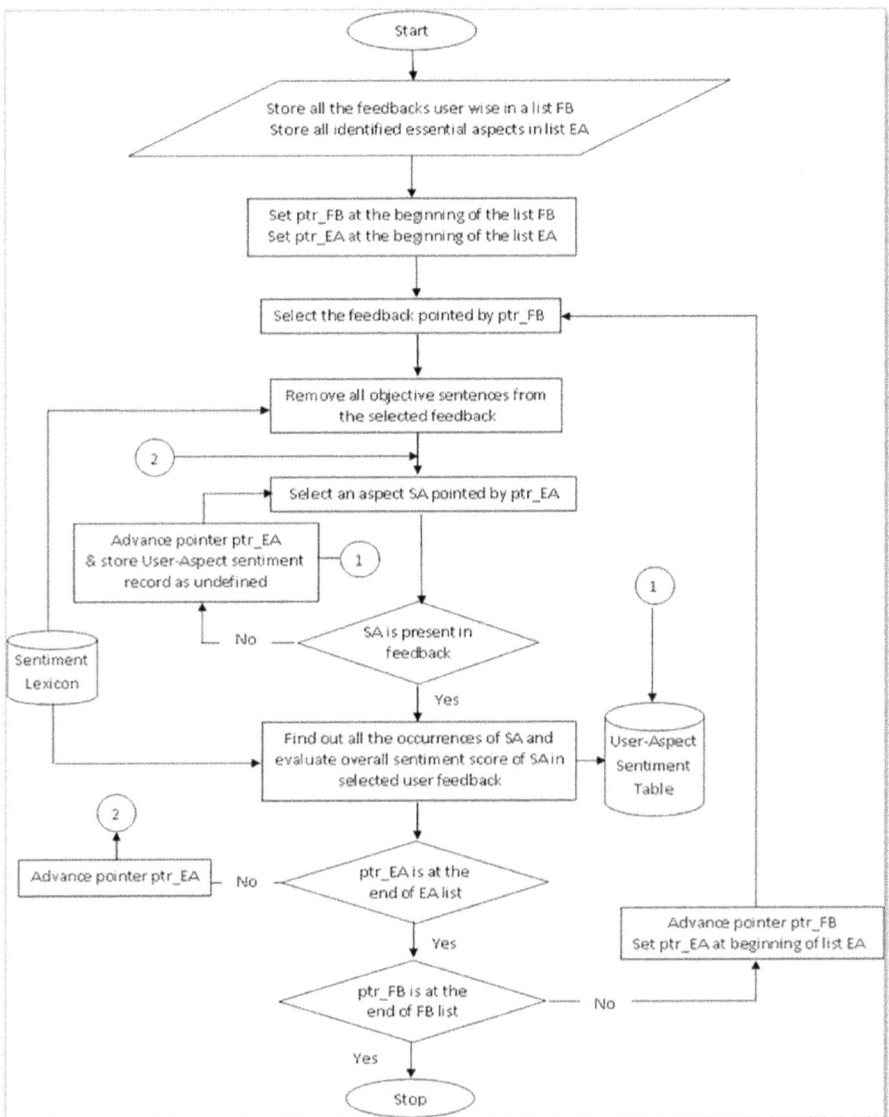

FIGURE 9.18   Flowchart of sentiment estimator module.

appear near to sentiment-bearing words. The flowchart of the sentiment estimator is shown in Figure 9.18.

Lexicons, which are available in the open domain, provide sentiment scores. Still, to get better sentiment scores, domain-specific sentiment lexicon has been developed. In this system, we have used SentiWordNet 3.0 [18], which is freely available in the public domain. The sentiment estimator module's output is the User Aspect Sentiment Record with respect to each feedback, as shown in Table 9.4.

TABLE 9.4

User Aspect Sentiment Record Format

	Aspect-1/ Feature-1	Aspect-2/ Feature-2	Aspect-3/ Feature-3	Aspect-3/ Feature-3	...	Aspect-m/ Feature-m
User-1	+1	−1		0	...	
User-2	0			-1	...	−1
User-3	−1		+1		...	
User-4		+1		−1	...	+1
User-5	+1		0			0
...	....	....	....		....	....
User-n		−1	−1	−1	....	

9.4.3.2 Sentiment Aggregator

Sentiment aggregator module aggregates the sentiment scores for each aspect to give a final sentiment score indicating the sentiments of all the users' feedbacks. Input to this module is User Aspect Sentiment Record computed by the sentiment estimator module. This module does not calculate the simple average of scores to compute the final sentiment score, and rather, it adds a bias term by multiplying the number of neutral sentiments by a very small factor $\log \dfrac{P+1}{N+1}$. The aggregate sentiment score of a feature 'a' (*feature_aggregate_score (a)*) is calculated using the formula given in equation (9.7). In this aggregation, no positive sentiments are multiplied by '+1', negative sentiments by '−1', and neutral sentiments are multiplied by a fractional biasness of positivity and negativity particular aspect 'a' to get the total sentiment score. The final aggregate sentiment score is evaluated by dividing by the sum of the total number of sentiments retrieved for that aspect from all feedbacks.

Let P, N, and *Neut* are the numbers of positive, negative, and neutral sentiments of a particular feature 'a' with respect to all feedbacks.

$$feature\_aggregate\_score\ (a) = \frac{P*(+1) + N*(-1) + \log\left(\dfrac{P+1}{N+1}\right)* Neut*(+1)}{P + N + Neut}$$

(9.7)

The value of *feature_aggregate_score (a)* is always between +1 and −1. These values are helpful in calculating the overall sentiment score of an item. Aggregate sentiment scores of some required aspects for a teacher are shown in Figure 9.19.

9.4.4 CUSTOMIZED EVALUATION

After calculating each aspect's final sentiment score and the weight of each aspect, a comprehensive overall sentiment score can be evaluated by simply taking the weighted average. However, all users do not have the same requirement. Expectations of each user from a product or a service are as per his or her need and utility. For example,

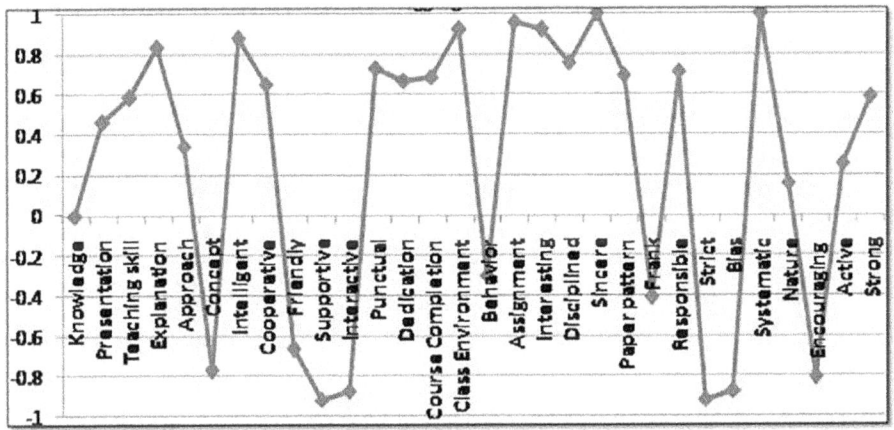

FIGURE 9.19 Aggregate sentiment score of some aspects in teacher domain.

mobile phones are essential for all in the current scenario, but the users expect different kinds of features of different domains. People of one domain may want long life of battery and low price, while people of other domains may look for the good-quality microphone, speaker, and camera. Therefore, it may be very useful to a user if he or she can get a customized sentiment score of an item/service, taking into consideration only user-selected aspects. Statistical feedback evaluation system allows us to perform a customized evaluation of items.

The key functions of this module are performance evaluation and performance comparison. With this module, we can measure effective performance in two ways, i.e., comprehensive performance evaluation and customized performance evaluation. In comprehensive performance evaluation, the item's overall effectiveness is computed by considering weights and aggregate sentiment scores of all essential aspects of the item. The comprehensive performance of an item 'x' is computed by the following formula shown in equation (9.8). Let there are n aspects, weights and aggregate sentiment scores are stored in lists W and ASS retrieved by aspect extraction and aspect weight estimation modules, respectively.

$$SP(X) = \left(\sum_{i=1}^{n} W[i] * ASS[i] \right) \Big/ n \tag{9.8}$$

Customized performance evaluation is similar, but aspects are chosen by the user. The user also has the facility to assign weights on its own also. This performance index helps users in choosing the most appropriate item according to his or her need and utility.

The customized evaluation module allows us to compare two or more items effectively. This module also generates few customized reports without the intervention of the user. Aspect summary report of laptop 'L1' is shown in Figure 9.20. The laptop brand comparison report is shown in Figure 9.21. In this report, three popular brands are compared for user-selected aspects.

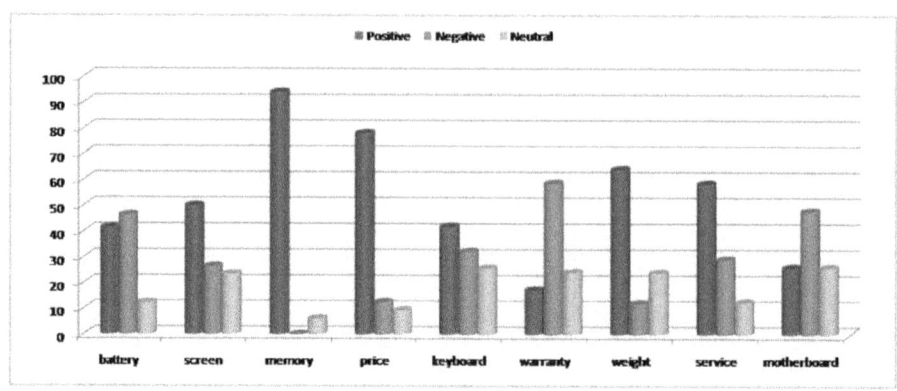

FIGURE 9.20 Individual laptop report of a laptop 'L1'.

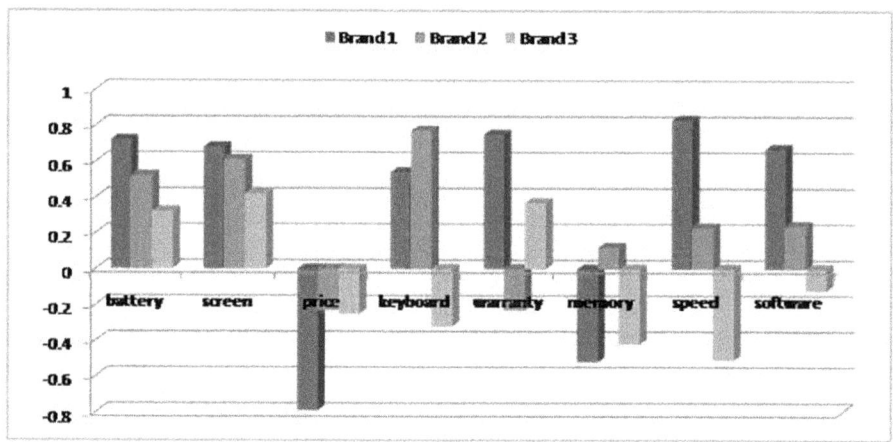

FIGURE 9.21 Comparison of laptop brands.

9.4.5 ASPECT-BASED QUESTIONNAIRE DESIGN

Commonly, questions of the questionnaire are decided by the senior members of the organization on the basis of their knowledge, experience, and assumptions, and all questions have equal importance. A traditional questionnaire's quality can be enhanced by incorporating the knowledge extracted by the statistical feedback evaluation system.

A questionnaire designed with the help of the statistical feedback evaluation system is called aspect-based questionnaire. In this questionnaire, questions related to all the identified aspects are included, and the weight is also assigned to each question. A sample aspect-based questionnaire designed for evaluating teacher performance using the planned method is mentioned in Table 9.5. Rows of the questionnaire contain questions related to important aspects of faculty and columns for different subjects. Set of questions are formed to know the feedback of faculty members in

TABLE 9.5

Statistical Questionnaire for Performance Evaluation of Faculty Members

S.No.	Questions	MTH – S101	PHY – S101	...	HSS – S101
1	The instructor was well prepared for his or her class and was able to deliver the lecture effectively		....		
2	The instructor was punctual in the class		....		
3	The instructor was able to answer the question raised in the class satisfactorily		....		
4	The quizzes and examinations covered the entire syllabus taught in the class		....		
5	The question papers were balanced to test the knowledge gain in the subject		....		
6	The instructor was unbiased in correcting the answer sheets		....		
7	All quizzes were held in time		....		
8	Proper assignments were given on each and every topic of the subject		....		
9	Assignments were graded on time		....		
10	The instructor was able to maintain discipline in the class		....		
11	Overall understanding and knowledge were gained in the course		....		

different subjects of a semester. The importance of questions is decided on the basis of aspects' weight estimated by the aspect weight estimation module. Similar kinds of questionnaires can be prepared and used by the product manufacturers, service providers, and employers to capture users' opinions on related products or service.

9.5   RESULT ANALYSIS AND DISCUSSION

The effectiveness of the proposed system depends on the correctness of the aspect extraction module and sentiment evaluation module. The accuracy of those modules has been estimated with three performance parameters, i.e., precision, recall, and accuracy. To estimate these parameters, the results of the system are compared with the results provided by human experts. For the analysis of teachers' feedbacks, feedbacks were collected from three different sources [25–27], and appropriate labels/tags (aspect terms) are assigned to approximately 500 sentences manually from each source. Similarly, to analyze laptop reviews, feedbacks were collected from a public repository [22,24]. In this repository, all sentences are pre tagged with the aspect terms. On the basis of results generated by the system and pre tagged results, all three performance parameters are estimated. Performances of aspect extraction and sentiment evaluation modules for teachers and laptops are shown in Tables 9.6 and 9.7.

Statistical feedback evaluation system allows the users to perform a customized evaluation, which helps the users find the most suitable items for them. With

TABLE 9.6

Correctness of Aspect Extraction and Sentiment Evaluation Module in Teachers' Feedback

	Average Precision			Average Recall			Average Accuracy		
	Source1	Source2	Source3	Source1	Source2	Source3	Source1	Source2	Source3
Aspect extraction	84.32	85.02	86.71	84.53	86.34	86.21	87.67	87.21	88.02
Sentiment evaluation	82.05	83.31	84.56	83.45	81.23	83.43	85.46	86.67	87.49

TABLE 9.7
Correctness of Sentiment Evaluation Module in Teachers' Feedback

	Average Precision	Average Recall	Average Accuracy
Aspect extraction	85.17	85.65	83.14
Sentiment evaluation	83.24	87.45	82.09

customized reports, product manufacturers and service providers can easily identify weak aspects of their product or service and improve quality. With the help of this system, a more meaningful and strong questionnaire can be generated, which can capture users' views also. In such a questionnaire, questions are dynamically extracted from the textual feedbacks, and weights are assigned to questions depending upon the importance given to aspects by the users. The proposed system is versatile and flexible and shall perform well in new scenarios.

It is observed that traditional teaching methodology has been changed due to COVID-19. Teachers are taking classes online, and the evaluation pattern is also changed where students are giving online examinations. To evaluate teachers' performance in the current scenario, a different kind of questionnaire is required. The proposed system can generate an environment-dependent questionnaire by taking the feedback of the students.

9.6 CONCLUSION

In this chapter, different feedback evaluation systems are discussed in detail with their merits and demerits. In this study, it is found that QBFESs are traditional and quite straightforward, but these systems are very static and restrictive. Existing textual feedback evaluation systems provide freedom to the users to express their opinion without any restrictions. A major problem of textual feedback systems is opinion aggregation due to the uneven structure of feedbacks. Projected statistical feedback evaluation system allows us to design an effective questionnaire using statistical knowledge of the text. In this questionnaire, questions and their weights are not pre decided. Statistical feedback evaluation system is helpful to the users and manufacturers in finding the appropriate item as per their choices.

9.7 FUTURE WORK

An automatic module is used in the proposed system to extract important aspects; sentiment evaluation and weight estimation representing the importance of extracted aspects are also done automatically. At present, questions for the modified questionnaire are being framed manually but that part can be automated easily by using simple AI techniques. Our future plan is also to explore other domains like tours, travels, hospitality, and employees for the organization. We are also trying to incorporate some domain-specific linguistic knowledge to enhance the performance of statistical feedback evaluation system.

REFERENCES

1. Hu, M. & Liu, B. (2004). Mining and summarizing customer reviews. In *Proceedings of the 10th ACM SIGKDD International Conference on Knowledge Discovery and Data Mining (KDD '04)*. ACM, New York, NY, pp. 168–177. doi: 10.1145/1014052.1014073.
2. Ghallab, A., Mohsen, A. &Ali, Y. (2020). Arabic sentiment analysis: A systematic literature review. *Applied Computational Intelligence and Soft Computing*, Article ID: 7403128. doi: 10.1155/2020/7403128.
3. Kumar, A. & Jain, R. (2015). Sentiment analysis and feedback evaluation. In *IEEE 3rd International Conference on MOOCs. Innovation and Technology in Education (MITE)*. IEEE, Amritsar, India.
4. Kumar, A. & Jain, R. (2020). Attribute extraction from textual feedbacks for effective opinion analysis. *Journal of Critical Reviews*, 7(11): 1706–1716. doi: 10.31838/jcr.07.11.276.
5. Neethu, M. S. & Rajasree, R. (2013). Sentiment analysis in twitter using machine learning techniques. In *Fourth International Conference on Computing, Communications and Networking Technologies (ICCCNT)*, Tiruchengode, pp. 1–5. doi: 10.1109/ICCCNT.2013.6726818.
6. Al-Smadi, M., Qawasmeh, O., Al-Ayyoub, M., Jararweh, Y. & Gupta, B. (2018). Deep Recurrent neural network vs. support vector machine for aspect-based sentiment analysis of Arabic hotels' reviews. *Journal of Computational Science*, 27: 386–393.
7. Poria, S., Cambria, E., Ku, L.-W., Gui, C. & Gelbukh, A. (2014). A rule-based approach to aspect extraction from product reviews. In *Workshop on Natural Language Processing for Social Media (Social NLP)*, Dublin, Ireland, pp. 28–37.
8. Maharani, W., Widyantoro, D. H. & Khodra, M. L. (2015). Aspect extraction in customer reviews using syntactic pattern. *Procedia Computer Science*, 59: 244–253.
9. Ilmania, A., Cahyawijaya, S. & Purwarianti, A. (2018). Aspect detection and sentiment classification using deep neural network for Indonesian aspect-based sentiment analysis. In *International Conference on Asian Language Processing (IALP)*, Bandung, Indonesia, pp. 62–67, doi: 10.1109/IALP.2018.8629181.
10. Jin, W., Ho, H. H. & Srihari, R. K. (2009). OpinionMiner: A novel machine learning system for web opinion mining and extraction. In *Proceedings of the 15th ACM SIGKDD International Conference on Knowledge Discovery and Data Mining (KDD '09')*. ACM, New York, NY, pp. 1195–1204. doi: 10.1145/1557019.1557148.
11. Shariaty, S. & Moghaddam, S. (2011). Fine-grained opinion mining using conditional random fields. In *Data Mining Workshops (ICDMW). IEEE 11th International Conference*. IEEE, Vancouver, BC, Canada, pp. 109–114.
12. Li, F., Han, C., Huang, M., Zhu, X., Xia, Y.-J., Zhang, S. & Yu, H. (2010). Structure-aware review mining and summarization. In *Proceedings of the 23rd International Conference on Computational Linguistics*, Beijing, China, pp. 653–661.
13. Nandal, N., Tanwar, R. & Pruthi, J. (2020). Machine learning based aspect level sentiment analysis for Amazon products. *Spatial Information Research*, 28: 601–607.
14. Wu, C., Wu, F., Wu, S., Yuan, Z. & Huang, Y. (2018). A hybrid unsupervised method for aspect term and opinion target extraction. *Knowledge-Based Systems*, 148: 66–73.
15. Alsaeedi, A. & Khan, M. Z. (2019). A study on sentiment analysis techniques of Twitter data. *International Journal of Advanced Computer Science and Applications*, 10 (2): 361–374.
16. Shah, A., Kothari, K., Thakkar, U. & Khara, S. (2020). User review classification and star rating prediction by sentimental analysis and machine learning classifiers. In: Tuba, M., Akashe, S. & Joshi, A. (eds). *Information and Communication Technology for Sustainable Development*. Advances in Intelligent Systems and Computing, vol. 933. Springer, Singapore. doi: 10.1007/978-981-13-7166-0_27.

17. Thet, T. T., Na, J.-C. & Khoo, C. S. G. (2010). Aspect-based sentiment analysis of movie reviews on discussion boards. *Journal of Information Science*, 36 (6); 823–848.

18. SentiWordNet. https://github.com/aesuli/SentiWordNet.

19. Cambria, E., Schuller, B., Xia, Y. & Havasi, C. (2013). New avenues in opinion mining and sentiment analysis. *IEEE Intelligent Systems*, 28 (2): 15–21. doi: 10.1109/MIS.2013.30.

20. Fink, A. (2016). *How to Conduct Surveys: A Step-by-Step Guide*, 6th ed., Sage, London.

21. Kumar, A. & Jain, R. (2016). Opinion sentiment analysis. *International Journal of Advances in Applied Sciences*, 5 (3): 128–136, ISSN: 2252-8814.

22. Bing Liu. (May 2012). *Sentiment Analysis and Opinion Mining*. Morgan & Claypool Publishers, San Rafael, CA.

23. SemEval (2014). Train Data. (Accessed: 24 May, 2020). https://metashare.ilsp.gr:8080/repository/browse/semeval-2014-absa-laptop-reviews-train-data/94748ff4624e11e38d18842b2b6a04d7ca9201ec33f34d74a8551626be122856/ [Online].

24. SemEval (2014). Trial Data. (Accessed: 24 May, 2020). https://alt.qcri.org/semeval2014/task4/data/uploads/laptops-trial.xml [Online].

25. Online American Platform for Teachers' Feedback. (Accessed: 23 March, 2020). https://www.ratemyprofessor.com [Online].

26. Online Indian Platform for Teachers' Feedback. (Accessed: 12 August, 2020). https://www.myfaveteacher.com [Online].

27. Textual feedbacks collected from 120 engineering students for 20 teachers of University Institute of Engineering and Technology, CSJM University, Kanpur.

28. Python. (Accessed: 20 March, 2020). https://www.python.org/ [Online].

29. NLTK. (Accessed: 20 March, 2020). https://www.nltk.org/ [Online]

30. Stanford Dependency Parser. (Accessed: 20 March, 2020). https://nlp.stanford.edu:8080/parser/index.jsp.

10 Emission of Herbal Woods to Deal with Pollution and Diseases
Pandemic-Based Threats

Rohit Rastogi
ABES Engineering College

Mamta Saxena
Ministry of Statistics, Govt. of India

D. K. Chaturvedi
DEI

Sheelu Sagar
Amity International Business School

CONTENTS

10.1 Introduction 184
    10.1.1 Scenario of Pollution and Need to Connect with Indian Culture ... 184
    10.1.2 Global Pollution Scenario 184
    10.1.3 Indian Crisis on Pollution and Worrying Stats 186
    10.1.4 Efforts Made to Curb Pollution World Wide 186
    10.1.5 Indian Ancient Vedic Sciences to Curb Pollution and
            Related Diseases 187
    10.1.6 The Yajna Science: A Boon to Human Race from
            Rishis and Munis 188
    10.1.7 The Science of Mantra Associated with Yajna and
            Its Scientific Effects 188
    10.1.8 Effect of Different Woods and Cow Dung Used in Yajna 189
    10.1.9 Use of Sensors and IoT to Record Experimental Data 189
    10.1.10 Analysis and Pattern Recognition by ML and AI 190
10.2 Literature Survey 191
    10.2.1 Gist 191
    10.2.2 Methodology Used in This Paper 191
    10.2.3 Instruments and Data Set Used 191

DOI: 10.1201/9781003138020-10

       10.2.4 The Future Scope Discussed 192
10.3   The Methodology and Protocols Followed... 192
10.4   Experimental Setup of an Experiment 193
       10.4.1 Airveda and Different Sensor-Based Instruments......................... 193
10.5   Results and Discussions.. 194
       10.5.1 Mango v/s Banyan (Bargad) 194
             10.5.1.1 Mango 194
             10.5.1.2 Bargad... 194
10.6   Applications of Yagya and Mantra Therapy in Pollution
       Control and its Significance.. 198
10.7   Future Research Perspectives 199
10.8   Novelty of Our Research 199
10.9   Recommendations 199
10.10 Conclusions... 199
References... 200

10.1 INTRODUCTION

10.1.1 SCENARIO OF POLLUTION AND NEED TO CONNECT WITH INDIAN CULTURE

As we look at the doings and actions of the Indian culture, it totally believes to take that much from the environment which it will regenerate or produce again easily. The traditional Indian cultural practices include Yajna or agnihotra in which offerings of various medicinal herbs and medicinal entities are made for the fire, which as a result purify the environment and decrease the amount of contaminants in the atmosphere. In Indian culture, rivers are given the importance of mothers and trees are considered as second figures, which are worshipped by the people of the country (Rastogi et al., 2019). In ancient Indian culture, people used to throw copper coins and some special types of coins made up of alloys that chemically purifies the water and reduce the amount of impurity in the water and trees are saved, as they are considered medicinal and equivalent to human life. Rainwater harvesting is also a part of such traditional practices of Indian culture. A special type of ecofriendly clothing known as khadi is widely used in India; it is very biodegradable and causes no harm to the environment; even the kitchens of India show how the country is ecofriendly by using the utensils made up of special type of clay, which are used for cooking. The water containers made up of clay worked as natural refrigerators due to the pores in them. In Indian culture, people use carry bags made up of jute and the wet curtains used at the windows are made up of jute, which makes the coming air cool (Dev, 2017).

10.1.2 GLOBAL POLLUTION SCENARIO

Table 10.1 shows the health impact of the human body by air pollutants and their source of emission, average time, and standard levels (Rastogi et al., 2020a). Air pollutant is a mixture of small particles in the air that can have an adverse effect on living things and ecosystems. These particles are in the form of liquid droplets, gaseous or solid particles

TABLE 10.1
Different Effects of Yagya on Environment in Studies by Researchers

Experiment	Effect on the Day of Yagya
Dr. Hafkine (Sharma et al., 2019) On burning of mixture of ghee and sugar in Hawan Samagri	Destroys the bacteria of different diseases and kills the germs of certain diseases
Trelle of France (Gupta, 2012) On burning of mango wood and jaggery and formaldehyde	Destroys the harmful bacteria
Tautilk (Gupta, 2012)	Destroyed germs of typhoid
National Botanical Research Institute (Nautial et al., 2007)	Reduces airborne bacteria to a large extents
Dr. Shirowich, a Russian scientist	Reduces the atomic radiation in a significant amount
Cow's ghee is burn into fire medicinal fumes emanating from agnihotra (Sharma et al., 2019)	These eradicate bacteria and other microorganisms
Dr. Kundanlal M.D. in allopathic medicine (Gupta, 2012) When 1 kg of mango wood was burnt with Hawan Samagri in open air	Bacterial count was reduced by 94%
Chander Shekhar Nautiyal, head of the division, studying plant microbe interactions (Nautiyal et al., 2007)	On the Yagya day, bacteria count was lowered the infections like tuberculosis (TB) and other viral infections First day after Yagya, bacteria count was lowered the infections like TB and other viral infections
The study was done by CPCB, a Govt. of India body (Sharma et al., 2019)	On the Yagya day, reduction of bacteria—79%, fungi—68%, total microflora (TMF)—69%, and pathogens—33% First day after Yagya, reduction of bacteria—55%, fungi—15%, and pathogens—79% Second day after Yagya, reduction of pathogens—79% and TMF—49% seventh day after Yagya, reduction of bacteria—93%, fungi—88%, and pathogens—93%

present in the air, like nitrogen dioxide, lead oxide, ground-level ozone, and polycyclic aromatic hydrocarbons. The standard level measured in microgram per cubic meter ($\mu g/m^3$) in the air is a unit of amount of chemical vapors, fumes, or dust in the ambient air (Figure 10.1) (Ghorani-Azam et al., 2016; Rastogi et al., 2020b).

World regional capital city in ranking of 2018 as sorted by average yearly PM 2.5 among countries from Asia and the Middle East were indexed and Delhi occupies most of the highest mixture and poor AQI. The second is Dhaka than the third ranking capital, Kabul (the world regional capital city ranking 2018), where PM 2.5 stands for particulate matter. PM 2.5 is dangerous and harmful to health. PM 2.5 is a mixture of liquid droplets and solid particles present in the air. When the number of these particles

Air pollutants	Major Source of Emission	Averaging Time	Standard level	Health impact target organs
Particle pollutants PM25	Smokes, Motor engines, Industrial activities	24h	35µg/m³	Respiratory and cardiovascular diseases, CNS and
PM10		24h	150µg/m³	reproductive dysfunctions, cancer
Ground-level ozone	Industrial activities, Vehicular exhaust	1h	0.12mg/m³	Respiratory and cardiovascular dysfunction, eye irritation
Carbon monoxide	Smokes, Motor engines, burning coal, oil and wood, Industrial activities	1h	35mg/m³	CNS and cardiovascular damages
Nitrogen dioxide	Feel-burning, Vehicular exhaust	1h	100µg/m³	Damage to liver, lung, spleen, and blood
Lead	Lead smelting, Industrial activities, leaded petrol	3months average	0.15µg/m³	CNS and hematologic dysfunctions, eye irritation
Sulfur dioxide	Fuel combustion, burning coal	1h	75µg/m³	Respiratory and CNS involvement, eye irritation
Polycyclic aromatic hydrocarbons	Fuel combustion, wood fires, motor engines	1 year	1ng/m³	Respiratory and CNS involvement, cancer

Air quality standards according to the European Union. PM₂.₅ is stand for PM of 2.5µ or less. PM₁₀ is stand for PM of 10 µ or more. PM=Particulate matter, CNS=Central nervous system.

FIGURE 10.1 Standard levels of criteria of air pollutants and their sources with health impact based on the United States Environmental Protection Agency. (Website: jmsjournal. net.)

increases and penetrates deeply into the lungs, it can affect your body like breathing problems, burning, or sensation in the eyes (Figure 10.2) (Rastogi et al., 2020c).

10.1.3 INDIAN CRISIS ON POLLUTION AND WORRYING STATS

The quality of air has become so poor in India that it is leading to death consequences for the larger section of the society. This severe air pollution affects millions of people especially in the densely populated region where the people are forced to bear dense and poisonous air for long time (Rastogi et al., 2020d). The air pollution in Delhi and its nearby areas reached the worst level on November 3 and 4, 2019, where PM 2.5 levels were shown by some indices at 407 and more than 500, respectively. In addition, as per a Washington Post report (paywall) which was published in November 2018, the lifespan of people in India is reduced by 5.3 years due to the air pollution. Hence, air pollution is the biggest challenge for humankind and some additional measures should be taken in order to control it (Ghosh & Parida, 2015; Rastogi et al., 2020e).

10.1.4 EFFORTS MADE TO CURB POLLUTION WORLD WIDE

Pollution can be in air or can be in water, but pollution exists in various forms. Pollution is the main reason for the presence of harmful contaminants in the biological ecosystem, which in return creates adverse and negative effects on the environment. Manufacturing operations, industrialization at global level, and the ever-increasing demand to increase the standard of living incline the graph of pollution ultimately resulting in life loss (Rastogi et al., 2020f; Apte & Salvi, 2016).

FIGURE 10.2   World regional capital city ranking 2018 (Website: IQAir.)

10.1.5   INDIAN ANCIENT VEDIC SCIENCES TO CURB POLLUTION AND RELATED DISEASES

Effects of Hawan on poisonous gases—Experiment shows the observations made by distinguished scientists as given in Table 10.1 that shows effects of Hawan on poisonous gases and disease-causing agents (Sharma et al., 2019) (Table 10.1).

The medicinal plants' woods, cow ghee, and sweet products used in Yagya activity volatilize easily and diffuse in the environment; the fumes generated in Yagya kill pathogens like bacteria; fungi; virus; and parasites like flies, ringworm, fleas, and dice.

These volatile substances further subjected to photochemical reaction in sunlight and undergo through decomposition, oxidation, reduction, etc. (Rastogi et al., 2020g).

10.1.6 THE YAJNA SCIENCE: A BOON TO HUMAN RACE FROM RISHIS AND MUNIS

One can easily understand the benefits of Yagya as per the details given in the figures below (Figure 10.3, Figure 10.4, Romana et al., 2020).

10.1.7 THE SCIENCE OF MANTRA ASSOCIATED WITH YAJNA AND ITS SCIENTIFIC EFFECTS

An Indian scientific process through which the balance of O_2 and CO_2 is maintained in the atmosphere is termed as Yajna. In the rituals performed during Yajna, besides, proper use of herbs and woods, application of mantra is important. Mantra is usually considered as a part of cultural dimension. Recent advancements in the field of science and medicine carry the healing role of various dimensions of old age therapies like mantras, which were originally considered as a part of the cultural traditions. Therefore, it is said that the Yajna is considered as a ritual that also has an effect on humans. This also tells us the importance of mantra in Yajna. Mantra results show us that it plays an important role in human mental capacity. In addition, positive effects of mantras are seen on the plant species (Rastogi et al., 2020h; Chauhan et al., 2015).

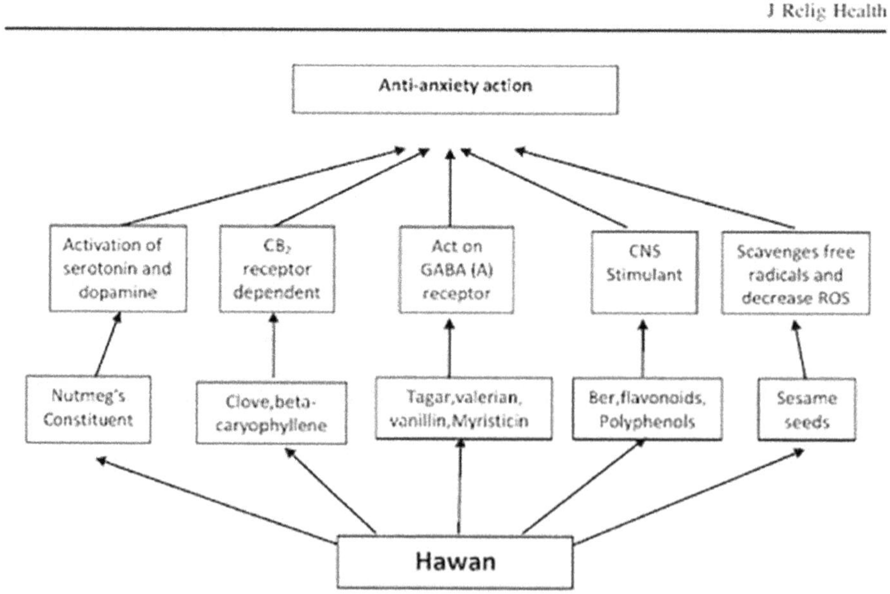

FIGURE 10.3 Diagrammatic representation of component of Hawan Samagri along with probable multiple mechanism of action (Romana et al., 2020).

FIGURE 10.4   Society residents chanting Vedic mantras and performing community Yagya. This Yagya was organized in Ward 9, Sector 29, Noida, NCR, UP, India, on World Environment Day June 5, 2017.

A study tested the effect on bacterial growth using smoke obtained from Yajna with complete procedure along with mantra chanting, and using the smoke obtained from Yajna without mantra chanting. However, the effect was magnificent when the smoke obtained from Yajna with complete procedure along with mantra chanting (Singh& Singh, 2018; Devender et al., 2019).

10.1.8   Effect of Different Woods and Cow Dung Used in Yajna

It is observed that six types of gases are released from the combustion of the special types of woods that directly reduce the bacteria. This whole process is a natural fumigation process in which these antibacterial fumes fill up the environment and disinfect it. The gases and volatile substances release from the combustion of the volatile oils present in these woods and materials further do some photochemical reactions (Rastogi et al., 2020i). The carbon dioxide released is also be reduced to some extent into formaldehyde due to which an amount of oxygen is also are added to the environment. Many experiments and researches regarding the checking of mineral content and medicinal content of these woods have taken place, and it is observed that the ashes obtained from the Yajna also act as a very good fertilizer because this also contains that mineral content in it. Cow dung cakes are also confirmed to have some antibacterial properties. Hence, these fumes are therapeutic in nature, naturally heal the atmosphere, and are beneficial for living creatures (Limaye, 2019) (Figure 10.5).

10.1.9   Use of Sensors and IoT to Record Experimental Data

Internet of Things (IoT) is a source that can connect hardware and software devices very easily. It is also known as the IoT. Although sensors were themselves smart enough but when we need to analyze the data, it would have become difficult without IoT. IoT is the answer to many problems be it crime or environment. As time will pass, it is expected that due to IoT being more and more advanced, less energy and

FIGURE 10.5   People in Indian and South Asian continent celebrating Holi festival (a mass Yajna process to purify atmosphere) with cow dung, herbal woods and spices, and new barley and grains.

less money will be spent in getting our tasks done (Rastogi et al., 2020i). We know that at this time specifically, the data is growing exponentially and it will continue in the coming years as well. Therefore, to control this data, to analyze it, and to store it, we need IoT. For this kind of data, traditional efforts would not bring much advantage as compared to the use of IoT devices (Figures 10.6 and 10.7) (How do Smart Devices Work: Sensors, IoT, Big Data and AI, November 04, 2018).

10.1.10   ANALYSIS AND PATTERN RECOGNITION BY ML AND AI

Whenever we talk about recognition of data, we talk about how regular the data is. By identifying the traits and characteristics of the data, we select the machine learning (ML) algorithm that will predict the results accurately. Recognizing these regularities plays an important role in building an efficient model. This is useful not only in the technological field but also in various fields such as e-commerce or medication (Rastogi et al., 2020j) (Pattern Recognition (Tutorial) and Machine Learning: An Introduction, April 29, 2020).

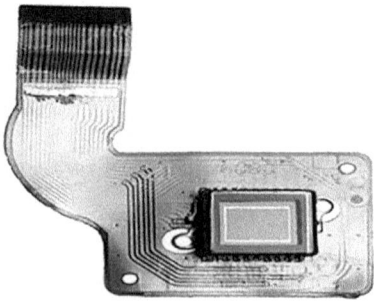

FIGURE 10.6   CCD image sensor for capturing images. (Website: Wikipedia.)

FIGURE 10.7 Infrared proximity sensor used for distance measurement. (Website: Flipkart.com.)

10.2 LITERATURE SURVEY

10.2.1 GIST

In this chapter, an attempt is made in order to determine whether the emissions from the gas industry are responsible for the hospitalizations in the Darling Downs, Queensland, Australia. Outdoor air pollution especially due to the emissions from the industries has resulted in the negative health impact on humans. Air pollution is responsible for numerous diseases that include respiratory and cardiac diseases, and it is responsible for cancer deaths. In addition, the air pollution has reduced the human life span and it is leading to premature deaths of humans. The hospitalization data was obtained from the Darling Downs Hospital and Health Service (DDHHS). In addition, the data is analyzed using linear regression analysis (McCarron, 2018).

10.2.2 METHODOLOGY USED IN THIS PAPER

In this research paper, the data was collected via different sources. The data on admissions to the hospitals was obtained from the DDHHS for the years 2006–2015 (McCarron, 2018).

10.2.3 INSTRUMENTS AND DATA SET USED

Here the DDHHS acute hospital admissions according to residence and year was collected. Using SPSS, linear regression analysis was performed. Linear regression is a statistical technique that is used to perform data analysis. It is used to find out the relationship between the dependent variables and one or more independent variables. The linear regression analysis was carried out on the hospital admission data, controlled for population versus time. The circulatory conditions increased from 0.87% in 2007 to 1.86% in 2014 while the respiratory admissions increased from 0.50% in 2007 to 1.10% in 2014 (McCarron, 2018; Oprea and Iliadis et al., 2011).

10.2.4 THE FUTURE SCOPE DISCUSSED

There has been a rapid increase in the health impairment of the people due to the rapidly escalating air pollution from the heavy industries. The controls to limit exposures are not effectual as there is a growth in the hospitalization due to the acute respiratory and circulatory conditions with the increase in toxic pollutants. The acute hospitalization data indicates a danger. Furthermore, a comprehensive investigation of the health impacts due to the unconventional gas industries in Australia can be made (McCarron, 2018; Cortés et al., 2000).

10.3 THE METHODOLOGY AND PROTOCOLS FOLLOWED

This experiment was accomplished in a room by taking into consideration proper methodology in the duration of February and March 2020 at one town of Uttar Pradesh, India, on various people who were ready to perform Yagya and chant mantras by following various protocols. The Yagya was performed daily for several days for 20 minutes. Various mantras were chanted including Vaidik Mantra of Shatkarma, Gayatri Mantra, Mahamrityunjay Mantra.

Firstly, the volunteers were asked to wake up early in the morning and pray to God for the happiness and prosperity for themselves and for their family members. Hawan Kund was prepared by dropping some sand in it and putting four or five pieces of wood used in it. Ghee was poured over the woods with some camphor in it (Ritchie and Roser, 2019).

At the beginning of this process, water was taken in the right hands of the volunteers and some mantras were pronounced and then they were required to drink that water. Fire was set up with a matchstick in order to start the process. Then, ghee was offered to the fire 108 times with a teaspoon along with repeating the mantras. Again, Hawan Samagri was offered while reciting the Vedic mantras. At the end of Hawan, coconut kept with ghee was offered into the fire while reciting some more mantras. Finally, something was donated after the Hawan as, according to the Vedic rule, it is said that Hawan remains incomplete without the donation.

The protocol and experiment glimpses are shown in Figure 10.8.

FIGURE 10.8   In Yajna besides burning material objects, also chanting and praying is done to make one self-pious and mighty. It purifies air and provides fresh content through respiration.

10.4   EXPERIMENTAL SETUP OF AN EXPERIMENT

The researcher developed artificial intelligence (AI) with ML system to study the effect of Yagya with the aims to measure and forecast air quality and pollution levels where serious air pollution problems prevailed in Delhi region. The live data captured by sensors were deployed at Lodi Complex area of Delhi region, the data collected with focus to test the algorithms. The researcher has tried to explore the feasibility of linking pollutant airborne particles or aerosols, present in the air to carbon control through Yagya. The specific shape and size of the Hawan Kund, specific selection of wood pieces, the specific amount of Hawan Samagri used were monitored for controlling chemical processing in the fire, sublimation, chemical components for the reduction of pollutant level in atmosphere. This research will help to develop an ancient technique to clean air in the crowded area of a city, which will help to improve the health of the population in the future. There are variety of Yajna for different purposes and different woods and different Mantras for different expected results are applied in them.

10.4.1   AIRVEDA AND DIFFERENT SENSOR-BASED INSTRUMENTS

The temperature and humidity can be accurately measured with the help of IoT temperature and humidity sensor (Rastogi et al., 2020k). The external sensors are placed inside the room and the indicator is mounted outside the room. These kinds of sensors are used at various places like data centers and IT server rooms. These sensors help the researchers to alert remotely by the means of short message service (SMS) or email. They also contain a cloud function that works with the help of a Web-based interface in order to view real-time data transmitted by the device. They provide reports in Excel/PDF format; also custom report designing is also possible as per the requirements of the user (Figure 10.9) (Rastogi et al., 2020l).

FIGURE 10.9   IoT-based sensors capturing the humidity and temperature data from atmospheric air in certain interval of time.

10.5   RESULTS AND DISCUSSIONS

10.5.1   MANGO V/S BANYAN (BARGAD)

10.5.1.1   Mango

The following are the time series graph of the Hawan(s) done using mango wood from November 13 to 17, 2019, every day. AQI, PM 2.5, PM 10, CO_2, temperature, and humidity were recorded. The Hawan was conducted between 6:30 and 6:50 a.m. Hence, a sharp change in values can be observed at 6:30 and 7:0 a.m. This change is of primary interest to us. In addition, it should be noted that there is no significant change in the temperature due to the Hawan (Figures 10.10 and 10.11).

10.5.1.2   Bargad

The following are the time series graph of the Hawan(s) done using Bargad wood on November 18 and 19, 2019. AQI, PM 2.5, PM 10, CO_2, temperature, and humidity were recorded. The Hawan was conducted between 6:30 and 6:50 a.m. Hence, a sharp change in values can be observed at 6:30 and 7:0 a.m. This change is of primary interest to us. In addition, it should be noted that there is no significant change in the temperature due to the Hawan. There is a '0' CO_2 value recorded between 16:00 and 20:30 on November 18, 2019. This can be attributed to the failure of recording instrument and it has been disregarded while using the data (Figures 10.12 and 10.13).

Comparative analysis is depicted in Figure 10.14).

The above graph is based on the comparison of the environmental conditions before the Hawan and after the Hawan. The difference mentioned above is in percentage. As it is clear from the graph that one the one hand, using mango wood

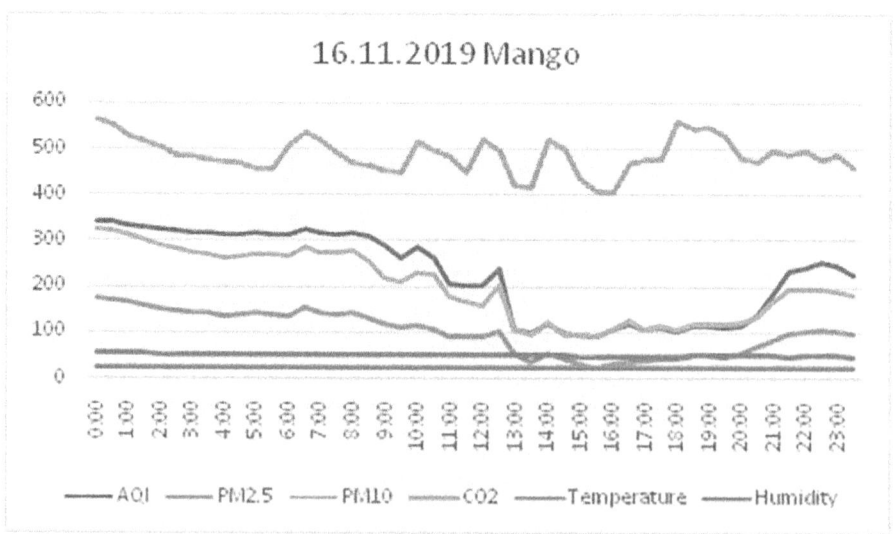

FIGURE 10.10   Measurement of different parameters of AQI on November 16, 2019. (Yagya was performed with mango wood at 6:30 a.m. with a 20-minute protocol.)

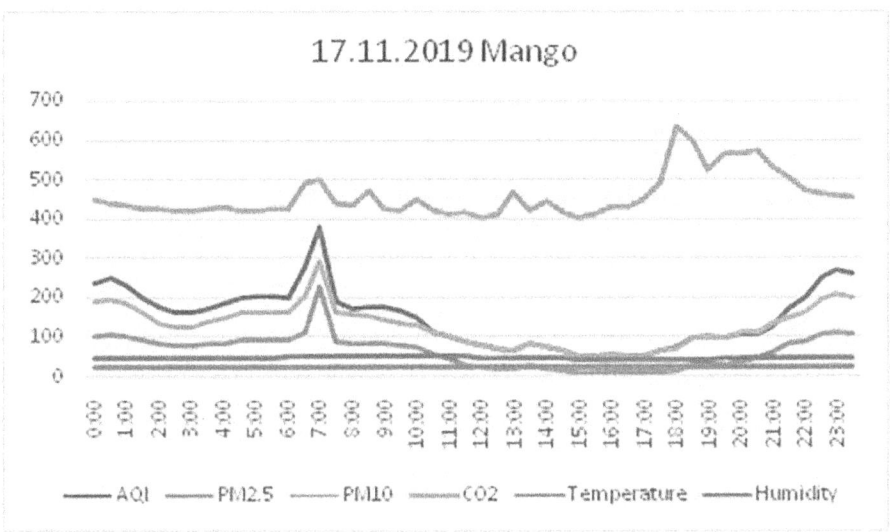

FIGURE 10.11 Measurement of different parameters of AQI on November 17, 2019. (Yagya was performed with mango wood at 6:30 a.m. with a 20-minute protocol.)

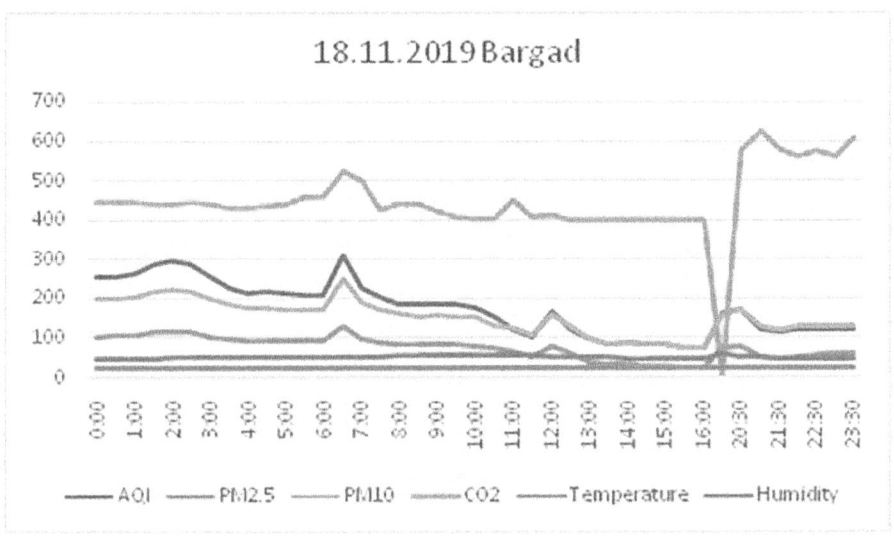

FIGURE 10.12 Measurement of different parameters of AQI on November 18, 2019. (Yagya was performed with banyan wood at 6:30 a.m. with a 20-minute protocol.)

brings an increase in AQI and PM 2.5 and decrease in PM 10 and CO_2 levels; on the other hand, use of bargad wood brings a significant decrease in AQI, PM 2.5, and CO_2 while an increase in PM 10.

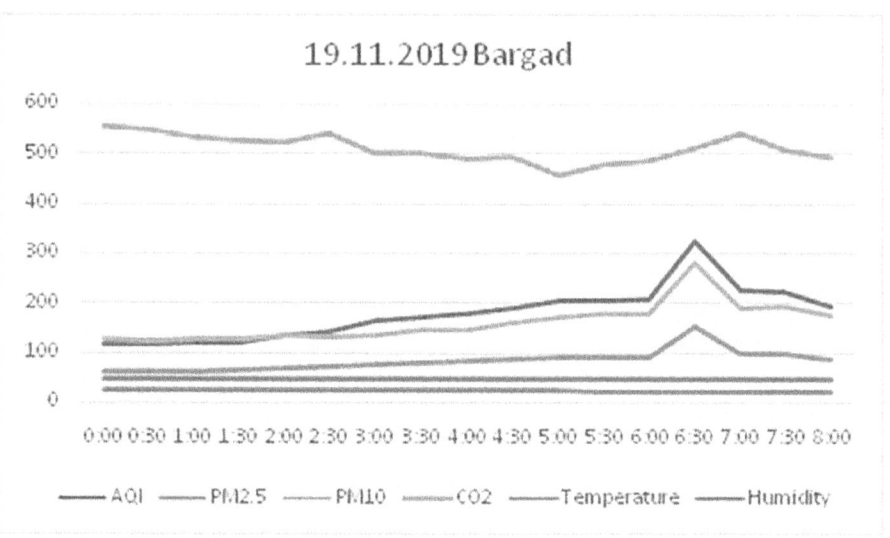

FIGURE 10.13   Measurement of different parameters of AQI on November 19, 2019. (Yagya was performed with banyan wood at 6:30 a.m. with a 20-minute protocol.)

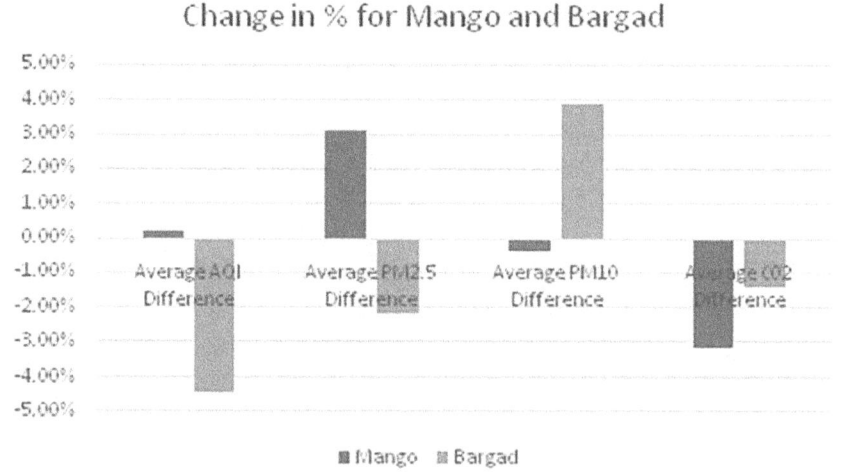

FIGURE 10.14   Comparative analysis of emission of different gaseous elements and AQI parameters through Yagya experiment done with mango and banyan woods. (Yagya was performed in morning 6:30 a.m. with a 20-minute protocol.)

10.5.1.2.1   From Data to Insights

If reader can refer to the working Excel sheet shared in Annex and can compare it with the original sheet, they will find some additional columns that have been added. These are namely AQI difference, PM 2.5 difference, PM 10 difference, and

CO_2 difference. The second sheet only contains the graphs copied from the first sheet for easy understandability, so it is less important. The third sheet contains the extracted data from the first sheet. These is of primary importance to us as our main goal is to identify the comparative analysis between mango and banyan woods' emission.

The above graph shows the average humidity, temperature, CO_2 levels, PM 10, PM 2.5, and AQI values observed when the Hawan done with mango and bargad woods, respectively. It can be clearly seen that the levels of CO_2, PM 10, PM 2.5, and AQI were very high when the Hawan was done with mango wood as compared to the days when it was done with bargad wood. In addition, it can be noticed that there is a small difference in the humidity levels of the both. Still, bargad is the winner here, although the natural weather conditions could also have contributed to the huge difference in pollution levels (Figure 10.15).

The above graph contains the average difference measured over time in a day when the Hawan(s) were conducted. This can be a significant set of features in determining the stability of environment when the Hawan(s) are conducted using mango and bargad woods. This graph shows that CO_2, PM 10, and PM 2.5 levels are less stable and show more fluctuations over the day when Hawan is done using mango wood; however the above-mentioned factors are comparatively more stable when the Hawan is done using bargad wood. However, when talking of the AQI, we can see a very huge instability when the Hawan was done using bargad wood. AQI was comparatively much more stable when the Hawan was done using mango wood (Figure 10.16).

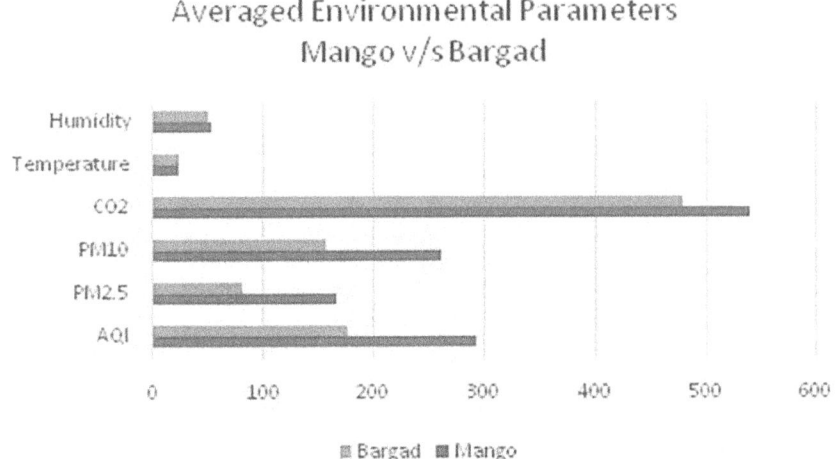

FIGURE 10.15 Comparative analysis of averaged environmental parameters in fume emission of different gaseous elements through Yagya experiment done with mango and banyan woods.

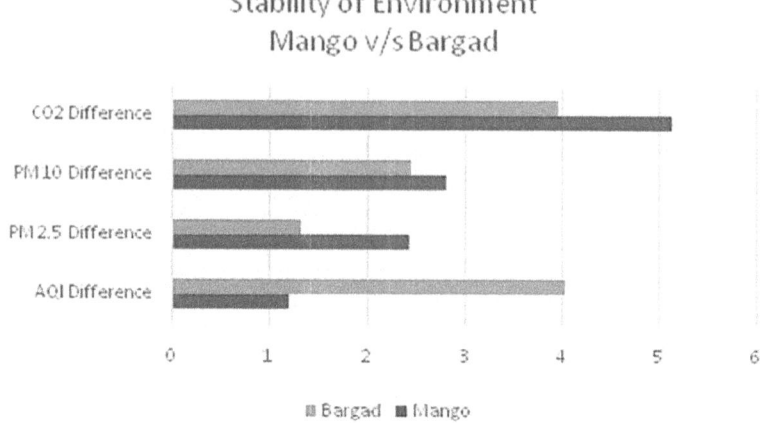

FIGURE 10.16 Study of the comparative analysis of stability of the environment by the emission of different gaseous elements and AQI parameters through Yagya experiment done with mango and banyan woods:

10.6 APPLICATIONS OF YAGYA AND MANTRA THERAPY IN POLLUTION CONTROL AND ITS SIGNIFICANCE

Yagya is a technique mentioned in the ancient texts including Vedas and Upanishads. When mantras are chanted along with this technique, it becomes more influential and powerful. This technique has various applications, as if it purifies the environment especially the polluted air (Rastogi et al., 2020m). Some research shows that Yagya helps in reducing air pollution, which is generated when the SO_2 and NO_2 levels get disturbed in the environment along with biological air pollutants such as microorganisms. There are various other applications of Yagya and mantra therapy (Figure 10.17). People are chanting mantras that help in depression treatment and sleeplessness (Saxena et al., 2018).

FIGURE 10.17 Collective recital of mantras helps in depression treatment and sleeplessness.

10.7   FUTURE RESEARCH PERSPECTIVES

Significant findings from earlier research have proved that Yagya is the simplest, cheapest, and most potential method to purify indoor and outdoor environment pollutants such as PM 2.5, PM 10, and CO_2 and has laid the foundation for further research also. Future researches and studies with statistical analyses will definitely throw more vision and enlighten further the significant aspects of Yagya and how well we can control pollution (Rastogi et al., 2020n).

10.8   NOVELTY OF OUR RESEARCH

According to Hinduism, while performing Hawan, we use the sound energy of mantras and thermal energy of fire, which purify the environment as well as the people around and drive away all negative energies around us. Mango wood burnt with Hawan Samagri releases formic aldehyde, a gas that kills harmful bacteria (Rastogi et al., 2020o). Air samples even after 24 hours showed that bacteria count goes lower by 94%. So, Hawan is helpful to prevent airborne infection. Between cosmic consciousness and human consciousness, supposedly the main link is fire. The pious fire mixed with herbal woods, medicinal spices and grains, and mixture of Sanskrit mantra produces the combination of heat, light and thermal energy for special effects in human biophysics structure and external nature.

10.9   RECOMMENDATIONS

Yajna with mantra has a great impact on the human body physically as well as mentally. To conduct the experiment, Yagya along with chanting Vedic mantras was performed in a fixed size room at different times on different days (Rastogi et al., 2020p). The data based on different parameters was collected with the help of software tools Airveda Instrument and IoT-based sensors. The Airveda tool was used for pollution data measurement. The different AQIs were measured with the help of this device. AI and ML were used in order to recognize the pattern of the data, so that the experiment is turned out to be fruitful in purifying the air.

10.10   CONCLUSIONS

The study indicates that the emission of volatile oils and gasses from herbal woods of banyan tree (Ficus benghalensis) and other ingredients of Hawan Samagri used in Yagya helps to reduce maximum environment pollutants such as PM 2.5, PM 10, and CO_2.

Therefore, Indian culture is full of such ecofriendly and balanced practices, which if used in long term, help in sustainable development and reduction in pollution for global scan. To overcome these threats, we have to recall the sayings and actions associated with our Indian culture that believes in the sustainable development and conservation of the environment because the sustainable development is the only long-term solution for these problems. Therefore, the regression to the roots of the Indian culture is what which is needed at this time.

REFERENCES

Apte, K., Salvi, S. (2016). 'Household air pollution and its effects on health'. *F1000Research*, 5, F1000 Faculty Rev-2593. doi:10.12688/f1000research.7552.1.

Cortés, U., Sànchez-Marrè, M., Ceccaroni, L. (2000). 'Artificial intelligence and environmental decision support systems'. *Applied Intelligence*, Vol. 13, No. 1, pp. 77–91.

Chauhan, P. (September 2015). 'An overview of air pollution in India at present scenario'. *International Journal of Research Graanthalayah*, Vol. 3, pp. 1–4, ISSN: 2350-0530.

Dev, M. (2017). 'Indian culture and lifestyle for environment conservation: A path towards Sustainable development'. *International Journal on Emerging Technologies*, Vol. 8, No. 1, pp. 256–260, ISSN: 2249-3255.

Devender, K. (2019). 'Air pollution mitigation through yajna: Vedic and modern views'. *Environment Conservation Journal*, Vol. 20, No. 3, pp. 57–60. https://environcj.in/volume-20-issue-3-20308/2019; doi:136953/ECJ.2019.20308.

Ghorani-Azam, A., Riahi-Zanjani, B., Balali-Mood, M. (2016). 'Effects of air pollution on human health and practical measures for prevention in Iran'. *Journal of Research in Medical Sciences*. doi:10.4103/1735-1995.189646; http://www.jmsjournal.net/text.asp?2016/21/1/65/189646.

Ghosh, D., Parida, P. (2015). 'Air pollution and India: Current scenario'. *International Journal of Current Research*, Vol. 7, No. 11, pp. 22194–22196.

Gupta, S. (2012). 'Hawan for cleansing the environment', October 17. Retrieved 26 February, 2019. Available from https://www.speakingtree.in/blog.

Limaye, V.G. (11 February 2019). 'Agnihotra (the everyday homa) and production of brassinosteroids: A scientific validation'. *International Journal of Modern Engineering Research*, Vol. 8, No. 2, pp. 41–51, ISSN: 2249-6645.

McCarron, G. (2018). 'Air Pollution and human health hazards: A compilation of air toxins acknowledged by the gas industry in Queensland's Darling Downs'. *International Journal of Environmental Studies*, Vol. 75, No. 1, pp. 171–185, ISSN: 1029-0400.

Nautiyal, C.S., Chauhan, P.S., Nene, Y.L. (3 December 2007). 'Medicine smoke reduces air born bacteria'. *Journal of Ethnopharmacology*, Vol. 114, No. 3, pp. 446–451. Epub 2007, August 28.

Oprea, M., Iliadis, L. (2011). 'An artificial intelligence-based environment quality analysis system', Iliadis, L., Jayne, C. (eds). *Engineering Applications of Neural Networks*. EANN 2011, AIAI 2011. IFIP Advances in Information and Communication Technology, vol. 363, Springer, Berlin, Heidelberg. doi:10.1007/978-3-642-23957-1_55.

Rani, R., Sharma, R. (2019). 'Ayurvedic management in Rajaswala Paricharya'. *World Journal of Pharmaceutical Research*, Vol. 8, No. 12, p. 427. https://www.wjpr.net.

Rastogi, R., Chaturvedi, D.K., Satya, S., Arora, N., Trivedi, P., Singh, A., Sharma, A., Singh, A. (19–20 January 2019). 'Intelligent analysis for personality detection on various indicators by clinical reliable psychological TTH and stress surveys'. Das, A.K., et al. (eds). In the *Proceedings of International Conference on Computational Intelligence in Pattern Recognition*, pp. 127–144, (CIPR 2019) at Indian Institute of Engineering Science and Technology, Shibpur, Springer Advances in Intelligent Systems and Computing (AISC) Series. doi:10.1007/978-981-13-9042-5_12.

Rastogi, R., Chaturvedi, D.K., Gupta, M., Singhal, P. (2020a). 'Intelligent mental health analyzer by biofeedback: App and analysis', Wickramasinghe, N. (eds). *Handbook of Research on Optimizing Healthcare Management Techniques*, p. 27, pp. 1–431, IGI Global, Hershey, PA. doi:10.4018/978-1-7998-1371-2.

Rastogi, R., Chaturvedi, D.K., Verma, H., Mishra, Y., Gupta, M. (2020b). 'Identifying better? Analytical trends to check subjects' medications using biofeedback therapies'. *International Journal of Applied Research on Public Health Management*, Vol. 5, No. 1, pp. 14–31. https://www.igi-global.com/article/identifying-better/240753.

Rastogi, R., Gupta, M., Chaturvedi, D.K. (2020c). 'Efficacy of study for correlation of TTH vs Age and gender factors using EMG biofeedback technique', *International Journal of Applied Research on Public Health Management (IJARPHM)*, Vol. 5, No. 1, pp. 49–66. doi:10.4018/IJARPHM.2020010104.

Rastogi, R., Chaturvedi, D.K., Satya, S., Arora, N., Gupta, M., Verma, H., Saini, H. (2020d). 'An optimized biofeedback EMG and GSR biofeedback therapy for chronic TTH on SF-36 scores of different MMBD modes on various medical symptoms'. Bhattacharya, S., et al. (eds). *Hybrid Machine Intelligence for Medical Image Analysis*, Studies in Computational Intelligence, vol. 841. doi:10.1007/978-981-13-8930-6_8.

Rastogi, R., Chaturvedi, D.K., Satya, S., Arora, N., Trivedi, P., Singh, A.K., Sharma, A.K., Singh, A. (2020e). 'Intelligent personality analysis on indicators in IoT-MMBD enabled environment'. Tanwar, S., Tyagi, S., Kumar, N. (eds). *Multimedia Big Data Computing for IoT Applications: Concepts, Paradigms, and Solutions*, pp. 185–215, Springer Nature, Singapore. doi:10.1007/978-981-13-8759-3_7.

Rastogi, R., Chaturvedi, D.K., Satya, S., Arora, N., Trivedi, P., Gupta, M., Singhal, P., Gulati, M. (2020f). 'MM Big Data applications: Statistical resultant analysis of psychosomatic survey on various human personality indicators', In *Proceedings of Second International Conference on Computational Intelligence 2018*. doi:10.1007/978-981-13-8222-2_25.

Rastogi, R., Chaturvedi, D.K., Gupta, M., Singhal, P. (10 May 2020g). 'Surveillance of Type –I & II diabetic subjects on physical characteristics: IoT and Big Data perspective in healthcare'. Alam, M., Shakil, K. A., Khan, S. (eds). *Internet of Things (IoT), Concept and Applications*. doi:10.1007/978-3-030-37468-6_23; ISBN: 978-3-030-37467-9.

Rastogi, R., Chaturvedi, D.K., Gupta, M., Sirohi, H., Gulati, M., Pratyusha, K. (2020h). 'Analytical observations between subjects' medications movement & medication scores correlation based on their gender and age using GSR biofeedback: Intelligent application in health care'. Vejar, M.A., Pozo, F. (eds). *Pattern Recognition Applications in Engineering*, pp. 229–257, IGI Global, Hershey, PA. doi:10.4018/978-1-7998-1839-7. ch010.

Rastogi, R., Chaturvedi, D.K., Gupta, M. (2020i). 'Exhibiting App and analysis for bio-feedback based mental health analyzer'. *Handbook of Research on Advancements of Artificial Intelligence in Healthcare Engineering*, p. 300. doi:10.4018/978-1-7998-2120-5.ch015.

Rastogi, R., Chaturvedi, D.K., Gupta, M. (2020j). 'Computational approach for personality detection on attributes: An IoT-MMBD enabled environment'. *Handbook of Research on Advancements of Artificial Intelligence in Healthcare Engineering*, p. 300. doi:10.4018/978-1-7998-2120-5.ch016.

Rastogi, R., Chaturvedi, D.K., Gupta, M. (2020k). 'Tension type headache: IoT and Fog applications in healthcare using different biofeedback'. *Handbook of Research on Advancements of Artificial Intelligence in Healthcare Engineering*, p. 300. doi:10.4018/978-1-7998-2120-5.ch017.

Rastogi, R., Chaturvedi, D.K. (2020l). 'Tension type headache: IoT applications to cure TTH using different biofeedback: A statistical approach in healthcare'. Taukeni, S.G. (ed). *Biopsychosocial Perspectives and Practices for Addressing Communicable and Non-Communicable Diseases*, IGI Global, Hershey, PA. doi:10.4018/978-1-7998-2139-7. ch010.

Rastogi, R., Saxena, M., Maheshwari, M., Garg, P., Gupta, M., Shrivastava, R., Rastogi, M., Gupta, H. (2020m). 'Yajna and mantra science bringing health and comfort to Indo-Asian Public: A Healthcare 4.0 approach and computational study'. Jain, V., Chatterjee, J. (eds). *Machine Learning with Health Care Perspective. Learning and Analytics in Intelligent Systems*, vol. 13, pp. 357–390, Springer, Cham. https://link.springer.com/chapter/10.10 07%2F978-3-030-40850-3_15.

Rastogi, R., Chaturvedi, D.K., Satya, S., Arora, N. (2020n). 'Intelligent heart disease prediction on physical and mental parameters: A ML based IoT and Big Data application and analysis'. Jain, V., Chatterjee, J. (eds). *Machine Learning with Health Care Perspective.* Learning and Analytics in Intelligent Systems, vol. 13, pp. 199–236, Springer, Cham, doi:10.1007/978-3-030-40850-3_10.

Rastogi, R., Chaturvedi, D.K., Singhal, P., Gupta, M. (2020o). 'Investigating diabetic subjects on their correlation with TTH and CAD: A statistical approach on experimental results'. Sandhu, K. (ed). *Opportunities and Challenges in Digital Healthcare Innovation.* doi:10.4018/978-1-7998-3274-4.ch012.

Rastogi, R., Chaturvedi, D.K., Singhal, P., Gupta, M. (2020p). 'Investigating correlation of tension type headache and diabetes: IoT perspective in health care'. Chakraborty, C. (ed). *IoTHT: Internet of Things for Healthcare Technologies*, Springer Nature, Singapore. doi:10.1007/978-981-15-4112-4_4.

Ritchie, H., Roser, M. (2019). 'Outdoor air pollution'. Published online at OurWorldInData.org. Retrieved from: https://ourworldindata.org/outdoor-air-pollution [Online Resource].

Romana, R.K., Sharma, A., Gupta, V., Kaur, R., Kumar, S., Bansal, P. (01 February 2020). 'Was Hawan designed to fight anxiety-scientific evidence?'. *Journal of Religion and Health*. Vol. 59, No. 1, NLM (Medline). doi:10.1007/s10943-016-0345-1.

Saxena, M., Kumar, B., Matharu, S. (2018). 'Impact of yagya on particulate matters'. *Interdisciplinary Journal of Yagya Research*, Vol. 1, pp. 1–8. doi:10.36018/ijyr.v1i1.5.

Sharma, P.K., Ayub, S., Tripathi, C.N., Anjavi, S., Dubev, S.K. (26 February 2019). 'AGNIHOTRA – A non conventional solution to air pollution'. *International Journal of Innovative Research in Science & Engineering*, Vol. 8, No. 12. ISSN (Online): 2347-3207. https://m.hindustantimes.com/columns.

Singh, R., Singh, S. (2018). 'Gayatri mantra chanting helps generate higher antimicrobial activity of yagya's smoke'. *Interdisciplinary Journal of Yagya Research*, Vol. 1, pp. 9–14. doi:10.36018/ijyr.v1i1.6; http://ijyr.dsvv.ac.in/index.php/ijyr/article/view/6.

11 Artificial Neural Networks
A Comprehensive Review

Neelam Nehra and Pardeep Sangwan
MSIT

Divya Kumar
IFTMU

CONTENTS

11.1 Introduction 204
11.2 Activation Function 206
    11.2.1 Linear Activation Function..207
    11.2.2 Nonlinear Activation Function 207
        11.2.2.1 Sigmoid (Logistic) Function 207
        11.2.2.2 Tanh Activation Function...208
        11.2.2.3 Rectified Linear Unit (ReLU) Function..............................208
11.3 Artificial Neural Network (ANN) 208
    11.3.1 Supervised Learning...209
    11.3.2 Unsupervised Learning 210
    11.3.3 Reinforcement Learning...210
11.4 Types of Artificial Neural Network..211
    11.4.1 Single-Layer Feedforward Neural Network 211
    11.4.2 Multilayer Feedforward Neural Networks...212
    11.4.3 Recursive Neural Network (RNN) 212
    11.4.4 Convolutional Layer Network (CNN)..214
    11.4.5 Backpropagation Neural Network 214
        11.4.5.1 Static Backpropagation 216
        11.4.5.2 Recurrent Backpropagation 216
11.5 Problems in Artificial Neural Networks...216
    11.5.1 Techniques to Avoid Overfitting When Neural Networks are
        Trained..218
11.6 Convergence of Neural Network 220
    11.6.1 Adaptive Convergence (or Just Convergence).....................................220
    11.6.2 Reactive Convergence..221
11.7 Key Features of the Error Surface 221

DOI: 10.1201/9781003138020-11

11.7.1 Local Minima 221
11.7.2 Flat Regions (Saddle Points) 222
11.7.3 High-Dimensional 223
11.8 Application of Artificial Neural Network...223
11.9 Conclusion 226
References..226

11.1 INTRODUCTION

Deep learning focuses on creating algorithms for machine learning (ML) that models high-level abstractions with many nonlinear data transformations. This technology works with the artificial neural network (ANN) systems. Learning algorithms are continuously being used by such ANNs, and the efficiency of training processes may be enhanced by increasing the quantity of data. The performance depends on higher data volumes. Since the neural network (NN) number increases with time, the training phase is called deep. The functioning of this process depends on both levels, known as the training and the testing phase. The training process involves marking and deciding the matching characteristics of large amounts of data and in the inferring phase, conclusions are drawn, and using their prior information, new unexposed data is labeled. Deep learning is a method that supports the machine to recognize with maximum precision of the complex tasks of perception. It is also referred to as a hierarchical and deep-structured learning, comprising of large layers that include nonlinear processing units for transformation purposes and feature extraction. The results of the previous layer are taken as the input by each subsequent layer (Figure 11.1) [1].

Artificial neural network (ANN) is the processing model for biological nervous system-motivated information, like information in brain process [2]. It consists of a large number of processing elements (neurons) that are strongly connected and worked to solve a specific problem. The basic elements of brain, the cells responsible for acquiring input response from an outside environment through dendrites, processing it and transmitting output through Axons, are known as biological neurons (also known as nerve cells).

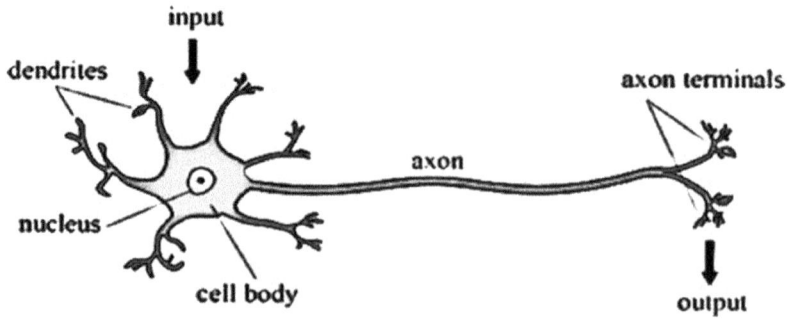

FIGURE 11.1 A biological neuron.

Dendrites: Every other neuron is enclosed by thin tube-shaped branches that are just like hair (extensions). They grow out across the body of the cell into a tree that recognizes received signs.

Axon: It is a tube-shaped part that is long and thin.

Synapse: These are the neurons that are bound to each other in the complicated system. It branches again as the axon approaches its final destination, termed as terminal arborization. Synapse is highly complicated and special structure which is in the end of the axon. When the relation between both neurons occurs at the synapses via other neuron synapses, dendrites get feedback. Over time, the soma processes these received signals and transforms the processed value into an output transmitted to further neurons via axon and synapses [3]. A NN with only one layer is called a perceptron. It provides only one response.

In Figure 11.2, x_1, x_2, and x_m represent the network's various inputs (independent variables). For a single observation with a connection weight or synapse, each of these inputs is multiplied. The weights are illustrated as wk_1, w_{k2},... W_{km} weights. The strength of a given node is indicated by its weight. The bias value is defined by b_k. The bias value of the activation function can be increased or decreased. In the basic terms to produce a result, those products are summarized and constantly fed into a transfer function (activation function), and then, this result is being sent out like an output. Mathematically, it can be given as follows:

$$x_1 \cdot w_{k1} + x_2 \cdot w_2 + x_m \cdot w_{km} = \sum x_i \cdot w_i.$$

The $(x_i \cdot w_i)$ activation function is now added (Table 11.1).

There is a relation between ANNs and biological neural networks (BNNs). The human brain is similar to NNs in the following two respects—learning acquires knowledge through a NN and the information of a NN is contained within the forces of interneuron communication known as "synaptic weights."

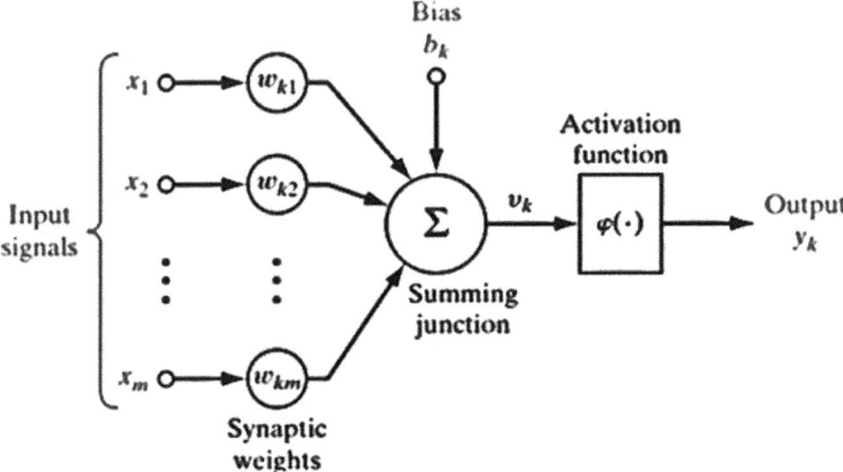

FIGURE 11.2   Perceptron.

TABLE 11.1

Major Difference between Artificial Neural Networks (ANNs) and Biological Neural Networks (BNNs)

Characteristics	ANNs	BNNs
Speed	Having information processing faster. The response time is in nanoseconds	Processing information is slower. The response time is in ms
Size and complexity	Small size and simple. However, it does not perform complicated tasks for pattern recognition	A more complicated and compact system of 1011 neurons and 1015 interconnections that are interconnected neurons
Processing	Serial processing	Massively parallel processing
Control mechanism	To track computing operations, there is a control system	No specific external control Computing function mechanism
Fault tolerance	Fault intolerant. When lost, in the case of system failure, information cannot be retrieved	Adaptable retention of information demonstrates that new information is introduced by changing the strengths of the interconnection without altering prior knowledge
Storage	Replaceable storage of information means the deletion of an old data can be applied to new data	Extremely complex and tightly packed interconnected neuron network, comprising 1011 neurons with 1015 interconnections

11.2   ACTIVATION FUNCTION

It is extremely essential for an ANN to understand something very complicated and make sense of it. Its main objective is to turn the input signal of the node into such output signal in ANN. The input to the next stack layer will be used for this output signal. The activation function decides whether, by evaluating the weighted sum and applying more bias to it, a neuron is activated or not. The objective is to introduce nonlinearity into a neuron's output. A linear (one-degree polynomial) function will simply be the output signal when we do not implement the activation function. A linear function is simple to solve, but has less power [4] and is restricted in complexity. Functions having degree more than one are nonlinear functions, which has a curvature. NN can be used to learn and explain these type of functions that maps input to output. The NN is called "approximators of universal functions." This essentially states that every function at all can be learned and computed. The activation functions can be categorized into two major types:

- Linear activation function and
- Nonlinear activation function.

11.2.1 LINEAR ACTIVATION FUNCTION

It generally takes inputs, each and every neuron multiplied by weights and generates a signal of the output which is proportional to the input. It also does not deal with either complexities or basic parameters of the traditional NN information fed in [5].

11.2.2 NONLINEAR ACTIVATION FUNCTION

Most frequently used function these days is nonlinear activation function. It gives the model a simple way to categorize or develop with additional data and to differentiate between outputs. The main terms required for nonlinear functions to be explained are as follows:

- **Derivative Function**: It is also referred to as a slope. The derivative of any function is the degree of variation of an output value corresponding with its input value.
- **Monotonic Function**: This function will be either absolutely nondecreasing or nonincreasing. Nonlinearity is desired for functions, as its motive in a NN is to build a nonlinear boundary of decision through nonlinear weight and input combinations. These functions are essentially categorized by their own curve or range.

11.2.2.1 Sigmoid (Logistic) Function

The shape of sigmoid function is in the shape of S (Figure 11.3).

The main purpose to choose the sigmoid function is because it actually occurs within (0–1). For models where we would have to predict the likelihood, it is also particularly used as an output. Since there is only the possibility of anything between the 0 and 1 scale, the best choice is sigmoid. That means it is possible to find the slope of the sigmoid curve at any two points. The function is monotonous, but it is not a derivative of the function. The logistic sigmoid function will cause the training time

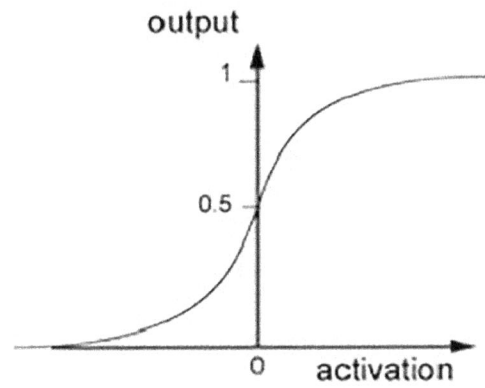

FIGURE 11.3   Sigmoid function.

to stick to a NN [5]. The more generalized function for logistic activation is softmax, and multiclass classification is also used.

11.2.2.2 Tanh Activation Function

Tanh is much better than sigmoid logistics. The tanh is *s*-shaped. The range of this tanh function is −1 to 1 (Figure 11.4).

The major benefit would be that the negative inputs are mapped highly negative, and the positive inputs are mapped near zero in the tanh graph. Function of this is differentiable. This function is monotonous although its derivative is not monotonous. For classification between two classes, tanh function is mainly used. In feed-forward neural network (FFNN), both the functions of tanh and the logistic sigmoid activation are used.

11.2.2.3 Rectified Linear Unit (ReLU) Function

The maximum used activation feature in modern era is the rectified linear unit (ReLU). Because that can be used for almost any network of convolutional neural network (CNN) (Figure 11.5).

The ReLU (from bottom) function is half rectified, $f(z) = z$, if z is greater than or equivalent to zero and $f(z) = 0$ should z be less than zero. In this case range: [zero to infinity], both are monotonic function as well as its derivative. However, the main concern is that most of the negative values typically are automatically zero that also decreases the model's ability to fit or train any of the data appropriately. This implies that they were given with some other input of negative value. This function immediately converts its value in the graph to zero.

11.3 ARTIFICIAL NEURAL NETWORK (ANN)

Artificial neural network is the successful ML algorithm that tries to mimic the way information is processed by the human brain. It offers a flexible way to approach regression and classification issues without specifying any relationship between the input and output variables explicitly. Neural networks are usually organized in three layers: one input layer, one or more hidden layers, and one output layer (Figure 11.6).

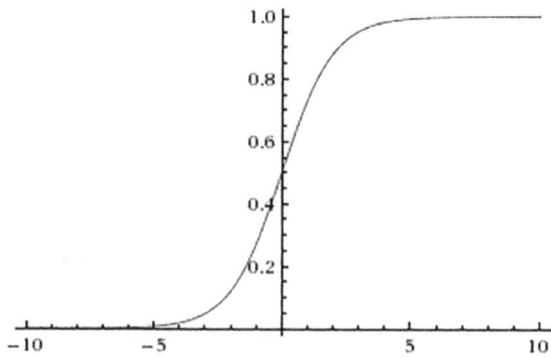

FIGURE 11.4   Tanh vs. logistic sigmoid activation function.

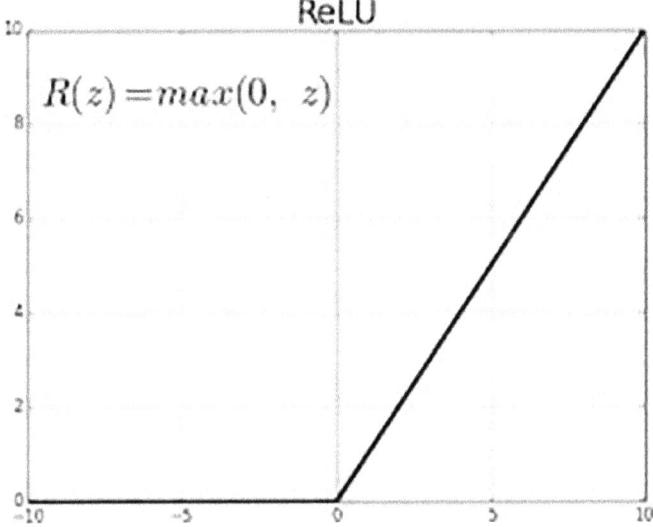

FIGURE 11.5 Rectified linear unit activation function.

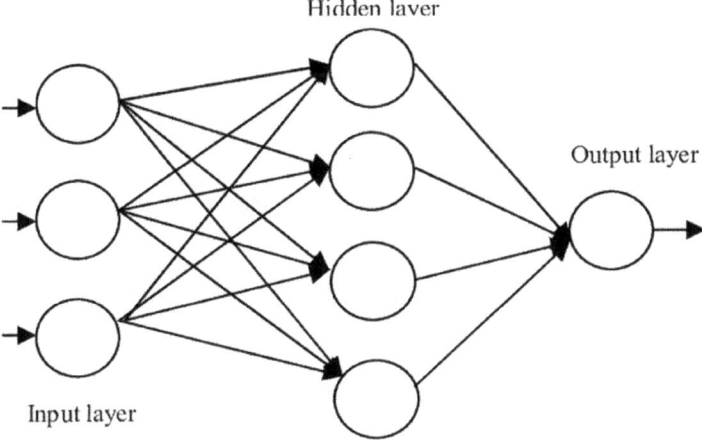

FIGURE 11.6 Schematic architecture of an ANN algorithm.

We are making the network deeper by increasing the number of hidden layers.

Learning Paradigms: The three main paradigms of learning are supervised learning, unsupervised learning, and reinforced learning. They relate to each one of a specific task of learning.

11.3.1 SUPERVISED LEARNING

Supervised learning can also train the machine with those data that are "labeled." However, it may be compared with learning that takes place when a supervisor or teacher is involved [6]. However, this algorithm is based on the labeled training data it allows to evaluate possible outcomes of unpredicted data. A group of highly experienced data scientists takes time and expertise from the data science model. Additionally, data scientists need to redesign models to ensure that the observations provided remain true before changes to their data.

Supervised learning helps to gather information and generate output of data from the previous one only using experience to help in optimizing efficiency parameters. This type of learning supports to overcome different kinds of computational issues in the real world.

11.3.2 UNSUPERVISED LEARNING

This is a ML system in which the model should not be supervised. Alternatively, to discover information by itself, you need to have the model. It primarily covers all unlabeled data. Unsupervised learning algorithms focus on performing processing tasks that are rather more complex than supervised learning [6].

Here are the main reasons why unsupervised learning should be used:

- All types of unknown patterns are found in data in unsupervised ML.
- Unsupervised approaches help you determine features that might be helpful to categorize [7].
- Only collecting unlabeled data from a computer is simpler and faster than labeled data involving manual intervention (Table 11.2) [8].

11.3.3 REINFORCEMENT LEARNING

It is the type of ML that addresses where software agents should perform action to achieve maximum cumulative reward concepts in a system. This learning method for the NN allows us to learn how to accomplish a specific goal or optimize a particular feature over many steps. For reinforcement learning [9], here are a few key terms used:

- **Agent**: It is a presumed individual who carries out behavior to obtain some reward in an area.
- **Reward** (R): An immediate response to just an agent when a particular task or action is performed by him or her.
- **State** (s): The existing environment-friendly situation refers to the state.
- **Policy** (π): It is a technique that the agent is using in deciding on the next move depending on the current situation.
- **Value** (V): as conflicting to the short-term reward, long-term return with discount is predicted.
- **Value Function**: It determines a state's value that is the whole reward number.
- **Model of the Environment**: This mimics the environment's behavior. It helps you to draw conclusions and also to decide how the environment should act.

TABLE 11.2
Difference between Supervised and Unsupervised Learning

Performance	Supervised Learning	Unsupervised Learning
Method	Input and output variables are given in the supervised model of learning	Only input data is provided in this model
Input signal	The data that is labeled is used to train algorithms	Algorithms are being used in contrast to nonlabeled data
Computational complexity	This technique of learning is the easier one	It is difficult in computational terms
Algorithms used	NN, logistic regression, classification tree support vector machine (SVM), linear regression, random forest	Unsupervised algorithms are made up of various categories: hierarchical clustering, cluster algorithms, K-means etc.
Use of data	Supervised model of learning makes use of data from training to learn a relation for both input and output	Output data is not used in unsupervised learning
Accuracy of results	A highly reliable and efficient approach	A less trustworthy and reliable method
Number of output	There is known number of output	There is unknown number of output
Major limitation	In supervised learning, it can be a real challenge to classify big data	You cannot get detailed information about the data sorting and the output is classified and unknown as data used in unsupervised learning

- **Model Dependent Techniques**: It is an approach that uses model-based approach for solving problems with reinforcement learning.
- **Q Value or Value of the Action** (Q): Value Q is almost similar to that of value. The only difference in the two is that, as a current operation, it takes one extra parameter.

11.4 TYPES OF ARTIFICIAL NEURAL NETWORK

Neural networks could be software-based (computer models) or hardware-based (physical components represent the neurons) that can use a variety of algorithms and topologies for learning.

11.4.1 SINGLE-LAYER FEEDFORWARD NEURAL NETWORK

The primary and simplest type of ANN designed was the FFNN. In such a type of network, information moves in a single direction from the input nodes to the hidden nodes (if any) and from the output nodes. The network has no cycles or loops. This NN may have the hidden layers, or may not. In a simplified way, it usually uses a classifying activation function to have a propagated front wave and no backpropagation (Figure 11.7).

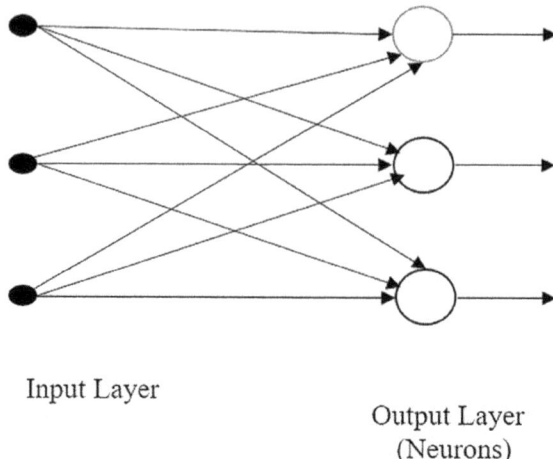

Input Layer

Output Layer
(Neurons)

FIGURE 11.7   Single-layer feedforward neural network.

11.4.2 MULTILAYER FEEDFORWARD NEURAL NETWORKS

Another class of the FFNN is characterized by the involvement of one or more
hidden layers, the computational nodes of which are referred to as hidden units or
hidden neurons. Hidden neurons have the function of interacting between both the
external input and the network output in some useful way. NNs is called deep NN
if it has more than one hidden layer. The network is allowed to acquire higher-order
statistics by adding one or more hidden layers. The source nodes in the network input
layer include the corresponding elements of the activation pattern (input vector) that
consists mainly of input signals to the second layer (i.e., the first hidden layer) of the
neurons (computation nodes) (Figure 11.8).

For the rest of the network, the second layer output signals are used as inputs
to the third layer, and so on. Normally, the neurons from each network layer have
only the output signals from the following layers as their inputs. The overall response
of the network to the pattern of activation given by the source nodes within the input
layer is the output neuron signal sequence in the output (final) layer of the network.
In the case of this unique hidden layer, the framework describes the architecture of a
multilayer FFNN in Figure [10].

11.4.3 RECURSIVE NEURAL NETWORK (RNN)

Recursive neural network (RNN) consists of deep neural network (DNN) generated
by repeatedly applying the same set of weights over such a structured input. It processes the natural language in training sequences and tree structures, mainly phras-
ing and sentencing consistent presentations depending on word embedding. Next,
RNNs were used for learning about distributed conceptual concepts, such as logical
terms. Since the 1990s, models and general systems have emerged in further works
(Figure 11.9).

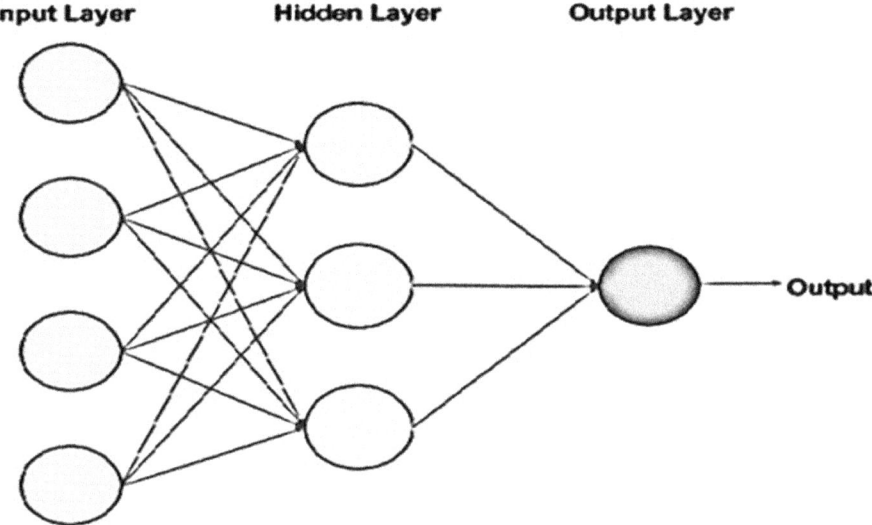

FIGURE 11.8   Feedforward neural network with one hidden layer and three neurons.

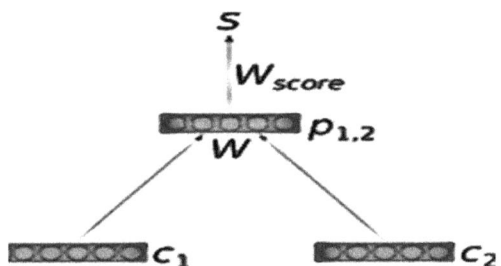

FIGURE 11.9   Architecture recursive neural network (RNN).

In the RNN, parents are formed by combining utilizing a weight matrix. The nonlinear activation function used in this architecture is tanh. The N-dimensional parent vector expression is given below:

$$P_{12} = \tanh\left[w(c_1; c_2)\right]$$

where
    W is a weight matrix.

Through some enhancements, this architecture is applied to effectively decode and to phrase natural language sentences in a syntactical way. Long short-term memory (LSTM) is a RNN architecture used in the field of deep learning. LSTM has feedback links, unlike normal FFNNs. It can form not just too specific data points (like images) but also whole data sequences (like voice or video). LSTM refers, for example, to tasks such as unsegment speech recognition, linked handwriting recognition, speech recognition, and network traffic interruption detection or intrusion detection systems (IDS). A major LSTM unit consists of a cell, an input gate, an output gate, and a forgotten gate. For all arbitrary time periods, the cell observes values, and then, the three gates regulate and control the flow of data in the cell and out of it. For classifying, processing, and drawing conclusions depending only on time slot data, LSTM networks are very well suited, completely indefinite duration between major events in a time series. This has been proposed to apply with the explosive and diminishing gradient issues that can recently be experienced when training conventional RNNs. Relative gap length insensitivity is an extra benefit of LSTM over the RNNs and many additional sequence learning techniques.

11.4.4 CONVOLUTIONAL LAYER NETWORK (CNN)

Convolutional neural networks are very similar to ordinary NNs: they only contain neurons with weights and biases that can be learned. Every neuron obtains few inputs, conducts a dot product, and performs a nonlinearity to it, alternatively. Throughout the network, a single differentiable score feature is still expressed: on one end, from the raw image pixels, to the class scores on the other [11]. They too also have a loss function on the very last (full-connected) layer as well (e.g., SVM/Softmax) (Figure 11.10).

Convolutional neural networks (CNNs) benefit from the fact that now the input comprises images, which are more sensibly restrict the architecture. In specific, the layers of a CNN have three-dimensional neurons unlike a regular NN: width, height, depth. (Keep in mind that the depth of the term or expression here now discusses to the third dimension of the amount of activation, not even just the depth of the entire NN, which may also mention to the total number of layers in the network.) As we can eventually note, the neurons in a layer are thus connected to a small area of the layer before it in a completely connected manner, rather than all the neurons.

To build CNN architectures, we utilize three major types of layers: pooling layer, convolutional layer, and fully connected layer [12] (basically like those seen in conventional NNs). These layers will be stacked to create a full CNN architecture.

11.4.5 BACKPROPAGATION NEURAL NETWORK

The basis of neural net training is backpropagation. It is the technique of fine-tuning the weights of a NN based on the error rate obtained during the previous epoch

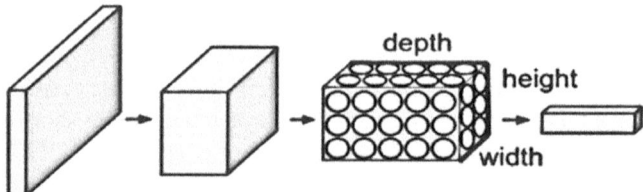

FIGURE 11.10 Dimensional volumes of the neurons.

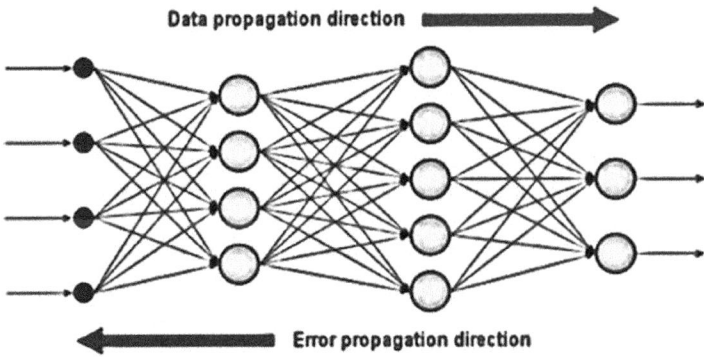

FIGURE 11.11   Backpropagation model.

(i.e., iteration). The error rates can be lowered by regular weight training and make the model robust by increasing its generalization [13]. Backpropagation is also known as "backward error propagation," a popular approach for training ANNs. This process helps to measure a loss function's gradient with respect to all the network weights [5]. The network uses its parameters to estimate about data. A loss function measures the networks. The error is propagated back and corrects the inaccurate parameters (Figure 11.11).

The backpropagation takes the error that a NN identifies with a mistake and this error can be used to change the NN parameters in the direction with few errors.

Gradient Descent: A slope at an angle that we can determine is a gradient. Like all other slopes, gradient descent can be displayed as an association between both variables: "y over x." In such situation, y is the error generated by the NN, and the parameter of NN is x. We can increase or decrease the error by adjusting the parameter. Therefore, y on x [14], the gradient tells us how much improvement we can expect. We need to use differential calculus to obtain this information that allows to calculate immediate rate of change that can be the case in which the tangent of an increasing slope representing the correlation of NN error. This error varies as change in parameter, and to shift all of the variables in less error direction. There are many parameters in a NN and what we are actually computing total error change, in which each one is the partial derivative of contribution.

The parameters of NN can be calculated by feeding input data sequentially. So, backpropagation determines the correlation among the NN error and the previous network layer parameters; after this, it provides the correlation in application of the chain rule of calculus between all the parameters of the last layer of the neural net, and those of the second-to-last layer, and so on [10].

Benefits of Backpropagation:

- Backpropagation can be programmed fast, simply, and effectively;
- Apart from the input numbers, there are no parameters to adjust;

- It is a flexible technique because previous network experience does not require in it;
- It is a standard method that works well in general.

Disadvantages of Using Backpropagation:
Backpropagation has a few problems associated that include paralysis of the network, slow convergence, and local minima.

- "Local minima" is known best. This occurs because the algorithm often changes the weights to minimize the error. But the error may need to increase briefly as a result for a more general downfall, that is the case, it will "stuck" the algorithm (because it cannot go uphill) and the error will not go downward it further.
- Network paralysis happens whenever weights are set to quite high values during training, large weights can lead many units to perform at high values, in a region wherein the activation function derivative is quite low.

A multilayer NN involves multiple presentations of repeated input patterns, to which the weights should be modified until the network can resolve through an optimal solution [14].

Two Types of Backpropagation Networks are:

- Static backpropagation
- Recurrent backpropagation

11.4.5.1   Static Backpropagation

Static backpropagation is some type of network for backpropagation that plots a static input to the static output. These networks may address static classification issues like those of optical character identification (OCR).

11.4.5.2   Recurrent Backpropagation

Recurrent backpropagation is the another method used for the training of fixed points. For example, NeuroSolutions is a software with that capability. In recurrent backpropagation, activation is fed forward when a set value has been achieved. The error is then evaluated and backward propagated. The primary differences between these two methods is in static backpropagation, the mapping is static, while in recurrent backpropagation it is not static. However, it is much more difficult to train a network that use fixed point learning than static backpropagation [10].

11.5   PROBLEMS IN ARTIFICIAL NEURAL NETWORKS

However, the NN algorithm still presents several challenges. First, the parameters of input must be independent and have very little relationship with one another. Otherwise, there are large errors in the results being trained. Secondly, the training method and the configuration of the NN often matter [15].

- **Overffitting**: During training of NNs, one of the problem that arise is called overfitting. When a model tries to forecast a too noisy pattern in the data, then overfitting occurs. This is the consequence of an overly complex model, with so many parameters. An overfitted model is inaccurate even though the trend does not match reality existing in the data. It can be determined when the model produces great results more on the data (training set) seen, although the data (test set) not seen does not work badly. The aim of a ML model is to categorize all information from both the problem domain and the training data effectively. The training set error is directed by a really low value, but when new data is given to the network [15], the error is high. The network's training examples have memorized but have not learned to categorize into unfamiliar situations. In fact, it is hard to identify that the model is being overfitted. It is not uncommon as we already have our qualified development model, and then, we start to realize that anything is mistaken. In reality, you can only ensure that everything works properly by approaching new data. During the training, however, we will try to replicate as much as possible the real conditions. t is best practice training to divide the dataset into three sections known as training, dev, and test set. Dev set is also known as crossvalidation or holdout that is used for tracking system progress and drawing conclusions for optimizing the model. While a test set at the end of the training process is set to measure the efficiency of the model, using totally different data helps to get an unbiased view of how much the algorithm operates. It is quite essential to ensure for crossvalidation and test set comes as of that similar source and accurately reflect the results we suppose to obtain in the future. Then, we can have a great deal of confidence here that the decisions we take throughout the learning process will enable us to come up with something better solution.
- **Under Fit Model**: It is a model that does not study the problem properly and then underperforms on the training dataset, and it does not function the same on a holdout sample [16].
- **Good Fit Model**: It is the model that understands the training dataset correctly and simplifies well into the dataset (Figure 11.12).

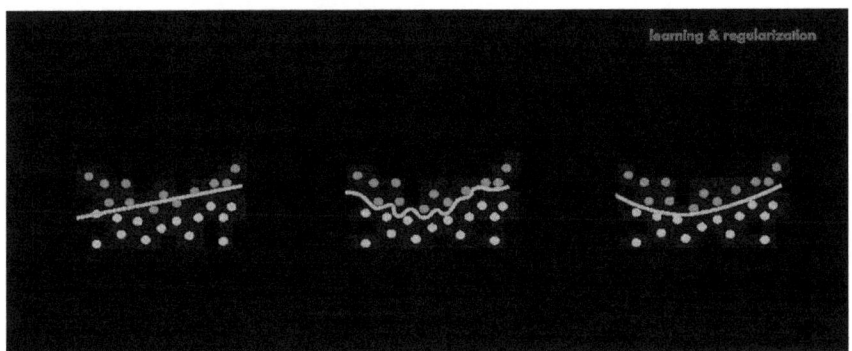

FIGURE 11.12 Underfitting and overfitting in artificial intelligence.

11.5.1 TECHNIQUES TO AVOID OVERFITTING WHEN NEURAL NETWORKS ARE TRAINED

The techniques to avoid overfitting are as follows:

- **Simplifying the Model**: To solve overfitting, the initial step is to actually reduce the model's complexity. We may simply delete layers or reduce the number of neurons to minimize the network in order to decrease complexity. As such, it is important to determine the input and output dimensions of the different layers involved in the NN. There is no basic rule about how much or how large the network must be reduced. Try to make it smaller when NN is overfitted.
- **Early Stopping**: When a model is trained with an iterative method such as gradient descent, then early stopping is a type of regularization like all NNs learn through the application of gradient descent [16]. Early stopping is essentially a strategy that applies to all issues. To better fit each iteration with the training data, this method updates the model. This increases the efficiency of the model on the test set data to a level. However, improving the fit of the model to the training data does lead to an increased generalization error after this point. Early stopping rules give instructions of how many iterations can be performed since the model continues to overfit (Figure 11.13) [17].
- **Use Regularization**: Regularization is a method for reducing the model's complexity. It does so by applying the loss feature to a penalty word. The commonest techniques are called regularization of L1 and L2. The penalty

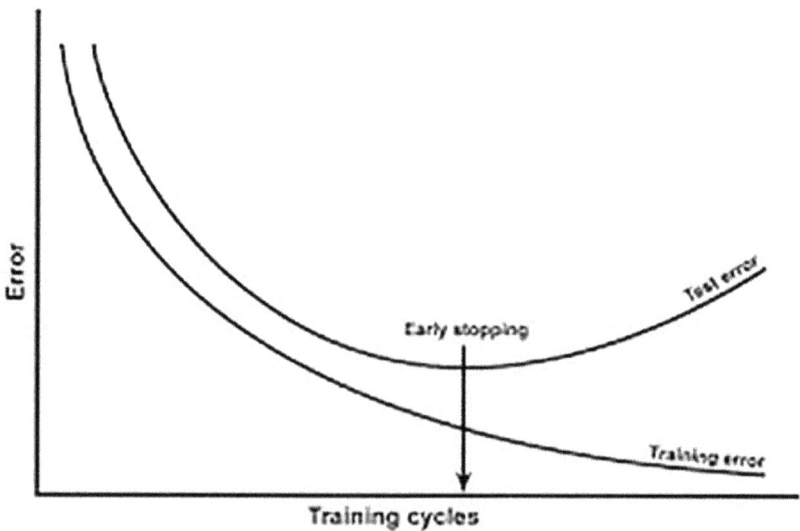

FIGURE 11.13   Overfitting in neural network.

L1 is aimed at decreasing the actual weight value. The penalty for L2 is intended to minimize the squared weight magnitude (Table 11.3).

So what is the best method to prevent overfitting? The answer depends on that if the data is too much complex to be accurately modeled, then L2 is a better option as it can learn the patterns embedded within the data. However, L1 is significantly better when the data is simple enough to be accurately modeled. For most of the computer vision problems, regularization with L2 almost always provides better performance. But L1 has an additional benefit of being robust to outliers. So the right option of regularization is dependent on the issue that we are attempting to solve.

- **Dropout**: Also a very common method of NN regularization is dropout. The situation is pretty clear—each unit of the NN (with the exception of those belonging to the output layer) will be provided in calculations and the probability p will be temporary ignored. The value of p is set to 0.5 as a default value and this hyperparameter p is known as "dropout rate." Alternatively, we randomly chose the neurons for each iteration that we lower according to the specified probability. Because of this, we operate in every situation with reduced NN. Figure 11.14 illustrates a model of a dropout NN. We can see how random second and fourth layer neurons are deactivated in each iteration (Figure 11.14).

TABLE 11.3

Difference between L1 Regularization and L2 Regularization

S.No.	L1 Regularization	L2 Regularization
1	L1 penalizes sum of absolute values of weight	L2 penalizes sum of square values of weight
2	L1 is robust to outliers	L2 is not robust to outliers
3	L1 provides a simple and interpretable model	L2 regularization will learn complex patterns of data

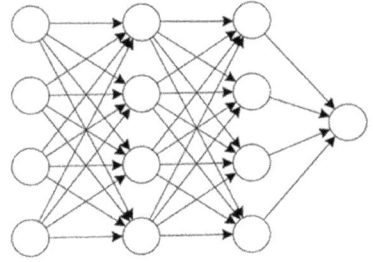
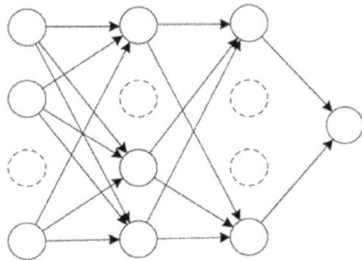

(a) Standard Neural Network   (b) Network after Dropout

FIGURE 11.14   Comparison between standard and dropout network.

This method's effectiveness is quite disturbing and counterintuitive. After all, the factory's productivity will not improve in the real world if its boss picks workers arbitrarily and sends them home every day. Now, let us aim at that question from a single neuron's perspective. Since any input value may be discarded randomly in each iteration, and the neuron attempts to assess the risk and selects none of the features. Consequently, the weight values are distributed more uniformly in the matrix. The model generally avoids a condition it will no longer apply the approach it suggests, since it no longer has knowledge that flows.

- **Use Data Augmentation**: Data augmentation is a technique that allows practitioners to increase the variety of available data for training models substantially, without actively collecting new data. For training large NNs, techniques for data augmentation such as cropping, padding, and horizontal flipping are widely used. In either the case of NNs, data augmentation essentially leads to an increase in data size that increases number of the images available in the dataset. A few of the successful techniques for image augmentation are adding noise, scaling flipping, rotation, shifting brightness, translation, etc. A lot of similar images can be produced by using data augmentation. It helps to increase the size of the dataset, thereby reducing overfitting. The cause is that the model is unable to overfit all the samples and is forced to generalize, as we add more data.

11.6 CONVERGENCE OF NEURAL NETWORK

Adaptive convergence is just "convergence." That is the traditional form of the term, as used in many current literature on ANNs. In supervised training, it describes a set of weights as they start to find (converge) the values required to produce the appropriate (trained) response [18].

Reactive convergence is the particular sense of the term convergence, used in Netlab. This explains convergence of signal propagation within networks, using signal feedback (i.e., "reactive feedback"). It has nothing to do with weight-value changes, or performance changes. Adaptive convergence simply represents weight convergence, while reactive convergence explains the convergence of signal values.

11.6.1 ADAPTIVE CONVERGENCE (OR JUST CONVERGENCE)

Convergence is commonly understood to mean traditional use when used without a prerequisite in the field of NNs. Convergence defines a progression toward a network state in the sense of traditional ANNs, where the network has learned to respond appropriately to a collection of training patterns within a certain margin of error. For example, a 10% convergence error means that a network has converged to a training set when it generates output responses within 10% of the expected output values for all input patterns in the trained repertoire. There is a type of adaptive convergence that is unique to Netlab, and there are multitemporal synapses. This use is further described below, with its own section (titled "Mixed-Mode Convergence Scenarios").

11.6.2   REACTIVE CONVERGENCE

The word convergence may also be used to define a mechanism that is entirely based on network-wide signal propagation (stimuli). This connotation has nothing to do with exercise, or shifts in weight values, unlike traditional ANN use. Typically, it will be qualified as reactive convergence when used in this manner, although it may not always be so qualified [18].

The word "reactive convergence" has no meaning in traditional feedforward-only networks, since there are no "convergence" dynamics that need to be represented in such a way without feedback.

The network may oscillate or be unstable for multiple iterations in NNs that employ reactive feedback before it settles on a given set of responds to a given set of inputs. In this case, when (or where) the oscillations (or "ringing") settle, the network will be said to converge, and the network generates some usable output. The performance is not expected to be static (steady) to converge, only to provide available, correct answers to a given encounter in the present moment.

Notice also that if it has not completely learned how to adapt to a given situation, the network does not automatically converge. The oscillations in this case may very well function as a type of trial and error for whatever learning processes may be used, but the unconverged feature of the stimulus. If it is defined in reactive terms, the definition of the reactive, propagating, signaling, and not of any changes in the connection weights is pure. The reactive term "convergence" connotation is an exact match for how it is used in computer design [17].

11.7   KEY FEATURES OF THE ERROR SURFACE

There have been several different types of nonconvex optimization issues yet then we really solve the particular type of problem while training the NN is difficult in particular. Using the characteristics of both landscape or surface of error, in order to provide a practical solution, we can define the difficulty the optimization algorithm can encounter and should overcome several features of optimizing NN weights that complicate the problem. However, the existence of flat regions, local minima, and the search space of high dimensionality are currently three commonly discussed features of an error landscape.

11.7.1   LOCAL MINIMA

Backpropagation can be very low with several local minima and flat regions, particularly for multilayered networks in which cost area is usually high-dimensional, nonquadratic, and nonconvex. This local minima comes from the fact that there are various regions in the error landscape In that the loss is considerably lower. These are valleys where, compared to the slopes and peaks in those valleys, solutions look perfect. The point is that, the valley has a reasonably high elevation in the wider view of the rest landscape and there may be better solutions (Figure 11.15) [18].

It is difficult to know exactly whether the optimization algorithm is in a valley or not, so starting the optimization process with lots of noise is a good practice,

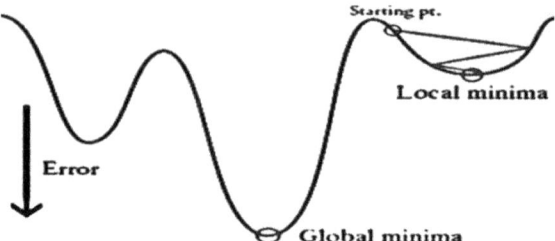

FIGURE 11.15   Local minima.

allowing thorough sampling of the landscape before finding a valley in which to fall. Alternatively, the lowermost point is called "global minima." There may be more than one global minima in NNs, and the issue is also that the distinction between local minima and global minima that does not make a substantial change.

This assumption would be that they consider a much more tractable set of weights rational enough and, in fact, it is more desirable than having ideal global weights or a decent set of weights. A typical approach to solving the local minima issue is just to repeatedly resume the search process with such a random initial weight (different starting points) and enable the optimization algorithm to identify alternatively local minima. This is known as "multiple restarts."

11.7.2   FLAT REGIONS (SADDLE POINTS)

It is the point in the landscape known as "flat regions or saddle points" where there is zero gradient. The main issue is that the optimization algorithm implies a zero gradient, which would not know which path to travel to increase the model. The existence of the saddle points, particularly areas in which there is flat error function that depends on iterative procedures to stick for longer duration, aims to mimic local minima (Figure 11.16) [18].

However, latest work may indicate that it may be less of a problem for local minima and flat regions than previously assumed. NNs join as well as exit a sequence of local minima. Do they simply travel as well as they begin to approach at different

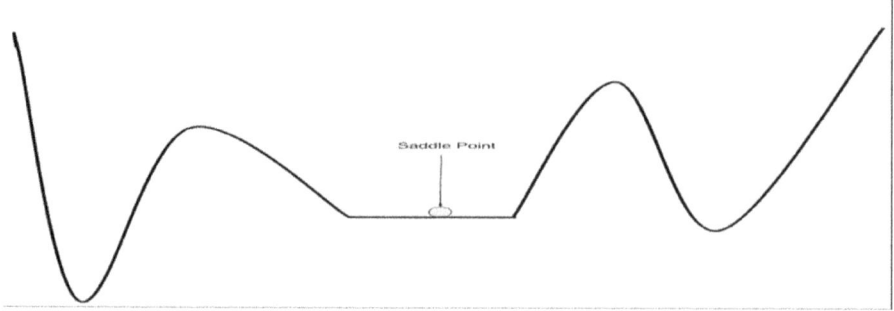

FIGURE 11.16   Saddle point on error surface.

speeds allow a number of saddle points? According to clear facts, the answer to all these questions is no.

11.7.3   HIGH-DIMENSIONAL

When training a NN, the optimization problem solved is using higher dimensional. Increasing weight in the network signifies alternative error surface parameter or dimension. DNN often has thousands of such parameters, essentially creating the navigable universe of the algorithm very high-dimensional in direct comparison with more compared to conventional ML algorithms. By adding each new dimension, the process of guiding a high-dimensional space is to increase the range between space points or hyperlength significantly. It is also known as the "curse of dimensionality." The number of potential distinct configurations of a set of variables grows exponentially as that of the number of variables increases.

Implications for the Training:
The extremely difficult nature of the optimization issues that need to be explained while using DNNs has consequences in practice while training the model. Broadly speaking, one of the powerful algorithms available is stochastic gradient descent, and that algorithm provides no guarantees. There is almost no ideal formula available that ensures that the network will converge it into good alternative approach or converge rapidly or there will be no convergence at all.

These implications are as given below:

- **Possibly Questionable Solution Quality**: This optimization approach will or does not find a proper solution, because of the misleading local minima, solutions can only be compared comparatively.
- **Possibly Long Training Time**: because of the iterative type of the search, it will take a long time for this form of optimization to find a suitable solution.
- **Possible Failure**: Due to the presence of flat regions, this optimization method can be failed to recover or unable to find a convenient resolution.

11.8   APPLICATION OF ARTIFICIAL NEURAL NETWORK

- **Image Processing and Character Recognition**: ANNs have the capacity to take in multiple inputs and process them to predict both hidden and complex, nonlinear relations. ANN plays a significant role in identifying character and image. In fraud detection (e.g., bank fraud) as well as national safety assessments, character recognition such as handwriting has many applications. Image recognition is a rapidly growing field with broad applications such as defence, social media facial recognition, cancer detection, satellite imagery processing, and agriculture. DNNs formed the base for "deep learning" and have now opening up all the interesting and revolutionary developments in speech recognition, computer vision, natural language processing—prominent examples being self-driving cars [19].

- **Forecasting**: In day-to-day business decisions (i.e., revenues, financial allocation among goods, and capacity utilization), in economic and monetary policy, and in finance and capital markets, there is a broad need for forecasts. More often than not, issues with forecasting are complicated, for example, forecasting stock prices is a difficult problem with many essential factors (some known, some unseen). With its ability to model and remove hidden features and relationships, ANNs, implemented in the correct way, can provide a robust alternative. ANN also does not place any constraint on input and residual distributions, unlike these traditional models [20].

- **Handwriting Recognition**: It is easy to use the FFNNs to identify handwritten types. The handwritten character bitmap shape is input, with the desired output being the appropriate letter or digit. These systems enable the network to be trained by the user, offering the system with handwritten patterns. The common handwriting recognition application areas are [21]: optical character recognition in the data entry and authentication of the signature on the bank cheque.

- **Traveling of Salesman**: The issue of traveling salesmen states to finding the straight potential route for traveling across all the cities in a specified area. To solve this problem, NNs can be used. ANN algorithm such as a genetic algorithm begins through the arbitrary position of the network to solve this issue. Every time, this algorithm selects a city randomly and nearest city is find out. This method thus continues a number of times. After each iteration, the form of the network changes and the network converges into a loop around most of the cities. Ring longitude is decreased by the algorithm used. We can overcome the problem of traveling in this way.

- **Image Compression**: A NN used it to compress images comprises the input and output layers of equal size. The intermediate layer is smaller. The network's compression ratio is input layer-to-intermediate layer ratio [22]. It is necessary to calculate the reference ratio for image compression using the formula given:

$$\text{Comparison Ratio} = \text{Input Layer}/\text{Intermediate Layer}$$

Data or image compression in NN is based on storage, encryption, and decryption of the actual image.

- **Stock Exchange Prediction**: The predictive precision of NNs has allowed them to be useful in stock market forecasting. In big corporations, making stock exchange predictions is natural. It uses criteria such as recent developments, the political climate, public sentiment, and the advice of economists. In business failure prediction, currency prediction, debt risk assessment, and credit acceptance, we can use NNs. In order to demand excellent 199.2% returns over a two-year period, companies such as MJ Futures use their NN prediction methods. With a simple NN at LBS Capital

Management, Dean Barr and Walter Loick have obtained excellent results. Just six financial metrics were used as inputs. That includes ADX, the current value of the S&P 500, as well as a net increase from 5 days ago over the entire S&P 500 value. The average directional change over the previous 18 days is demonstrated by ADX [23].

- **Linguistic Parsing and Answering Question**: Answering question systems reply various kinds of questions automatically in as well as questions about meanings, multilingual issues, biographical queries, etc. Use of NNs enables high answering query networks to be developed.
- **Paraphrase Detection**: Paraphrase recognition decides when two sentences are of the same significance. For answering question systems, this task is especially important, because there are several ways of asking a similar question [24].
- **Language Generation and Multidocument Summarization**: Automatic report writing is often used to produce natural language, creating texts depends on statistical analysis of retail transactions, summarizing electronic health accounts, creating written climate predictions from satellite data [25].
- **Neural Networks in Medicine**: The main objective of research in medical diagnostics is to create more precise, cost-effective, and convenient-to-use clinician support systems, procedures, and techniques. Given this latest technique, completely new diagnostic instruments in the field of ECG, EEG, and macroscopic and microscopic image analysis systems could be developed in the coming years [26].
- **Forensic Speaker Recognition**: ANN is being used in forensic science. Whenever a suspect leaves his/her speech as proof, it might be captured in camera by telephone or by voice. Some offenders can try to change their voice so they will not be recognized. Therefore, forensics' recognition of speakers is more complicated. When proper speech information is available from the defendant and accused to extract speaker attribute from speech data and sample comparison is performed. Automatic speaker recognition systems' feature is extracted and requires extraction of the feature in order to perform a meaningful comparison [27].
- **Speech Recognition**: Speech recognition has several uses, including home automation, mobile phone, interactive assist, hands-free gaming, video games, and many more. Scientists describe how and when to introduce CNNs in a unique way to speech recognition in CNNs for speech recognition, CNN specifically introduces certain types of speech variation, such as varying speaking speeds [28].
- **Spell Checking**: Many text editors let users test if there are spelling errors in their text. NNs are now present in many spell testing methods. A new system for the identification of misspelled words was introduced in personalized spell checking using neural networks. This method is based on interpretations of the particular corrections a typist is making [29]. This overcomes several of the shortcoming conventional methods of spellchecking.

11.9 CONCLUSION

The main emphasis of this chapter is on NN representations and defining suitable problems for NN learning. The rest of the chapter covers numerous substitute designs for the primitive units making up an ANN such as perceptron units, sigmoid units, and linear units. This chapter also covers the learning algorithms for training single units. Later, the backpropagation algorithm for multilayer perceptron training is described in detail. Also, the general issues such as the representational capabilities of ANNs, overfitting problems, and substitutes to the backpropagation algorithm are also explained. Lastly, the concepts of convergence, local minima, generalization, overfitting, and stopping criterion are covered in detail.

REFERENCES

1. Grossi, E., & Buscema, M. (2007). Introduction to artificial neural networks. *European Journal of Gastroenterology & Hepatology*, 19(12), 1046–1054.
2. Krenker, A., Bester, J., & Kos, A. (2011). Introduction to the artificial neural networks. In *Artificial Neural Networks – Methodological Advances and Biomedical Applications*. IntechOpen, Croatia.
3. Dongare, A. D., Kharde, R. R., & Kachare, A. D. (2012). Introduction to artificial neural network. *International Journal of Engineering and Innovative Technology*, 2(1), 189–194.
4. Karlik, B., & Olgac, A. V. (2011). Performance analysis of various activation functions in generalized MLP architectures of neural networks. *International Journal of Artificial Intelligence and Expert Systems*, 1(4), 111–122.
5. Sibi, P., Jones, S. A., & Siddarth, P. (2013). Analysis of different activation functions using back propagation neural networks. *Journal of Theoretical and Applied Information Technology*, 47(3), 1264–1268.
6. Zou, J., Han, Y., & So, S. S. (2008). Overview of artificial neural networks. In *Artificial Neural Networks* (pp. 14–22). Humana Press, New York.
7. Love, B. C. (2002). Comparing supervised and unsupervised category learning. *Psychonomic Bulletin & Review*, 9(4), 829–835.
8. Huang, G., Song, S., Gupta, J. N., & Wu, C. (2014). Semi-supervised and unsupervised extreme learning machines. *IEEE Transactions on Cybernetics*, 44(12), 2405–2417.
9. Wiering, M., & Van Otterlo, M. (2012). Reinforcement learning. *Adaptation, Learning, and Optimization*, 12, 3.
10. Mandischer, M. (2002). A comparison of evolution strategies and backpropagation for neural network training. *Neurocomputing*, 42(1–4), 87–117.
11. Bouvrie, J. (2006). Notes on convolutional neural networks.
12. O'Shea, K., & Nash, R. (2015). An introduction to convolutional neural networks. arXiv preprint arXiv:1511.08458.
13. Li, J., Cheng, J. H., Shi, J. Y., & Huang, F. (2012). Brief introduction of back propagation (BP) neural network algorithm and its improvement. In Jin, D. & Lin, S. (eds). *Advances in Computer Science and Information Engineering* (pp. 553–558). Springer, Berlin, Heidelberg.
14. Wang, L., Zeng, Y., & Chen, T. (2015). Back propagation neural network with adaptive differential evolution algorithm for time series forecasting. *Expert Systems with Applications*, 42(2), 855–863.
15. Lawrence, S., Giles, C. L., & Tsoi, A. C. (1997, July). Lessons in neural network training: Overfitting may be harder than expected. In *AAAI/IAAI* (pp. 540–545). AAAI Press, Menlo Park, CA.

16. Koehrsen, W. (2018). Overfitting vs. underfitting: A complete example. Towards data science. https://towardsdatascience.com/overfitting-vs-underfitting-a-complete-example-d05dd7e19765.
17. Srivastava, N., Hinton, G., Krizhevsky, A., Sutskever, I., & Salakhutdinov, R. (2014). Dropout: A simple way to prevent neural networks from overfitting. *The Journal of Machine Learning Research*, 15(1), 1929–1958.
18. Atakulreka, A., & Sutivong, D. (2007, December). Avoiding local minima in feedforward neural networks by simultaneous learning. In *Australasian Joint Conference on Artificial Intelligence* (pp. 100–109). Springer, Berlin, Heidelberg.
19. Quraishi, M. I., Choudhury, J. P., & De, M. (2012, March). Image recognition and processing using artificial neural network. In *2012 1st International Conference on Recent Advances in Information Technology (RAIT)* (pp. 95–100). IEEE, Dhanbad, India.
20. Zhang, G., Patuwo, B. E., & Hu, M. Y. (1998). Forecasting with artificial neural networks: The state of the art. *International Journal of Forecasting*, 14(1), 35–62.
21. Kulik, S. D. (2015). Neural network model of artificial intelligence for handwriting recognition. *Journal of Theoretical & Applied Information Technology*, 73(2), 202–211.
22. Alshehri, S. A. (2016). Neural network technique for image compression. *IET Image Processing*, 10(3), 222–226.
23. Chong, E., Han, C., & Park, F. C. (2017). Deep learning networks for stock market analysis and prediction: Methodology, data representations, and case studies. *Expert Systems with Applications*, 83, 187–205.
24. Agarwal, B., Ramampiaro, H., Langseth, H., & Ruocco, M. (2018). A deep network model for paraphrase detection in short text messages. *Information Processing & Management*, 54(6), 922–937.
25. Gao, Y., Meyer, C. M., Mesgar, M., & Gurevych, I. (2019). Reward learning for efficient reinforcement learning in extractive document summarisation. arXiv preprint arXiv:1907.12894.
26. Amato, F., López, A., Peña-Méndez, E. M., Vaňhara, P., Hampl, A., & Havel, J. (2013). Artificial neural networks in medical diagnosis. *Journal of Applied Biomedicine*, 11(2), 47–58.
27. Künzel, H. J. (1994). Current approaches to forensic speaker recognition. In *ESCA Workshop on Automatic Speaker Recognition, Identification and Verification* (pp. 135–141).
28. Graves, A., Mohamed, A. R., & Hinton, G. (2013, May). Speech recognition with deep recurrent neural networks. In *2013 IEEE International Conference on Acoustics, Speech and Signal Processing* (pp. 6645–6649). IEEE, Vancouver, BC, Canada.
29. Hutson, M. (2017). AI glossary: Artificial intelligence, in so many words. *Science*, 357, 19.

12 A Case Study on Machine Learning to Predict the Students' Result in Higher Education

Tejashree U. Sawant and Urmila R. Pol
Shivaji University

CONTENTS

12.1 Introduction 229
    12.1.1 Literature Review 231
12.2 Proposed Model..232
    12.2.1 Participants and Datasets...233
    12.2.2 Data Retrieval 235
    12.2.3 Data Preprocessing 235
12.3 Result and Discussion 236
    12.3.1 Model Evaluation Metrics...236
    12.3.2 Decision Tree Classification 236
    12.3.3 KNN Classification...237
    12.3.4 Random Forest Tree Classification..237
    12.3.5 X-Gradient Boosting Tree Classification...238
12.4 Comparative Results for Different Classification Models 238
12.5 Conclusion and Future Scope 239
References..240

12.1 INTRODUCTION

Education in India is a turning point in the life of students in respect of the career. During such academic year, there are many factors which affect the students' result; some of them include demographic, academic, and socioeconomic factors. Focusing on such determinants can ultimately help to improve the performance of students. This is the need to study different factors affecting the students' performance and build a system which can predict the last semester result of postgraduate (PG) students so as to help them to improve and take actions accordingly before the final examination. This will gradually enhance the higher education performance. In the said aspect, this chapter is to design and develop a machine learning (ML) classification model, considering different variables. The proposed ML classification model to predict the final semester result of PG courses will be useful to students for taking appropriate decisions for the improvement of the result in the highly competitive world today.

Without doubt, this will help the society owing to its linkage with academic enhancement of student and consequently development of the nation. In this chapter, we intend to specify quality education in a university and develop an ML model for enhancement of quality education in terms to predicting last semester students' results and take necessary actions before it is too late. The author tried to study the main factors that are highly correlated with the dataset using ML with ML approach algorithms such as k-nearest neighbor (KNN), random forest, decision tree, and XGBoost.

There exists a broad variety of ML related work, in which many fascinating approaches and methods are brought forward to meet the goals of knowledge discovery, decision-making, and recommendations. A wide-ranging analysis of the literature on academic achievement of students and their estimation was conducted using performance models. Nevertheless, it was noted that a study on the factors affecting the students' academic performance at the university level and prediction of students' result using specific ML algorithms were carried out in restricted research investigations.

Nowadays, the biggest challenge is to make the university stronger in making education processes healthier, more effective, and more reliable. ML is considered a suitable technology that can provide unique insight into the faculty, student, employee, and other staffs.

The use of ML will be beneficial to resolve the disparities of information in the higher education sector. Hence, the secret patterns, correlations, and anomalies found with the methods are used to boost performance and efficiency in processes. Such reforms will benefit to the higher education system, such as optimizing the educational effectiveness, reducing dropout rates for students, optimizing student promotion rates, building up student retention rates, maximizing student efficiency, and maximizing learning. Educational patterns in higher education using the ML technique are presented in Table 12.1.

To accomplish the above-mentioned progress in quality, we need machine language in education platform that can provide with the knowledge and insights needed in the higher education context to take appropriate decisions.

TABLE 12.1

Patterns Generated through Educational ML Classifications

ML Technique	Generated Educational Patterns in Higher Education
Classification and prediction	Predicting students' learning outcomes in PG courses
	Predicting the percentage of students' academic performance
	Predicting the factors influencing the students' academic performance
	Predicting students' behavior and attitude toward learning
	Predicting final semester students' results

12.1.1 LITERATURE REVIEW

A comprehensive analysis gives the authors with such a summary of the higher education research with artificial intelligence. Of the 2656 articles which were originally listed for the period from 2007 to 2018, 146 articles are included in the final analysis, based on simple evaluation parameters for acceptance and rejection. The Artificial Intelligence in Education (AIED) theme within the Personalization strand of Technology Enhanced Learning (TEL) is concerned with exploring the ways in which the work conducted under TEL within and across projects can contribute to the (inter)discipline of AIED. The empirical results indicate that several of the functional areas within AIED studies originated through computer science and that the much more commonly used quantitative methods were in observational research (Zawacki-Richter et al., 2019). Authors plan to systematically review the published research focusing on implementation of artificial intelligence and ML. Both are seen as the motivating factor for smart industry transformation with introducing Industry 4.0 (Cioffi et al., 2020). Authors applied an ML approach to forecast the students' final grades depending on historical grade results. The idea referred to the historical academic knowledge available to students (Buenaño-Fernández, Gil, & Luján-Mora, 2019). Authors have proposed ways of using models to efficiently conduct instructional sequencing: (1) using different learning models to set up rules and policies that are consistent in how students actually learn and (2) integrating learning models based on psychological theory with data-driven approaches. Aims at combining human and artificial intelligence will further boost efforts to enhance the semiautomated teaching of students (Doroudi, 2019). Authors in their research identified features students who started in the Tennessee Board of Regents (TBR) program, moved after their first year to some other college, got graduated, or with credentials. Variables have been implemented to the predictive tasks for classifying few students before their departure (Whitlock, 2018). In their research, authors explored the potential of learning analytics in knowing learning skills in collective learning settings, with the goal of enhancing teaching and learning. More specifically, research questions include the following: how to accurately predict students' success through the learning analytics and social network analysis (SNA) using contextual, theory-based metrics and how to use SNA to assess collective online learning, direct, and evaluate data-driven intervention. The analysis methodology followed a systematic process of gathering, planning, and analyzing data. Students' data were collected using custom plug-ins and database queries from the online learning management system (Saqr, 2018). Authors also suggested a method that will use the decision tree methodology to predict the success of a student. Depending on the present performance of the student and other observable past characteristics, the end result can be expected to be graded as good or poor performers (Vyas & Gulwani, 2017). Authors in their paper explores the phenomena of the emergence of the use of artificial intelligence in teaching and learning in higher education. They investigate educational implications of emerging technologies on the way students learn and how institutions teach and evolve. Recent technological advancements and the increasing speed of adopting new technologies in higher education was explored in order to predict the future nature of higher education in a world where artificial intelligence is part of the fabric of universities. They pinpoint some challenges for institutions of higher education and student learning in the adoption of these technologies for teaching, learning, student support, and administration and explore further directions for research. (Popenici & Kerr, 2017). The authors, in their report, performed data analytics to cover potentially secret knowledge; highlight fascinating data stories; and investigate student learning behavior and learning success in an active learning environment, including collaborative learning in a flipped classroom model (Sener, 2017). Authors present the students' learning outcome prediction approach which uses four different kinds of characteristics, notably, family expenses, parental income, and personal knowledge. It also adapts a subclass feature selection method to determine the most influential predictors for forecasting academic success among students (Daud et al., 2017). Data science is used for acquiring knowledge, discovering the hidden information, and even applying ML techniques to the academic dataset. Academic data includes identification and external labels. The final semester marks are calculated based on the measured performance of each student (Kalaivani & Nalini, 2017). A decision table (DT) was derived from a decision tree or rule-based classifier to boost the consistency of academic performance for alcohol-consuming students at educational institutions. Predicting adolescents' addiction to alcohol by using demographic, family, and other student-related data, different classifiers performed that helped identify the best classifier for predicting the performance of students consuming alcohol with the help of data science methods. The classifiers sequential minimal optimization, bagging, Reduced Error Pruning (REP) tree, and DT were used to diagnose students' results (Pal & Chaurasia, 2017). Researchers used data mining methods to predict the final year students' outcome using gained marks and in which other attributes such as socioeconomic or demographic was not considered for the study (Asif, Hina, & Haque, 2017). Authors discussed the implications of big data in education, focusing on data created by the student's writing as an example. Writing was chosen because it poses specific difficulties, highlighting the variety of processes for collecting and evaluating learning evidence in the age of computer- mediated instruction and assessment, as well as the challenges (Cope & Kalantzis, 2016). The authors suggested a model for forecasting students' success in an academic organization. The algorithm used is neural networks, a technique in ML. However, to assess which of these are associated with students' success, the value of several different attributes or "features" is considered.

The results of an experiment eventually follow, demonstrating the strength of ML in such an application (Havan Agrawal, 2015). Authors have stated that diverse classification methods help the model to predict students' achievement. In addition to its common attributes in an online learning environment, login details such as log count, length of every session, etc. are used. Authors in the study have added features useful in knowing students learning to the existing list (Akçapınar, Altun, & Aşkar, 2015).

12.2 PROPOSED MODEL

Although many studies are being carried out on the prediction of student performance, very few studies focus on investing how the performance of students evolves during their course of study. Most of the approaches have used only the factors such as demographic and academic marks as their basis of prediction. Very little work is done with ML. The proposed conceptual development model builds in the context of higher education.

12.2.1 PARTICIPANTS AND DATASETS

The data collection segment focuses on gathering data. The proposed framework provides that data will be collected through the use of traditional methods and computational methods. Some of the examples of these traditional methods of gathering data include surveys, interviews, questionnaires, checklist, observations, focus groups, etc. These methods can also be automated, instead of being paper based. As with other studies in educational ML, the researcher has found the dataset for university PG students.

The research focuses on the ML model design and implementation using the students' semester outcome and features related to demographic, financial, and graduation students from different PG courses. A finite population formula has been used for determining the sample size under the study researcher.

The estimated sample size is 355 respondents. As a result, the researcher has based 355 respondents, randomly selected from various the PG departments from the Shivaji University campus. The researcher collected data through a standardized questionnaire for the selected sample. The questionnaire and a few 1-on-1 interviews with students were the tools used for data collection. It has been ensured in the data collection process that only students who have no problem when sharing their personal data are personally interviewed and chosen to be the respondent to this research study. The original raw data collected and further stored in the dataset comprises 2 personal attributes, 10 academic and 12 academic behavioral attributes of students, and 6 student social attributes with all past students' success records in 8 attributes as shown in Table 12.2.

TABLE 12.2
Dataset Features with Possible Values

Features	Description	Possible Values
Gender	Male or female	{Male, female}
Location	Student city	{Urban, rural}
Time taken to travel to the department (in hours)	Time to reach to the university	{Less than 1 hour, 1–2 hours, 2–3 hours, 3–4 hours, more than 4 hours}
Medium of school	The language of instructions in classes: English and Marathi	{English language, Marathi language, other}
Teaching method	The teaching methodology which is used in PG classes	{Understandable, not understandable, neutral}
Teaching or learning gap	The number of days off from the respective student's course calendar	{Yes, no}
Exam environment	Does the student feel stressed in the exam center/room during examination which can impact the student's performance	{Very much stressful, neutral, relaxed}
Teacher motivation	Did the student get proper/timely motivation from the teacher to study	{Yes, all the time, only sometimes, not at all}
Expert motivation	Did the student get any motivation from other expert to study/learn	{Yes, all the time, only sometimes, not at all}
Inappropriate syllabus	Exam syllabus and curriculum syllabus comparison	{Less appropriate, average, more appropriate}

(Continued)

TABLE 12.2 (*Continued*)
Dataset Features with Possible Values

Features	Description	Possible Values
Cultural or festival holidays	Does cultural or festival holidays impact the student performance?	{Yes, no}
Incentive programs	Is the student getting any incentives?	{Yes, no}
Study hours	How many hours students spend in study?	{Less than 1 hour, 1–2 hours, 2–3 hours, 3–4 hours, more than 4 hours}
Class attentiveness	How well is the student's attentiveness during class?	{Yes, no}
Referring previous papers	Has the student referred previous papers to comprehend his/her study for exams?	{Yes, no}
Time management	How is the time management skill of the student?	{Good, poor}
Communication skill	How is the communication skill of the student?	{Good, poor}
Extra tutions	Has the student taken any tuitions other than classes?	{Yes, no}
Class attendance	How well the student attended classes during the course?	{Above 70, between 40–70, below 40}
Own study notes	Does the student have study notes of his/her own? If yes, how well prepared notes one has?	{Yes, all the time, only sometimes, not at all}
Time spent on social media	How much time the students spend on social media like Facebook, Instagram, WhatsApp, etc.	{Yes, no}
Use of online course material	Has the student used any online material to comprehend his study?	{Yes, no}
Health issue	Health during the course	{Yes, sometimes, no}
Exam stress	What level of stress the student feels during exams?	{Yes, no}
Family literacy	How literate the parents are?	{Literate, illiterate}
Family motivation	Did the student get motivation from family members to study	{Yes, all the time, only sometimes, not at all}
Friends motivation	Did the student get motivation from the friend circle	{Yes, all the time, only sometimes, not at all}
Discussion with friends	How often study is carried out by discussing with friends	{Yes, sometimes, not at all}
Family responsiblities	How responsible a family member of the student is in his family?	{Yes, sometimes, not at all}
Financial status	What is the financial status of the student?	{Low, average, high}
Overall PG performance	How well the student has performed overall in the PG course?	{Good, average, poor}
SSC	Marks in 10th examination	{0–100}
HSC	Marks in 12th examination	{0–100}
UG	Marks in graduation examination	{0–100}
Semester 1	Marks in semester 1 of PG	{0–100}
Semester 2	Marks in semester 2 of PG	{0–100}
Semester 3	Marks in semester 3 of PG	{0–100}
Semester 4	Marks in semester 4 of PG	{0–100}

HSC, Higher Secondary Certificate; PG, Postgraduate; SSC, Secondary School Certificate; UG, Undergraduate.

12.2.2 DATA RETRIEVAL

The researcher leverages the mechanism for pandas to retrieve data from the comma-separated value format. The following section shows the information to be accessed and viewed (Figure 12.1).

12.2.3 DATA PREPROCESSING

Data preprocessing for ML tackles one of the most relevant problems within the well-known data process information discovery. It is possible that data taken directly from the source may have inconsistencies, errors, or, most importantly, it is not ready to be considered for an ML process.

Data preprocessing involves techniques for data reduction, which aim to reduce data complexity and detect or delete unnecessary and noisy elements from the data.

Data preprocessing includes operations that can arrange the data in a proper manner to better understand the method of ML. Because data is a major asset in many companies, inaccurate data can be risky. Incorrect data will reduce the efficacy of the ads, thus reducing revenue and performance.

ML is the extraction process of knowledge which involves investigating the hidden patterns from large datasets with analyzing results in various domains. The knowledge from datasets helps for decision-making and also improves organizational efficiency. ML is a standardized technique of obtaining insights from large repositories by means of different techniques. Popular techniques are clustering, classification, association rule mining, etc.

Today, most challenging work that the education system faces is to provide and use up-to-date information for sustainable growth, tracking student performance, and evaluating efficiency. PG education is the backbone of the innovation and competitiveness in higher education.

The ML classification task allows the university to keep track of those students who need immediate attention. For the development of the ML model, the researcher

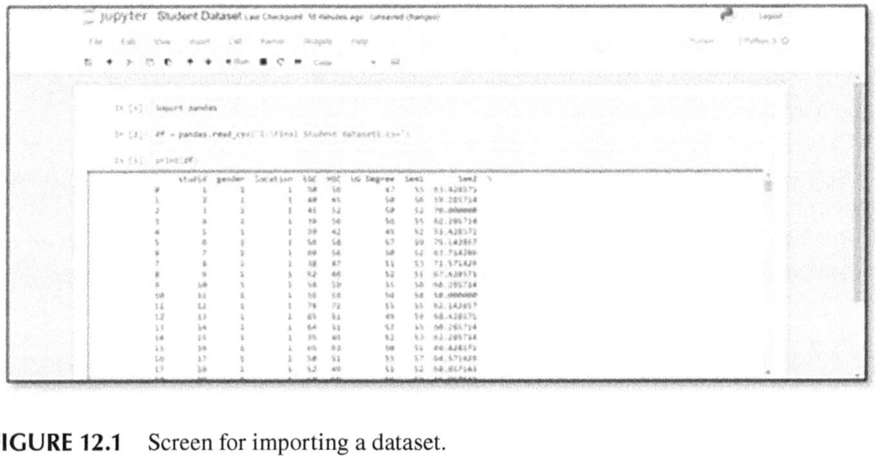

FIGURE 12.1   Screen for importing a dataset.

has applied different classification algorithms on the student dataset, namely, decision tree, KNN classification, random forest classification, and XGBoost classification algorithms.

Model evaluation is an important phase in the development of classification model. This phase assists to select the better performing model from the list of developed classification model. After the model evaluation is performed, the researcher will compare the developed ML model metrics for selection of the best model which solves the problem. With the help of the developed model, different educational patterns can be drawn, i.e., discovery of knowledge from the student dataset. The researcher has decided to predict the students' last semester result based on the past data using the student dataset, thus solving the statement of the problem of the study.

12.3   RESULT AND DISCUSSION

The researcher studied various classification algorithms and concluded that decision tree, KNN classification, random forest classifier, and XGBoost multiclassification are most popular algorithms for development of an ML model. Therefore, the researcher has applied selected classification algorithms. Performance evaluation has been carried out by the researcher along with the performance of the XGBoost multiclassification, decision tree, KNN classification, and random forest classifier methods. Scikit-learn provides a wide range of classification algorithms which are unified/consistent for predicting. The raw set of features after the feature selection process is further trained and tested using decision tree, KNN classification, and XGBoost classification algorithms. The researcher for the present study has used the scikit-learn library and XGBoost 0.82 library for fitting out the training dataset.

12.3.1   MODEL EVALUATION METRICS

Evaluation metrics explain the performance of a model. An important aspect of evaluation metrics is their capability to discriminate among model results. It is crucial to check the accuracy of your model prior to computing predicted values.

Classification Accuracy: This is what is usually meant when the term accuracy is used. It is the ratio of the number of correct predictions to the total number of input samples. Accuracy for the matrix can be calculated by taking average of the values lying across the "main diagonal".

Precision: It is the number of correct positive results divided by the number of positive results predicted by the classifier. Precision is a good measure to determine, when the costs of false positive are high.

Recall: It is the number of correct positive results divided by the number of all relevant samples (all samples that should have been identified as positive).

12.3.2   DECISION TREE CLASSIFICATION

The researcher has trained and tested our best performing feature set on different hyper parameters. The researcher recorded the five best performing sets of features and recorded results for each label class of all decision tree classifiers as shown in the below (Table 12.3).

12.3.3 KNN CLASSIFICATION

The researcher has trained and tested our best performing feature set on different hyper parameters. The researcher recorded the five best performing sets of features and recorded results of KNN tree classifiers as shown in the above (Table 12.4).

12.3.4 RANDOM FOREST TREE CLASSIFICATION

It is an ensemble classification algorithm which means the researcher uses a group of classifiers instead of one classifier for prediction outcome to fit in the training

TABLE 12.3

Top 5 Accuracy, Precision, and Recall Metric Values on the Test and Train Dataset with a Different Set of Hyperparameters and Trained Using Decision Tree ML Classification Algorithm

	On Training Data			On Testing Data		
Parameter	Accuracy (%)	Precision (%)	Recall (%)	Accuracy (%)	Precision (%)	Recall (%)
Set-I	74.24	49.65	46.37	74.14	48.76	45.25
Set-II	75.06	50.41	46.49	73.32	50.84	45.53
Set-III	**78.24%**	**52.25**	**48.37%**	**77.14**	**51.56**	**47.25**
Set-IV	73.19	46.54	43.83	72.92	44.45	42.60
Set-V	62.12	34.81	32.45	61.32	33.20	31.56

The bold values has the highest accuracy on the training and testing data.

TABLE 12.4

Top 5 Accuracy, Precision, and Recall Metric Values on the Test and Train Dataset with a Different Set of Hyperparameters and Trained Using the KNN Classification Algorithm

	On Training Data			On Testing Data		
Parameter	Accuracy (%)	Precision (%)	Recall (%)	Accuracy (%)	Precision (%)	Recall (%)
Set-I	55.49	32.21	31.17	54.63	30.42	30.25
Set-II	58.52	38.62	37.16	57.23	37.01	36.74
Set-III	61.37	42.03	42.20	60.19	41.41	41.52
Set-IV	62.16	47.28	46.31	61.82	46.12	45.22
Set-V	59.21	40.23	32.45	58.04	39.32	31.39

dataset. The target based on the majority voting method. It has different advantages over decision trees: it reduces the possibility of over fitting as well because the prediction is based on multiple classifiers or set of trees rather than one classifier where the researcher only has only one set of features.

The majority voting concept is very similar to the voting system among political parties. Every classifier will vote for one and only one target class. The target class getting the desired number of votes is considered as the predicted target class.

The researcher has trained and tested our best performing feature set on different hyper parameters. The researcher recorded the five best performing sets of features, and the results of all random forest tree classifiers are as shown in the above plot (Table 12.5).

12.3.5 X-GRADIENT BOOSTING TREE CLASSIFICATION

XGBoost (extreme gradient boosting) uses the gradient boosting framework at its base and belongs to a family of boosting algorithms.

The researcher has trained and tested our best performing feature set on different hyper parameters. The researcher recorded the five best performing sets of features and has recorded results of XGBoost multiclassification as shown in the above plot (Table 12.6).

12.4 COMPARATIVE RESULTS FOR DIFFERENT CLASSIFICATION MODELS

Table 12.7 shows values evaluated for the performance of the developed model. The precision and recall of the random forest classifier were 31.80% and 23.45%, respectively. The accuracy of the random forest classifier showed to be 60.09%. The performance of KNN was observed to be 47.28% in average precision and 46.31% for average recall and 62.16% for average accuracy. Decision tree classification was identified with a precision of 52.25%, a recall value of 48.37%, and 78.24% of average accuracy. The average accuracy for XGBoost multiclassification was 75.25%.

TABLE 12.5

Top 5 Accuracy, Precision, and Recall Metric Values on the Test and Train Dataset with a Different Set of Hyperparameters and Trained Using the Random Forest Classification Algorithm

	On Training Data			On Testing Data		
Parameter	Accuracy (%)	Precision (%)	Recall (%)	Accuracy (%)	Precision (%)	Recall (%)
Set-I	55.24	23.65	18.37	54.14	21.52	17.20
Set-II	60.09	31.80	23.45	59.68	30.41	19.05
Set-III	59.10	28.10	21.30	58.58	27.05	21.10
Set-IV	57.05	26.90	19.42	56.30	25.10	19.24
Set-V	56.03	24.50	18.30	55.26	23.07	18.10

XGBoost multiclassification was identified with a precision and recall value of 60.74% and 60.32%, respectively. The performance of XGBoost multiclassification performance showed as the best as other classification algorithms.

The researcher has hyper tuned each of the classification models, and the researcher selects best results, which provides best recall on a given set of features. As seen in the above table, XGBoost performs well.

12.5 CONCLUSION AND FUTURE SCOPE

The educational dataset with an ML approach to predict the students' final semester result has always been a crucial field of research. Predicting the academic result for further improvisation is one of the approaches that seek to track the students' performance and predict students at risk of learning pathways. Different classification algorithms have applied on the student dataset, and results are presented. The researcher noticed that XGBoost multiclassification gives best accuracy than the decision tree, KNN classification, and random forest classifier model.

TABLE 12.6

Top 5 Accuracy, Precision, and Recall Metric Values on the Test and Train Dataset with a Different Set of Hyperparameters and Trained Using the Extreme XGBoost Classification Algorithm

	On Training Data			On Testing Data		
Parameter	Accuracy (%)	Precision (%)	Recall (%)	Accuracy (%)	Precision (%)	Recall (%)
Set-I	71.08	55.30	55.21	70.40	54.42	54.10
Set-II	73.10	58.40	58.16	72.50	57.00	57.18
Set-III	75.25	60.74	60.32	75.15	59.72	59.42
Set-IV	72.04	56.80	56.32	71.20	55.04	55.72
Set-V	69.02	53.60	52.20	68.16	52.06	52.20

TABLE 12.7

Accuracy, Precision, and Recall Metric Values on the Test and Train Dataset of Different Classification Algorithms

		On Training Data			On Testing Data		
Parameter	Feature Set	Accuracy (%)	Precision (%)	Recall (%)	Accuracy (%)	Precision (%)	Recall (%)
DT classifier	Label-I	78.24	52.25	48.37	77.14	51.56	47.25
KNN classifier	Label-I	62.16	47.28	46.31	61.82	46.12	45.22
RF classifier	Label-I	60.09	31.80	23.45	59.68	30.41	19.05
XGBT classifier	Label-I	75.25	60.74	60.32	75.15	59.72	59.42

XGBoost, eXtreme Gradient Boosting; RF, Random Forest.

The whole study concentrated on the usage of educational data mining (EDM) with the ML approach. The study can be extended in diverse ways. The said work has indeed been studied for one university, but this could be applied to many different universities in India for future work.

The datasets of the PG course was considered for the present study for the prediction of the final semester result. Furthermore, this problem may be explored for the dataset from multiple programs at various levels. The involvement of several other programs in the model might provide a new viewpoint to the university for greater understanding of the prediction patterns.

This study focused on the last semester result prediction, students' different factors, and examination results for each semester. Using classification algorithms, decision tree classification, KNN classification, random forest classifier, and XGBoost multiclassification were applied on the dataset in the thesis for best results. But in the future research study, one can apply various appropriate EDM techniques on the student dataset for diverse research questions.

REFERENCES

Akçapınar, G., Altun, A., & Aşkar, P. (2015). Modeling students' academic performance based on their interactions in an online learning environment. *Elementary Education Online*, 14, 815–824.

Asif, R., Hina, S., & Haque, S. I. (2017). Predicting student academic performance using data mining methods. *International Journal of Computer Science and Network Security*, 17(5), 187–191.

Buenaño-Fernández, D., Gil, D., & Luján-Mora, S. (2019). Application of machine learning in predicting performance for computer engineering students: A case study. *Sustainability*, 11(10), 2833.

Cioffi, R., Travaglioni, M., Piscitelli, G., Petrillo, A., & De Felice, F. (2020). Artificial intelligence and machine learning applications in smart production: Progress, trends, and directions. *Sustainability*, 12(2), 492.

Cope, B., & Kalantzis, M. (2016). Big Data comes to school: Implications for learning, assessment, and research. *AERA Open*, 2, 1–19.

Daud, A., Aljohani, N. R., Abbasi, R. A., Lytras, M. D., Abbas, F., & Alowibdi, J. S. (2017, April). Predicting student performance using advanced learning analytics. *Proceedings of the 26th international conference on World Wide Web companion* (pp. 415–421). Perth, Australia.

Doroudi, S. (2019). Integrating human and machine intelligence for enhanced curriculum design. PhD diss., Air Force Research Laboratory.

Havan Agrawal, H. M. (2015). Student performance prediction using machine learning. *International Journal of Engineering Research & Technology*, 4, 111–113.

Kalaivani, S., & Nalini, S. (2017). Analyzing student's academic performance based on data mining approach. *International Journal of Innovative Research in Computer Science & Technology*, 5, 194–197.

Mostow, J., Beck, J., Cen, H., Cuneo, A., Gouvea, E., & Heiner, C. (2005). An educational data mining tool to browse tutor-student interactions: Time will tell! *Proceedings of the Workshop on Educational Data Mining* (pp. 15–22). Pittsburgh, PA: AAAI Press.

Pal, S., & Chaurasia, V. (2017). Performance analysis of students consuming alcohol using data mining techniques. *3rd International Conference on "Latest Innovations in Science, Engineering and Management"* (pp. 24–36). Panjim, Goa: The International Centre Goa.

Popenici, S., & Kerr, S. (2017). Exploring the impact of artificial intelligence on teaching and learning in higher education. *Research and Practice in Technology Enhanced Learning*, 12(1), 1–13.

Saqr, M. (2018). *Using Learning Analytics to Understand and Support Collaborative Learning.* Kista: Stockholm University.

Sener, A. C. (2017). A data science pipeline for educational data: A case study using learning catalytics in the active learning classroom.

Vyas, M. S. & Gulwani, G. (2017). Predicting student's performance using CART approach in data science. *2017 International conference of Electronics, Communication and Aerospace Technology (ICECA).* Coimbatore, India: IEEE.

Whitlock, J. L. (2018). Using data science and predictive analytics to understand 4-year University Student Churn. Electronic Theses and Dissertations. Paper 3356. https://dc.etsu.edu/etd/3356.

Zawacki-Richter, O., Marín, V. I., Bond, M., & Gouverneur, F. (2019). Systematic review of research on artificial intelligence applications in higher education – Where are the educators? *International Journal of Educational Technology in Higher Education*, 16, 39.

13 Data Analytic Approach for Assessment Status of Awareness of Tuberculosis in Nigeria

*Ishola Dada Muraina, Rafeeah Rufai Madaki,
and Aisha Umar Suleiman*
Yusuf Maitama Sule University

CONTENTS

13.1 Introduction 243
13.2 Related Works..244
13.3 Materials and Methods 245
    13.3.1 Population and Sample 245
    13.3.2 Tools and Designing 245
    13.3.3 Task Procedures...246
    13.3.4 Data Analysis and Results 247
13.4 Results and Discussion 247
13.5 Conclusions..248
Acknowledgements..248
References..248

13.1 INTRODUCTION

The spread of infectious diseases has been seen as a crucial issue that requires drastic attention toward its eradication across the world. Thus, its effect has been reported in almost all the continents of the world, with inclusion of Africa [1,2]. The previous study has stressed that public health workers in some of sub-Saharan Africa countries have intensified efforts to curb the spread of infectious diseases so as to protect their image in front of the international communities and to be viewed as healthy countries [3]. Hence, many of the multinational companies have seen opportunities in investing in healthcare industries as efforts to support governments for combatting the spread of diseases [3], specifically, infectious ones. The recent development is the additional 15% of contribution to the fight against tuberculosis (TB), malaria, and human immunodeficiency virus/acquired immunodeficiency syndrome as announced by The President of France, Emmanuel Macron [4].

DOI: 10.1201/9781003138020-13

The World Health Organization had simply described infectious diseases as that capable of spreading directly or indirectly from one person to another and are caused by pathogenic microorganisms, viruses, bacteria, parasites, or fungi [5]. In Nigeria, of many infectious or communicable diseases, TB is one of the visible diseases among the lower class of the population. This might be due to the low income of the affected people and the available healthcare facilities. Moreover, the prevalence rate of TB in Africa rates Nigeria as the first with 616 estimated cases per a populace size of 100,000 people [6]. In the study called "Nigeria Tuberculosis Fact Sheet" conducted by the United States of America (USA) Embassy in Nigeria [7], of 36 states including the Federal Capital Territory, Lagos, Kano, Oyo, Benue, and Kaduna states were seen to be the most prevalence states in hierarchy with TB.

On the other hand, it has been argued that pulmonary TB is highly prevalent in Kaduna state compared to the previous study by the USA Embassy in Nigeria [7]. The sudden change in the prevalence rate of TB vis-à-vis the Kaduna state could be due to the impact of awareness given to the populace by the policy makers in the healthcare industry. This variation in the prevalence level of TB reveals that enough awareness has not been provided by the people in charge as the disease still persists. Previous studies have stressed that giving awareness on infectious diseases has a broad impact on its prevalence rate toward its mitigation or total eradication [8,9]. Meanwhile, it has been argued that the fight against TB can be won by providing adequate education to both the secured and vulnerable groups [10]. Thus, this study focuses on the assessment status of the awareness of TB in Nigeria toward its mitigation or eradication.

Furthermore, analytical techniques have previously been argued as one of the approaches to address different issues in healthcare informatics, such as visual analytics and predictive analytics [11]. This is as a result of its ability to explore databases or repositories of some historical data of patients. Therefore, this study uses the data analytic approach to assess the awareness status of some participants on TB in their respective communities.

13.2 RELATED WORKS

Studies have argued that curtailing of infectious diseases spread could be achieved through the modeling of different tools together with synergy with both streaming and historical data [12–14]. The ultimate step of modeling is to engage the analysis of processes and methods toward obtaining the meaning of the hidden phenomenon and produce answers to the questions under research [15,16]. Meanwhile, the approaches or methods used in analytics could be in the form of supervised and unsupervised learning approach of data training [17,18]. This has been used by many organizations and industries in revealing the hidden pattern of some issues and predict future occurrence in their respective actions. Moreover, adequate codes to explore a pool of data [19,20] have been suggested for use in the process of analytical techniques in infectious diseases studies [21,22].

Different methods for treating TB and its associated symptoms have been proposed by many researchers, but the choice is based on the patients' physical conditions and other factors [23–25]. The study of Hanaoka et al. [23] devised a method for treating airway stenosis, a symptom secondary to bronchial TB causing dyspnea and recurrent infection through balloon dilation with laser cauterization, as opposed to the use of surgery and radical treatment. This is due to the fact that some patients may not be able to undergo the latter in good condition, without developing more complications. They argued that the surgical procedure may not be preferable to some patients owing to its high risk or severity of the patient's symptoms. However, a comprehensive report on the long-term prognosis of noninvasive therapy is still lacking, while the optimal treatment method remains questionable in some areas [24].

Moreover, a case of pelvic TB has been recommended to be treated using any other method other than surgery owing to the exhibiting symptoms of pelvic pains, fatigue, and dry cough [6]. The study has stressed that curbing the rate of TB transmission requires people to be educated on the preventive measures and counselling in the ways to prevent other households from being infected [10]. In addition, the study of Liu et al. [8] analyzed the level of cyber awareness with respect to emerging infectious diseases in Zhejiang, China. This implies that internet has made information readily available to people, specifically in the area of obtaining information about an outbreak of infectious diseases in any part of the world. Meanwhile, the rate of prevalence of infectious diseases had been categorized as high, medium, and low [26], thus tested in measuring differences in the levels of publicity of the epidemic.

13.3 MATERIALS AND METHODS

The research methodology for this study is in line with the objective of the study. A design approach in the form of an experiment was adopted during the study toward determining the level of awareness of some participants on TB in their communities.

13.3.1 POPULATION AND SAMPLE

The study engaged some undergraduate students from the computer science department of Yusuf Maitama Sule University (Formerly, Northwest University), Kano, for participation in the data collection processes. All the participants are studying computer science and information and communication technology programs; a total of 148 students were invited to participate in the data gathering processes.

13.3.2 TOOLS AND DESIGNING

A simple database was designed the using MySQL package to store the responses of participants with respect to their awareness on TB in their respective locations; an interface was also designed for participants through which they can submit their opinion about their awareness on TB. Both the database and interface were run on the Windows 8.1 operating system during the processes.

The study adopted four attributes, such as the perceptual awareness scale (PAS), confidence ratings (CRs), postdecision wagering (PDW), and feeling of warmth (FW), from Timmermans and Cleeremans [27] to measure the awareness of participants about the prevalence of TB in their communities. Previous studies have argued that the PAS has to do with the rating of issues that everyone believes one has seen or has knowledge about without being compensated [28,29]. Besides, the CR has to do with the act of testing people's confidence on the contemporary or bothered issues in their domains [30]. On the other hand, the PDW ensures the elimination of uncertainty in making decision on an issue [31], while the FW rates people's feelings of warmth on what they saw or believe to have happened [24]. Hence, all the four attributes, PAS, CR, PDW and FW, were used with respect to the assessment of the awareness status of TB by the participants in this study.

13.3.3 TASK PROCEDURES

A 5-minute video clip on TB was provided to all the participants for viewing to before their access to the interface for data collections as shown in the task and analytical procedures in Figure 13.1.

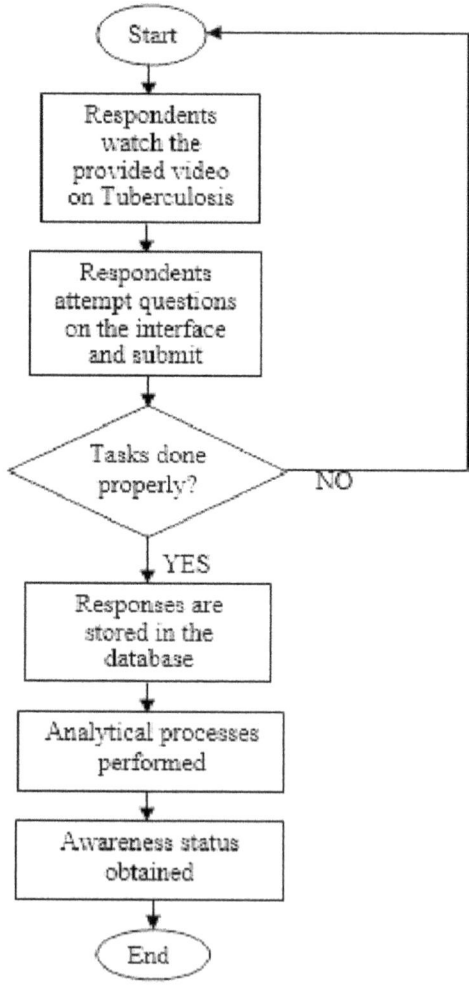

FIGURE 13.1   Task and analytical procedures.

Below is a flowchart diagram representing tasks of activities for the analytical approach in assessing the awareness status of TB.

All the participants were divided into three groups, and different times were fixed for their participation in the experiment toward data collection. Meanwhile, the same task was given to all the participants who were watching the five-minute video clip on TB. Therefore, participants were given a maximum time of 10 minutes to attend to questions displayed on the provided interface and submit the questions, and if there is an error in the process of attempting the questions, the process will begin again until the questions are correctly attempted. Hence, the responses were later stored in the database for onward analytical processes.

13.3.4 DATA ANALYSIS AND RESULTS

The 148 cases gathered in the database were imported into the data analytic software Tableau Prep 2018.1.2. Therefore, the data was screened and freed from all forms of errors before the analytical stage. The frequency of responses of cases as stored in the database was performed for all the three groups that participated in the data collection processes toward assessing the status of their awareness on TB.

13.4 RESULTS AND DISCUSSION

Table 1 presents the results of the analytical experiment of frequency of 148 stored cases in the database with respect to the assessment status of awareness of TB in Nigeria.

As shown in Table 13.1, the participants were divided into three different groups, with Group 1 containing 33.11% of males and 8.78% of females. There were 33.78% of males and 8.11% of females in Group 2, and Group 3 had 11.49% of males and 4.73% of females as participants. This shows the diversity of participants who were involved in the data collection processes. Thus, their opinions could represent some of the communities with respect to the objective of this study. Meanwhile, almost 80% of the participants claimed that they have low awareness on TB and its associated issues. This implies that the priority given to the awareness of TB is not yet enough toward sensitizing the societies and bringing an end to this infectious disease

TABLE 13.1
Results of the Analytical Experiment of the Participants' Awareness Status on TB

Assessment Attributes	Group 1		Group 2		Group 3	
	Male	Female	Male	Female	Male	Female
PAS	Medium	Low	Low	Low	Low	Medium
CR	Low	Low	Low	Low	Low	Medium
PDW	Low	Low	Low	Low	Low	Low
FW	Low	Medium	Low	Low	Medium	Low

in Nigeria. However, about 20% of the participants believed that the received awareness on TB is moderate which is not enough to mitigate the spread of TB in their communities.

13.5 CONCLUSIONS

The analytical technique used in this study assists in assessing the awareness status of people on TB toward its mitigation and serves as a contribution to the field of health informatics. Indeed, a majority of participants claimed that they had low awareness on TB and its associated issues in their communities. As the participants were from the Kano state, a strategic location in the northern part of Nigeria, the result of the experiment can represent major opinions of northern residents. Besides, the study intends to extend its scope beyond the analytical technique to visual analytics in the future so as to further explore the participants' responses on their awareness on TB in Nigeria.

ACKNOWLEDGEMENTS

This research was funded by Yusuf Maitama Sule University (Formerly, Northwest University), Kano, Nigeria, to the Smart Research Group of the Computer Science Department, Faculty of Science.

REFERENCES

1. Li, L., Xu, W., Wagner, A.L., Dong, X., Yin, J., Zhang, Y. & Boulton, M.L. (2019). Evaluation of health education interventions on Chinese factory workers knowledge, practices and behaviors related to infectious diseases. *Journal of Infectious and Public Health*, 12, pp. 70–76.
2. Omodior, O., Luetke, M.C. & Nelson, E.J. (2018). Mosquito-borne infectious disease, risk-perceptions, and personal protective behaviour among U.S international travelers. *Journal of Preventive Medicine Reports*, 12, pp. 336–342.
3. Smith, K.M., Machalaba, C.C., Seifman, R., Feferholtz, Y. & Karesh, W.B. (2019). Infectious disease and economics: the case for considering multi-sectoral impacts. *One Health*, 7, pp. 1–6.
4. RFI: the global fund hits target, raises $14 Billion to fight HIV, tuberculosis and malaria, Available at: http://en.rfi.fr/africa/20191010-france-pledges-15-percent-increase-its-total-contribution-fight-tb-aids-and-malaria, accessed October 11, 2019.
5. WHO: infectious diseases, Available at: https://www.who.int/topics/infectious_diseases/en/, accessed October 5, 2019.
6. Kooffreh, M.E., Offor, J.B., Ekerette, E.E. & Udom, U.I. (2016). Prevalence of tuberculosis in Calabar, Nigeria: a case study of patients attending the outpatients department of Dr. Lawrence Henshaw memorial hospital, Calabar. *Saudi Journal for Health Sciences*, 5 (3), pp. 130–133.
7. USA Embassy Nigeria (2012). Tuberculosis highlights, Available at: https://photos.state.gov/libraries/nigeria/487468/pdfs/JanuaryTuberculosisFactSheet.pdf, accessed October 6, 2019.
8. Liu, B., Wang, Z., Qi, X., Zhang, X. & Chen. H. (2015). Assessing cyber-user awareness of an emerging infectious diseases: evidence from human infectious with avian influenza a H7N9 in Zhejiang, China. *International Journal of Infectious Diseases*, 40, pp. 34–36.

9. Misra, A.K., Sharma, A. & Shukla, J.B. (2011). Modelling and analysis of effects of awareness programs by media on the spread of infectious diseases. *Mathematics and Computer Modelling*, 53, pp. 1221–1228.
10. Rakhmawati, W., Nilmanat, K. & Hatthakit, U. (2019). Moving from fear to realization: family engagement in tuberculosis prevention in children living in tuberculosis Sudanese households in Indonesia. *International Journal of Nursing Sciences*, 6, pp. 272–277.
11. Oliva, S.Z. & Felipe, J.C. (2018). Optimizing public healthcare management through a data warehousing analytic framework. *International Federation of Automatic Control*, 51 (27), pp. 407–412.
12. Baguelin, M., Medley, G.F., Nightingale, E.S., O'Reilly, K.M., Rees, E.M., Waterlow, N.R. & Wagner, M. (2020). Tooling-up for infectious disease transmission modelling. *Epidemic*, 32, p. 100395.
13. Lucas, T.C.D., Pollington, T.M., Davis, E.L. & Hollingsworth, T.D. (2020). Responding modelling: unit testing for infectious disease epidemiology. *Epidemic*, 33, p. 100425.
14. Elveback, L.R., Fox, J.P., Ackerman, E., Langworthy, A., Boyd, M. & Gatewood, L. (1976). An influenza simulation model for immunization studies. *American Journal of Epidemiology*, 103, pp. 152–165.
15. Wong, Z.S.Y., Zhou, J. & Zhang, Q. (2018). Artificial Intelligence for infectious disease big data analytics. *Infectious, Disease & Health*, 24 (1), pp. 1–5.
16. Han, B.A., Schmidt, J.P., Bowden, S. E. & Drake, J.M. (2015). Rodent reservoirs of future zoonotic diseases. *Proceedings of the National Academy of Sciences of the United States of America*, 112 (22), pp. 7039–7044.
17. Kononenko I. (2001). Machine learning for medical diagnosis: history, state of the art and perspective. *Artificial Intelligence in Medicine*, 23 (1), pp. 89–109.
18. Foster, K.R., Koprowski, R., Skufca, J.D. (2014). Machine learning, medical diagnosis, and biomedical engineering research-commentary. *BioMedical Engineering Online*, 13, p. 94.
19. Ferguson, N.M. (2020). MRC Centre for global infectious disease analysis. *Covid-sim*, Retrieved from https://github.com/mrc-ide/covid-sim, December 18, 2020.
20. Chatzilena, A., Van Leeuwen, E., Ratmann, O., Baguelin, M. & Demiris, N. (2019). Contemporary statistical inference for infectious disease models using Stan. *Epidemics*, 29, p. 100367.
21. Eglen, S.J. (2020). CODECHECK certificate for paper: Report 9: impact of non- pharmaceutical interventions (NPIs) to reduce COVID-19 mortality and healthcare demand.
22. Endo, A., Van Leeuwen, E. & Baguelin, M. (2019). Introduction to particle Markov-chain Monte Carlo for disease dynamics modellers. *Epidemics*, 29, p. 100363.
23. Hanaoka, J., Ohuchib, M., Kakua, R., Okamotoa, K. & Ohshioa, Y. (2019). Bronchoscopic balloon dilatation combined with laser cauterization of high and long segmental tracheal stenosis secondary to endobronchial tuberculosis: a case report. *Respiratory Medicine Case Report*, 28, pp. 1–5.
24. Kurasawa, T. (2010). Another type of tuberculosis: endobronchial tuberculosis. *Kekkaku*, 85, pp. 805–808.
25. Mathur, R., Vatliya, M. & Chitmilla, S. (2009). Tuberculosis of breast: a case report. *West London Medical Journal*, 1, pp. 37–43.
26. CINIC (2015). *The Statistical Report of China's Internet Development Status*. People's Republic of China; Ministry of Information, Beijing.
27. Timmermans, B. & Cleeremans, A. (2015). How can we measure awareness? An overview of current methods. In M. Overgaard (Ed.). *Behavioural Methods in Consciousness Research*, Chapter 3, pp. 21–46, Oxford University Press, Oxford.
28. Wierzchon, M., Asanowicz, D., Paulewicz, B. & Cleeremans, A. (2012). Subjective measures of consciousness artificial grammar learning task. *Consciousness and Cognition*, 21 (3), pp. 1141–1153.

29. Ramsoy, T.Z. & Overgaard, M. (2004). Introspection and subliminal perception. *Phenomenology and the Cognitive Sciences*, 3 (1), pp. 1–23.
30. Dienes, Z., Altmann, G.T.M., Kwan, L. & Goode, A. (1995). Unconsciousness knowledge of artificial grammars is applied strategically. *Journal of Experimental Psychology: Learning, Memory, and Cognition*, 21 (5), pp. 1322–1338.
31. Persaud, N., McLeod, P. & Cowey, A. (2007). Post-decision wagering objectivity measures awareness. *Nature Neuroscience*, 10 (2), pp. 257–261.

14 Active Learning from an Imbalanced Dataset
A Study Conducted on the Depression, Anxiety, and Stress Dataset

Umme Salma M. and Amala Ann K. A.
CHRIST (Deemed to be University)

CONTENTS

14.1 Introduction 252
14.2 Literature Survey 252
14.3 Problem Statement..254
14.4 Necessity of Defining the Problem/Research Gap 255
14.5 Objectives 255
    14.5.1 Primary Objective..255
    14.5.2 Secondary Objective..255
14.6 Dataset 255
    14.6.1 Data Collection 255
    14.6.2 Data Description 256
    14.6.3 Data Preprocessing 257
    14.6.4 Exploratory Data Analysis...257
        14.6.4.1 Analysis of DASS 257
        14.6.4.2 Analysis of the TIPI Test 259
        14.6.4.3 Analysis of Time Taken by the Users to
                 Complete the Survey.. 260
        14.6.4.4 Analysis of the Validity-Check List and their
                 Relationship with the Education Information................... 260
14.7 Implementation Design 260
    14.7.1 Class Imbalance..260
    14.7.2 SMOTE...262
    14.7.3 Model Building...262
    14.7.4 Evaluation Metric 263
14.8 Results and Conclusion...263
References..264

DOI: 10.1201/9781003138020-14

14.1   INTRODUCTION

Depression can cause huge loads of side effects which can neither be neglected nor excused. Not only the children but also the grown-ups can face problems in recognizing psychological changes in them. The American Psychological Association (APA) reports that people older than the middleage suffering from depression have more consequences with respect to amnesia and response time period when compared with the younger people suffering from depression. The stress hormones are released in the brain of a person suffering from long-term panic attacks and anxiety. The increased stress hormones can lead to various changes in body such as depression, dizziness, headache, etc. In reality, stress can have a negative impact on one's physical, mental, and emotional health and in turn on one's conduct. These issues, whenever left untreated, can further lead to major complications such as coronary illness, corpulence, diabetes, and hypertension.

Therefore, this study focuses on the effect of these disorders using the depression, anxiety, and stress (DAS) scale (DASS). The approach is a multiclass classification in imbalanced data. The principal objective of adjusting classes is to either expand the number of rows in the minority class or diminish the row count of the majority class [22,27,28]. This is cleared out to get around an identical number of cases for both the classes. The technique adopted is the synthetic minority oversampling technique (SMOTE). This strategy is followed to avoid overfitting that happens when the most dataset is applied to correct reproductions of minority cases. For instance, a subset of data is taken from the minority class; at that point, new manufactured equivalent examples are produced [31]. These manufactured occurrences are then added to the first dataset. The new dataset is used to build the classification models. Section 14.1 of the implementation focuses on the data collection and preprocessing steps; Section 14.2 is about imbalanced data and the classification strategies; Section 14.3 discusses primarily SMOTE; and then the evaluation metrics of imbalanced data is compared with that of the balanced data.

14.2   LITERATURE SURVEY

"Deep learning (DL) in mental health outcome research: a scoping review" is a review article that makes use of state-of-the-art DL algorithms to investigate the uses of DL techniques in the field of mental health. The articles are pertinent to four social events, diagnosis and prognosis upheld clinical information; investigation of hereditary qualities and genomics information for understanding the state of the mind; visual and vocal articulation information examination for disease identification; and assessment of danger of dysfunctional behavior using online media data [14].

The authors in the article "Improving Diagnosis of Depression with XGBoost Machine Learning Model and a Large Biomarkers Dutch Dataset (n = 11,081)" bring out some significant insights into the diagnosis of depression. The aim was to investigate the identification of depression cases in the Dutch resident dataset having 11,081 instances. The artificial intelligence (AI) model dependent on the unbalanced dataset gives the outcome toward the majority class. The model will reliably envision the case as no slump case except for it is an occasion of melancholy [13]. The article presents different strategies, such as rose package, under sampling, oversampling etc., to work with the imbalanced data, and finally "Extreme Gradient Boosting" (XGBoost) is trained on each sample to classify the mental disease cases from healthy cases. The accuracy, $F1$-score, precision, and recall obtained from the classification report for the model was around 0.90. The study, although robust given the XGBoost models with different samples and therefore the amount of knowledge used, should be replicated in clinical test settings [13].

In the article entitled "Predicting Anxiety, Depression and Stress in Modern Life using Machine Learning Algorithms", forecasts of depression, insanity, and stress were made using AI calculations. For this, data was gathered from used and jobless people across various societies and networks through the depression, anxiety, and stress scale-21 (DASS21) poll. Nervousness, gloom, and stress were foreseen as occurring on five degrees of seriousness using five distinctive AI calculations. Subsequent to applying the different strategies, the imbalanced classes were observed. The authors selected F-measure to check the performance of the classification models they have used [12].

In the article entitled "Assessment of Anxiety, Depression and Stress using Machine Learning Models", the authors used eight algorithms to predict five different levels of severity of hysteria, depression, and stress. The calculations are assembled into four classifications: probabilistic, closest neighbor, neural organization, and tree-based. A half-and-half arrangement calculation was additionally applied for the forecast of changed seriousness level uneasiness, discouragement, and stress. Comparable procedures were similarly applied to an alternate dataset, DASS21. The forecast precision found by using the half breed calculation was more noteworthy than by using single calculations; however, the absolute best exactness was found by using the outspread premise work network that goes under the classification of neural organization. Notwithstanding, the results of the random forest are one hundred percentage for anxiety in DASS21 [7].

"Classifying Depression in Imbalanced Datasets using an Autoencoder-Based Anomaly Detection Approach", is an approach that makes use of the pipeline concept in machine learning. The first step includes making use of autoencoders to project the mobility attribute of the majority class. The second step involves training the autoencoder using one class support vector machine (SVM) to perform binary classification. This method was implemented on real-time data. The real-time data contained student information and was used to classify whether the student is depressed or not. The area under the curve (AUC) was around 0.92 [4].

According to the authors of the article, "XA-BiLSTM: A Deep Learning Approach for Depression Detection in Imbalanced Data", early depression discovery using web-based media information through deep learning models can assist with shifting daily routine directions and spare lives. Yet, the accuracy of these models was not fulfilling because of these real-world imbalanced data disseminations. The methods that have been adopted are X-A-BiLSTM and Attention-BiLSTM neural network. The analysis was carried out on the Reddit self-reported depression diagnosis (RSDD) [1].

The article entitled "Machine learning in major depression: From classification to treatment outcome prediction", discusses the commonly used machine learning classification models and prediction models used for brain imaging and that can provide a summary of studies, specifically for medical field. Here also, the pipeline approach was used. A significant change was observed in the pipeline after preprocessing. The classification was carried out using the SVM and leave one out cross-validation was used for cross-validation. The future scope includes multicross classification for a giant sample size and deep learning, multimodal magnetic resonance imaging in major depressive disorder (MDD), and enormous dataset classification [2].

The authors of the study "Feature Selection and Imbalanced Data Handling for Depression Detection", discuss different attribute selection methods and also highlight the effect of resampling the data on imbalanced data. The results obtained by ten-fold cross validations of random forests (RFs) and support vector machines (SVMs) with a Gaussian kernel (radial basis function, RBF) provided unbiased and were estimated by the AUC. Here, the performance of RF beats SVM for the depression classification task. The authors discuss that in the future study, different classification techniques will be considered to strengthen the performance of the model. Also, another idea is to use deep neural networks for classifying clinical data on depression [9].

The importance of a confusion matrix in an imbalanced dataset was highlighted by the authors of the article "Confusion-Matrix-Based Kernel Logistic Regression for Imbalanced Data Classification". The authors use the confusion matrix–based kernel logistic regression (CM-KLOGR) approach for classifying the imbalanced data. The CM-KLOGR mainly focused to improve the harmonic mean from the confusion matrix. They used positive predicted values, negative predicted values, specificity, and sensitivity to accomplish the task. It was observed that the CM-KLOGR outflanked the kernel logistic regression and SVM with or without testing, for various datasets under various data splits [10].

The article "Handling imbalanced datasets: A review" describes various techniques used in the literature for handling imbalance dataset problems such as re-examining, arbitrary oversampling with substitution; irregular undersampling; coordinated oversampling; coordinated undersampling; oversampling with educated and age regarding the accompanying tests; and a combination of all the above resampling methods. It is been concluded that the reduced classification performance on the imbalanced data is because of the minority class being poorly represented and suppressed by the majority class because of the overlapping of classes, and therefore, the class is divided into further clusters during preprocessing. For larger data indexes, the impact of those confusing variables is diminished by all accounts diminished, on the grounds, that the minority class is best addressed by a greater number of models [6].

14.3 PROBLEM STATEMENT

Depression or despondency is a psychological disorder which can cause aftermath in the central nervous system in one way or the other. Depression affects physical, mental, and emotional health of a human being, and if left untreated and/or ignored, it can lead to serious health issues such as cardiovascular disease, high blood pressure, diabetes, obesity, etc.

The proposed study focuses on the analysis of depression, anxiety, and stress in accordance with the requirement of the DASS scale requirement. The approach is treated as a multiclass classification in imbalanced data [11,15]. The principal objective of adjusting classes is to either escalate the count of the minority class or lessen the count of the majority class. The technique adopted to achieve this objective is SMOTE [8]. This strategy is followed to avoid overfitting. Thereafter, the study not only talks about the balancing techniques [17] but also reveals the strategy for altering existing classification methods to improve the model's performance. One such strategy is using XGBoost.

14.4   NECESSITY OF DEFINING THE PROBLEM/RESEARCH GAP

The necessity of defining the problem primarily lies on the analysis of depression, anxiety, and stress on a large dataset such as DASS-42 that has 39975 instances and 172 columns. Many studies are conducted on this dataset [23–25,30]. Classifying the imbalanced data will be our prime motive. To classify an imbalanced data, resampling should be carried out using SMOTE, which is our second objective. Finally, we will be comparing the performance of various classification techniques for classifying multiclass targets [16, 18] and checking their performance using the F-score and AUC.

14.5   OBJECTIVES

14.5.1   PRIMARY OBJECTIVE

The primary objective of this article is to study the effect and correlation between depression, anxiety, and stress.

14.5.2   SECONDARY OBJECTIVE

The secondary objective is to analyze the performance of machine learning models on an imbalanced dataset using SMOTE and apply classification techniques unlike the conventional techniques and study the difference in performance using the evaluation metrics, particularly in a multiclass classification.

14.6   DATASET

14.6.1   DATA COLLECTION

The data used in this research was contributed by the Psychology Foundation of Australia and was collected from an online repository [29]. The dataset contains the information gathered by the contributors following the DASS criteria. An overview was accommodating to anybody, and individuals were persuaded to expect it to ask customized results. At the most noteworthy point, they also got the selection to end a quick research survey. This dataset comes from those who agreed to finish the research survey and answered "yes" to the questions provided. The DASS questionnaire contains a set of three self-report scales which can be used to measure the most vital negative emotions such as depression, anxiety, and stress [29].

14.6.2 DATA DESCRIPTION

The dataset contains 39,975 instances and 172 columns and three classes, namely, stress, depression, and anxiety. Each of the three classes of this dataset contains a set of 14 items and is partitioned into 2–5 subsets. The scale of depression focuses on gathering answers related to dysphoria, misery, self-deprecation, depreciation of life, latency, and anhedonia. The scale of anxiety focuses on getting answers related to overexcitement, aftermath of anxiousness, nervousness, skeletal muscle impacts, etc. The scale of stress focuses on finding answers related to levels of chronic arousal such as temperament issue, over receptiveness, getting easily emotionally hurt, etc. [29]. Across the three classes (DAS), a common scale of 1–4 was fixed, in which 1 indicates weak impact and 4 indicate strong impact. The sample screenshot of the questions within the DASS is given in Table 14.1.

The DASS data consists of 42 behavioral questions given in a random order for every new participant. These 42 questions had each section A, E, and I. The response is stored in variable A (e.g., Q1A). Also, the time taken to answer that question, in milliseconds, was recorded was the time taken in milliseconds to answer that question (E), and question's position within the survey was denoted as (I).

The next section of the data frame consisted of generic demographics survey with many various questions. This was a quick measure of the ten-item personality inventory (TIPI) test. There were ten columns from TIPI1 to TIPI10. Each of those had categories of both positive and negative or organized and disorganized behavior. The score varied from 1 to 7, from weak to strong agreement.

The next section called *"In the grid below, check all the words whose definitions you're sure you know"* was used to get multiple responses from the subject by clicking the checklist. The columns were from VCL01 (Varnish Configuration Language) to VCL16 that consisted of random words. A value of 1 is checked, and 0 means unchecked. The words at VCL12, VCL09, and VCL06 are not genuine words and will be used as a legitimacy check.

The last section of the data frame consisted of private questions about education, age, gender, religion, orientation, race, etc. Times for various circumstances were also included in the survey such as intro-elapse, the time spent on the introduction/

TABLE 14.1

DASS-42 Sample Questions [29]

Q1	I found myself getting upset by quite trivial things [29]
Q2	I was aware of dryness of my mouth [29]
Q3	I couldn't seem to experience any positive feeling at all [29]
Q4	I experienced breathing difficulty (e.g., excessively rapid breathing, breathlessness in the absence of physical exertion) [29]
Q5	I just couldn't seem to get going [29]

landing page (in seconds); test-elapse,: the time spent on all the DASS questions (should be a bit like the time elapsed on all the individual questions combined); survey-elapse, the time spent answering the rest of the demographic and survey questions.

14.6.3 DATA PREPROCESSING

There are 172 columns and 39,975 instances in the dataset. Out of which, all the E values, which represented the position of the questions, from 42 questions are ignored as they are insignificant in accordance with the ground truth information. The insignificance was also confirmed by checking the correlation value with respect to the target label. On the same account, I-recording time is also being ignored. The A values for the first behavioral questions are taken as a subset. Out of these subsets, entries below 3 are again discarded. The data had no missing values, so we saved ourselves from dealing with them.

14.6.4 EXPLORATORY DATA ANALYSIS

14.6.4.1 Analysis of DASS

The 42 questions had indications of symptoms of either depression, anxiety, and stress. The questions were categorized (as shown in Table 14.2) in such a way that only the scores greater than 2 out of 4 across each of the disorders were considered. Then the total sum of scores from 1 to 4 (counting the 3s and 4s primarily) is calculated across each entry number. The range of the sum varied from a minimum of 35 to a maximum of 42. Taking into consideration this score value, the rows that had a score greater than 35 are extracted and saved. There were 3985 rows out of 39,975. Thus 9.631% of the total entries have serious issues with either stress, depression, anxiety, or all the three. The symptoms of stress, anxiety, and depression are given in Table 14.2, and the target column predicted from the features is given in Table 14.3. The data shows a distribution of 64% of depression, 35.3% of anxiety, and 0.7% of stress (displayed in Figure 14.1).

TABLE 14.2
Symptoms of DAS [29]

Depression Scale	Anxiety Scale	Stress Scale
1. Self-disparaging	1. Apprehensive, panicky	1. Over-aroused, tense
2. Dispirited, gloomy, blue	2. Trembly, shaky	2. Unable to relax
3. Convinced that life has no meaning or value	3. Aware of dryness of the mouth, breathing challenges, beating of the heart, dampness of the palms	3. Touchy, easily upset
4. Pessimistic about the future	4. Worried about performance and possible loss of control	4. Irritable
5. Unable to experience enjoyment or satisfaction		5. Easily startled
6. Unable to become interested or involved		6. Nervy, jumpy, fidgety

---

TABLE 14.3
Target Column Predicted from the Features

Sl. No	Labels Predicted
0	Depression
1	Anxiety
2	Depression
...	Anxiety
39,774	Depression

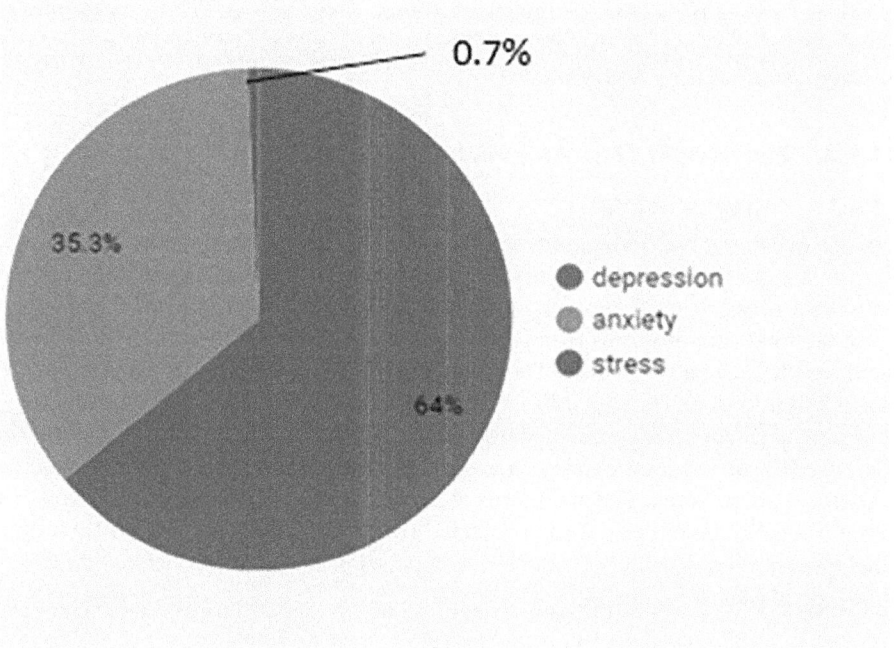

FIGURE 14.1   Distribution of DAS across the data.

It is evident from Figure 14.1 that the distribution of depression is the most and that of stress is the least. Thus, there exists a direct imbalance relationship between depression and stress and almost balanced correlation between depression and anxiety.

The Table 14.4 provides the scores from 36 to 42, and their corresponding frequency distribution respectively, i.e., around 618 of 39,975 are sufferers from serious disorders. Hence, with every 1.545% increase in the symptoms for depression, there is always equal chances of an increase in anxiety and less than half the probability of suffering from stress.

TABLE 14.4

Frequency Distribution of the Scores of DAS

DAS Score	Count
36	671
37	632
38	564
39	544
40	487
41	469
42	618

14.6.4.2   Analysis of the TIPI Test

The instance of symptoms below the score 5 is eliminated from a maximum score of 7, and the total sum of all 5s, 6s, and 7s across each column is calculated. The range of this sum came out as 5–10.

From Table 14.5, we can conclude that 9175 of the totals had score 5; 9619 had score 6; and so on. Now, the TIPI category of questions had a combination of organized and disorganized behaviors, and its details are provided in Table 14.6.

TABLE 14.5

Sample TIPI Score

Tipscore	Rate
5	9175
6	9619
7	6811
8	3390
9	1321
10	475

TABLE 14.6

Negative and Positive Behavior from TIPI

TIPI No.	Organized/Positive Behavior	TIP No.	Disorganized/Negative Behavior
TIPI1	Extraverted, enthusiastic	TIPI2	Critical, quarrelsome
TIPI3	Dependable, self-disciplined	TIPI4	Anxious, easily upset
TIPI5	Open to new experiences, complex	TIPI6	Reserved, quiet
TIPI7	Sympathetic, warm	TIPI8	Disorganized, careless
TIPI9	Calm, emotionally stable	TIPI10	Conventional, uncreative

Based on Table 14.6, we categorize the total score (i.e., the total number of entries who have rate scales more than 5 but less than 7) for positive and negative behaviors.

14.6.4.3 Analysis of Time Taken by the Users to Complete the Survey

Survey Elapse: It is the time taken to answer the demographic and survey questions other than the behavioral questions.

The time was recorded in seconds. The percentage of people who have finished answering the questions in not more than 20 seconds was observed to be 0.28% of 39,975 entries.

Test Elapse: The time spent on all the DASS questions (equivalent to the time elapsed on all the individual questions combined).

The results of the test elapse showed that there were only 17 instances, i.e., 0.042% of the total data.

The inference that is derived from the data is that less than 1% of the total entries have completed the survey. It may be probably because the questions, disorders, and symptoms were familiar to them, unlike others who must have spent more time in understanding the questions.

14.6.4.4 Analysis of the Validity-Check List and their Relationship with the Education Information

The total sum of all checked-1 values across each row is determined. The range falls between 0 and 16. Out of 16, the entries that have marked more than ten checked lists are then printed. It was observed that only 4692 patients were having the sufficient knowledge about validity words being asked. When it comes to the education, 1 indicates less than secondary school, 2 indicates high school, 3 indicates university certificate, and 4 indicates graduate degree. When we focused on the graduate degree, there comes around 13.1% of the total who falls into patient categories such as psychology, law, and medicine and engineering; 58.7% includes random patients other than mentioned ones. The rest opted for **prefer not to say** their educational details which accounted for 28.4% as shown in Figure 14.2.

14.7 IMPLEMENTATION DESIGN

14.7.1 CLASS IMBALANCE

Figure 14.3 indicates the system architecture of the proposed system. According the architecture, the first step involves data collection. We have collected the dataset from online repository [27]. The preprocessing includes discarding useless attributes. The next step is to balance the dataset using SMOTE. The visualization of data points after applying SMOTE is given in Figure 14.4. The resampling step is followed by exploratory data analysis where we examine the exploratory details of data. Various models such as SVM, random forest, XGBoost, etc. were implemented. After working with various models, finally we settled with the linear SVM that outperformed the other models. The details of the models and their result are given in Table 14.7. The result for the SVM was displayed as confusion matrix and F-score, and the AUC is used as the evaluation measure. While using a dataset with profoundly unequal

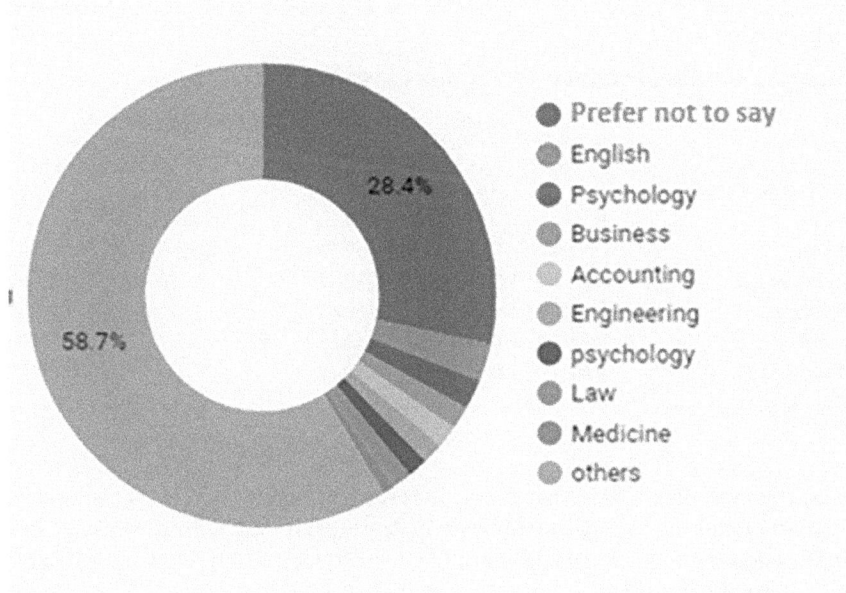

FIGURE 14.2   Statistics of qualifications shown by the patients as per the report.

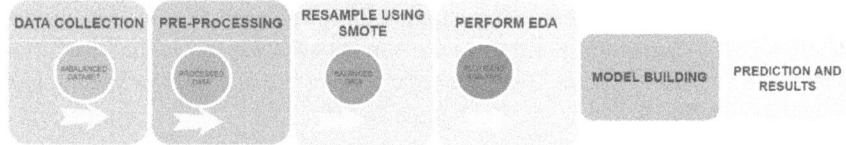

FIGURE 14.3   System architecture.

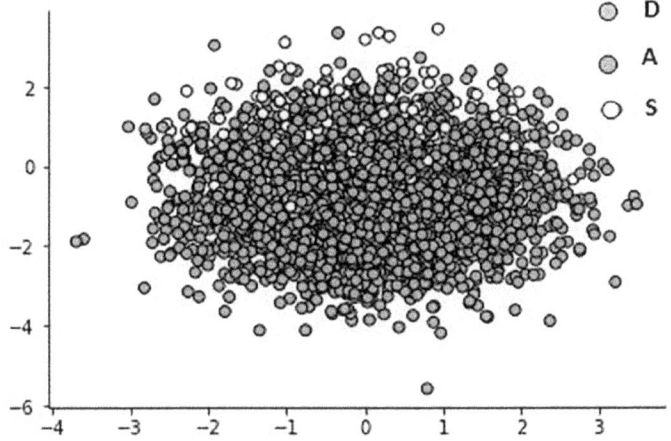

FIGURE 14.4   SMOTE algorithm applied for the DAS dataset.

TABLE 14.7

Comparing the Performance of Various Classification Models

Model	Precision	Recall	Weighted *F1*	Train Time (seconds)	AUROC
SVM linear	1	1	1	4.7414	0.98
Random forest classifier	0.97	0.97	0.97	3.3867	0.96
SVM rbf	0.99	0.99	0.99	21.54	0.95
SVM poly	0.99	0.99	0.99	11.06	0.95
Decision tree	0.99	0.99	0.99	0.1624	0.95
Naïve Bayes	0.68	0.61	0.689	1.49	0.66
XGBoost	0.61	0.62	0.624	3.110	0.65
SVM sigmoid	0.52	0.52	0.52	74.233	0.58

classes, the classifier will consistently "foresee" the first basic class without playing out any examination of the analysis of features, and it will, at present, have a high precision rate, which is clearly not the correct one [20,21]. Hence, SMOTE was applied to fix the issue.

14.7.2   SMOTE

This strategy produces engineered information for the minority class. SMOTE works by discretionarily choosing a point from the minority class and enlisting the k-closest neighbors for the chosen point. The synthetic focuses are added between the picked point and its neighbors.

SMOTE works in the accompanying four stages:

1. The minority class in the data is chosen as the input vector.
2. k-closest neighbors are to be found.
3. Pick one of these neighbors and recognize a synthetic point wherever on the line joining the point reasonable and its picked neighbor.
4. Continue the procedure until the data is balanced.

In this case, as the DASS suffers from class imbalance, with the first class having 25,559 instances, second class with 14,199, and third class with 17 only, we have adopted the method of SMOTE. After upsampling with SMOTE, the three classes had a size of 17,900 each. One main advantage of upsampling over downsampling is that it outperforms the latter without any much information loss. Once the class imbalance has been removed, we split the dataframe into two arbitrary groups, 70% for training and 30% for testing [5].

14.7.3   Model Building

The performance of models built on the balanced data after applying SMOTE are summarized in Table 14.7.

14.7.4 EVALUATION METRIC

Metrics that can give better understanding in case of imbalanced data classification are as follows [3,19]:

Confusion Matrix: A table demonstrating right forecasts and sorts of inaccurate predictions.

Precision: The quantity of genuine positives partitioned by every single positive forecast. Precision is likewise called positive predictive value. It is a proportion of a classifier's precision. Low precision demonstrates a high number of bogus positives.

Recall: The quantity of true positives isolated by the quantity of positive qualities in the test information. The recall is in like manner called Sensitivity or the True Positive Rate. It is an extent of a classifier's zenith. A low recall demonstrates a high number of bogus negatives.

F1 Score: The weighted normal of exactness and review.

Area Under ROC (Receiver Operating Characteristic) Curve (AUROC): AUROC indicates the probability of your model- particular observations from two classes.

14.8 RESULTS AND CONCLUSION

The relationship between the three disorders, like depression, anxiety, and stress, shows that 9.631% of the total entries have serious issues with either depression, anxiety, or stress, individually or as a combination or all the three. The resampling strategy of SMOTE with upsampling has outperformed the rest without any information loss.

The models built on the features namely {"age", "depression", "anxiety", "stress", "married", "familysize", "gender", "education", "vsum", "source", "urban", "gender.1", "engnat", "age.1", "screensize", "uniquenetworklocation", "hand", "religion", "orientation", "race", "voted", "married.1", "familysize.1", "tipscore"} with the label "target" shows the classification metrics as shown in Table 14.7.

The training time for each of the algorithms are recorded at the beginning and the end of the training, and the total time taken by the algorithm to train the features is determined by subtracting the stopping time from the starting time in seconds. The least time taken to train the model is by decision trees, with 0.1624 seconds, and the most time taken is by the SVM kernel-sigmoid. The order of higher F1-score of all the models are given as below:

SVM linear > SVM rbf > SVM poly > Decision tree >

Random forest classifier > Naïve Bayes > XGBoost > SVM sigmoid.

Comparing the time taken and the F1-score obtained from the algorithms, we can conclude that SVM polynomial (time taken = 11.06 seconds, F1-score = 0.99), decision tree (time = 0.1624 seconds, F1-score = 0.99), and random forest classifier (time = 3.3867 seconds, F1-score = 0.97) have outperformed rest of the algorithms.

The confusion matrix for SVM with kernel polynomial is given in Table 14.8, and other evaluation measure details are given in Table 14.9.

TABLE 14.8
Confusion Matrix for SVM with Kernel Polynomial

4262	121	18
0	4511	27
0	0	4482

TABLE 14.9
Evaluation Measures for Polynomial Kernel Based SVM Classifier

Precision	Recall	F1-score	Support
1.00	0.97	0.98	4401
0.97	0.99	0.98	4538
0.99	1.00	1.00	4482

From Table 14.9, it is obvious that Class 1 has precision of 1.00, Class 2 has precision of 0.97, and Class 3 has a precision of 0.99 which implies the number of correctly classified among the classes are having a support of 4401, 4538, and 4482, respectively.

The study is led on a survey data of general mental inquiries, and the reactions for every passage are grouped into depression, anxiety, or stress dependent on an assessed limit score. Thereafter, the study not only talks about the balancing techniques but also reveals the strategy for modifying existing grouping calculations to make them fitting for imbalanced data sets. The methodology used is a multiclass grouping in imbalanced information. The crucial objective of changing classes is to either increase the count of the minority class or reduce the count of the majority part class. This is performed to get a roughly similar number of occasions for the two classes. The method embraced is SMOTE. This procedure is followed to evade overfitting which happens when precise imitations of minority occurrences are added to the fundamental dataset. A subset of information is taken from the minority class for instance, and afterward, new manufactured comparable examples are made. These engineered cases are then added to the first dataset. The new dataset is used as an example to prepare the order models such as SVM and Naïve Bayes. It is observed that even though SVM-linear can be considered as an ideal fit, SVM and decision tree performs well here. The future extents of the study is to actualize deep learning models and look at the assessment measurements and to furthermore execute the thoughts on an electroencephalogram (EEG) information.

REFERENCES

1. Cong, Q., Feng, Z., Li, F., Xiang, Y., Rao, G., & Tao, C. (2018, December). XA-BiLSTM: A deep learning approach for depression detection in imbalanced data. In *2018 IEEE International Conference on Bioinformatics and Biomedicine (BIBM)* (pp. 1624–1627). IEEE, Madrid, Spain.

2. Gao, S., Calhoun, V. D., & Sui, J. (2018). Machine learning in major depression: From classification to treatment outcome prediction. *CNS Neuroscience & Therapeutics*, *24*(11), pp. 1037–1052.

3. Gautheron, L., Habrard, A., Morvant, E., & Sebban, M. (2019, November). Metric learning from imbalanced data. In *2019 IEEE 31st International Conference on Tools with Artificial Intelligence (ICTAI)* (pp. 923–930). IEEE, Portland, OR.

4. Gerych, W., Agu, E., & Rundensteiner, E. (2019, January). Classifying depression in imbalanced datasets using an autoencoder-based anomaly detection approach. In *2019 IEEE 13th International Conference on Semantic Computing (ICSC)* (pp. 124–127). IEEE, Newport Beach, CA.

5. Guo, J., Wan, X., Lin, H., Li, P., Liu, G., & He, Y. (2017, June). An active learning method based on mistake sampling for large scale imbalanced classification. In *2017 International Conference on Service Systems and Service Management* (pp. 1–6). IEEE, Dalian, China.

6. Kotsiantis, S., Kanellopoulos, D., & Pintelas, P. (2006). Handling imbalanced datasets: A review. *GESTS International Transactions on Computer Science and Engineering*, *30*(1), pp. 25–36.

7. Kumar, P., Garg, S., & Garg, A. (2020). Assessment of anxiety, depression and stress using machine learning models. *Procedia Computer Science*, *171*, pp. 1989–1998.

8. Li, K., Zhang, W., Lu, Q., & Fang, X. (2014, October). An improved SMOTE imbalanced data classification method based on support degree. In *2014 International Conference on Identification, Information and Knowledge in the Internet of Things* (pp. 34–38). IEEE, Beijing, China.

9. Mousavian, M., Chen, J., & Greening, S. (2018, December). Feature selection and imbalanced data handling for depression detection. In *International Conference on Brain Informatics* (pp. 349–358). Springer, Cham.

10. Ohsaki, M., Wang, P., Matsuda, K., Katagiri, S., Watanabe, H., & Ralescu, A. (2017). Confusion-matrix-based kernel logistic regression for imbalanced data classification. *IEEE Transactions on Knowledge and Data Engineering*, *29*(9), pp. 1806–1819.

11. Pristyanto, Y., Pratama, I., & Nugraha, A. F. (2018, March). Data level approach for imbalanced class handling on educational data mining multiclass classification. In *2018 International Conference on Information and Communications Technology (ICOIACT)* (pp. 310–314). IEEE, Yogyakarta, Indonesia.

12. Priya, A., Garg, S., & Tigga, N. P. (2020). Predicting anxiety, depression and stress in modern life using machine learning algorithms. *Procedia Computer Science*, *167*, pp. 1258–1267.

13. Sharma, A., & Verbeke, W. J. (2020). Improving diagnosis of depression with XGBOOST machine learning model and a large biomarkers Dutch dataset ($n = 11,081$). *Frontiers in Big Data*, *3*, 15.

14. Su, C., Xu, Z., Pathak, J., & Wang, F. (2020). Deep learning in mental health outcome research: A scoping review. *Translational Psychiatry*, *10*(1), pp. 1–26.

15. Yang, B., Zhai, Y., Qu, W., & An, B. (2010, October). The problem of classification in imbalanced data sets in knowledge discovery. In *2010 International Conference on Computer Application and System Modeling (ICCASM 2010)* (Vol. 9, p. V9-658). IEEE, Taiyuan, China.

16. Yu, H., Yang, X., Zheng, S., & Sun, C. (2018). Active learning from imbalanced data: A solution of online weighted extreme learning machine. *IEEE Transactions on Neural Networks and Learning Systems*, *30*(4), pp. 1088–1103.

17. Yuan, Z., & Zhao, P. (2019, May). An improved ensemble learning for imbalanced data classification. In *2019 IEEE 8th Joint International Information Technology and Artificial Intelligence Conference (ITAIC)* (pp. 408–411). IEEE, Chongqing, China.

18. Zhang, R., Li, L., Zhang, Y., & Bu, C. (2018, November). Imbalanced networked multi-label classification with active learning. In *2018 IEEE International Conference on Big Knowledge (ICBK)* (pp. 290–297). IEEE, Singapore.
19. Chawla, N. V., Bowyer, K. W., Hall, L. O., & Kegelmeyer, W. P. (2002). SMOTE: Synthetic minority over-sampling technique. *Journal of Artificial Intelligence Research*, *16*, pp. 321–357.
20. Dealing with imbalanced data. https://towardsdatascience.com/methods-for-dealing-with-imbalanced-data-5b761be45a18.
21. Lemaître, G., Nogueira, F., & Aridas, C. K. (2017). Imbalanced-learn: A python toolbox to tackle the curse of imbalanced datasets in machine learning. *The Journal of Machine Learning Research*, *18*(1), pp. 559–563.
22. Aggarwal, U., Popescu, A., & Hudelot, C. (2020). Active learning for imbalanced datasets. In *Proceedings of the IEEE/CVF Winter Conference on Applications of Computer Vision* (pp. 1428–1437). Snowmass Village, CO.
23. Belouadah, E., Popescu, A., Aggarwal, U., & Saci, L. (2020, August). Active class incremental learning for imbalanced datasets. In *European Conference on Computer Vision* (pp. 146–162). Springer, Cham.
24. Ertekin, S., Huang, J., Bottou, L., & Giles, L. (2007, November). Learning on the border: active learning in imbalanced data classification. In *Proceedings of the Sixteenth ACM Conference on Information and Knowledge Management* (pp. 127–136). ACM, Lisbon, Portugal.
25. Ramyachitra, D., & Manikandan, P. (2014). Imbalanced dataset classification and solutions: A review. *International Journal of Computing and Business Research (IJCBR)*, *5*(4), pp. 1–29.
26. Leevy, J. L., Khoshgoftaar, T. M., Bauder, R. A., & Seliya, N. (2018). A survey on addressing high-class imbalance in big data. *Journal of Big Data*, *5*(1), pp. 1–30.
27. Dataset. http://www2.psy.unsw.edu.au/dass/, accessed on 15–06–2020.
28. Antony, M. M., Bieling, P. J., Cox, B. J., Enns, M. W., & Swinson, R. P. (1998). Psychometric properties of the 42-item and 21-item versions of the depression anxiety stress scales in clinical groups and a community sample. *Psychological Assessment*, *10*(2), pp. 176–181.
29. Hanifah, F. S., Wijayanto, H., & Kurnia, A. (2015). Smotebagging algorithm for imbalanced dataset in logistic regression analysis (Case: Credit of bank x). *Applied Mathematical Sciences*, *9*(138), pp. 6857–6865.
30. Fernández, A., García, S., Galar, M., Prati, R. C., Krawczyk, B., & Herrera, F. (2018). *Learning from Imbalanced Data Sets*, Vol. 11. Berlin: Springer.

15 Classification of the Magnetic Resonance Imaging of the Brain Tumor Using the Residual Neural Network Framework

Tina and Sanjay Kumar Dubey
Amity University

CONTENTS

15.1 Introduction 267
15.2 Literature Review 269
15.3 Architecture of Resnet Medical Imaging Modalities..........................270
15.4 Stages for Implementation of the Resnet Framework.........................272
    15.4.1 Preprocessing...272
    15.4.2 Training the Network..272
    15.4.3 Segmentation 272
    15.4.4 Focal Loss Function...273
15.5 Results and Discussions..273
15.6 Conclusions and Future Scope..275
References...276

15.1 INTRODUCTION

Digitized medical imaging modalities handle the management of obtaining computerized image modalities of the parts of the body using medical probes attached to a digital computer. The course of transformation of a medical image into a computerized structure aims to produce the enhanced image to accomplish operation for the extraction of meaningful statistics about the region of interest. These digital images require to undergo continual dispensation for removal of abnormalities in the given anatomical structure of the organ and, to this, digitalized images endure several procedures for elimination of deficits [1] to attain better visual representation. It is desirable to acquire the fine-tuned image with its respective higher resolution

DOI: 10.1201/9781003138020-15

which will drive the image toward better classification of the particular section for complex accuracy leading to localization of the tumor in the part of the brain in magnetic resource imaging (MRI). When the medical image is captured using a digitized medical device, the defined resolution of the image gets degraded because of deformed geometric sequence, high noise, low-contrast images, or occurrence of imaging artifacts, and to end, the spatial resolution of the image gets lower. The objective for the formation of a higher resolution medical image from the respective low-resolution input image is to eradicate the noisy sequences in corresponding higher dimensional input image using the upsampling progression for the improvement in clinical trials. Along with that, it will support to overwhelm the hardware restrictions of medical imaging procedures through reformation of high-resolution images using super-resolution processing methods [2]. MRI of the brain was a well thought-out application of medical imaging which is there to assess the structure of the brain through which abnormalities, disorders, and the damaged region in the structure can be illustrated. The super-resolution processing procedures on the three-dimensional MRI help in achieving the fine scale structural study for further illustration in accurate clinical diagnosis. This system can be formulated in a mathematical notation using the provided equation such as $Y = F(X) + Ns$ where Y is higher resolution data or image, $F(X)$ is the function which will link the single input MRI image with the output function, and Ns are the possible noisy sequences in the input image [3]. One of the elucidations to this resolution problem is to use the inverse operator in the supervised way, but the challenge here is to estimate the mapping of the operator Y to another operator X as mentioned in the above equation.

A class of feedforward and deep artificial network termed as the convolutional neural network (CNN) has gained tremendous upshot in the field of image classification, super-resolution, image registration, etc. With the present medical imaging processing, several drawbacks are perceived, such as the use of added max pooling layers which leads to reduction in localization accuracy. The second drawback includes slow computation power because of redundancy issue due to overlapped patches. Other foremost problems due to the low-quality medical images are optical blurring, inhomogeneous brightness, poor contrast, etc. The medical probe needs to run for respectively single patch for the part of the organ, and this led to overlapping of various pixels in the image. To resolve these issues, i.e., to remove the noisy sequences or overlapped patches from the three-dimensional MRI, super-resolution convolution neural network framework provides the mapping context which will be potentially helpful in the formation of higher resolution image [4]. The deep learning–based approaches for the improvised version or reconstruction of higher image resolution from low image resolution can be categorized into two methods which comprise external database and self-learning. The former methodology anticipated by Choi et al. [5] applies the self-resemblance learning of images by using only a single example-based input image without any training image from the outsource. The sparse representation and collective frame structure variation allows to resolve the delinquent of characteristics for the image structure with identical and unlike sizes. Freeman et al. [6] segmented the blocks which are extracted through low-resolution images into various multiple subblocks via extracting the communiqué among high- and low-resolution medical input images, and the outcome improved edge processing effect. The latter method of the learning is based on external database, and it is built on sample learning, which is attained by means of nearest neighbor quest in the external medical image database [11]. The study proposed by Yang et al. [7] implements the sparse illustration to address noisy sequences in the medical input image along with a significant decrease in the computational complexity. Deep CNN and machine learning algorithms are becoming extensively distinguished, especially in relation to image super resolution, medical image classification and enhancement, and visual recognition.

In this article, we will propose a methodology for the manner of CNN with the framework of the residential network (ResNet) which extracts the representations in a semantic manner as of the embedding vector and minimizes dimensionalities of the input matrix for the input image. The mechanism of the attention parameter in the ResNet distinguishes the focus on the output, and the short- and long-term memory layers reminisces extracted features. Also, the problem of vindication of class imbalances in data can be resolved via applying the enhanced focal loss function [8]. The function of encoders in the architecture provides the minimization of the reconstruction error in the initial phases of input layer values and output layer values in the respective network. It can be defined as a two-layered deep CNN which will lessen the error by learned representations while reconstruction of the output image. The CNN approach of deep learning initiations is the meaningful representation for the structure of medical imaging data. The technique will propose the construction of segmentation of a tumor with predicted mask of the MRI using the CNN with the implementation of the ResNet in the CNN framework.

15.2 LITERATURE REVIEW

The purpose of the diagnosis is to sight the tumor detection, classification, and segmentation in the input MRI image modalities, which plays a decisive part as researchers need to advance a roadmap to confront every single task one at a time and link these tasks to the whole applied system. The automatic segmentation is considered as the initial phase in classification of the MRI image. RedMon et al. [9] implement a full-resolution convolutional neural network for image segmentation even though methods are further innovative in relation to automation and semiautomation with sustaining outcomes on diagnosis. G. Litjens et al. [10] propose the model in two phases, where in its first phase, design and its implementation of a model proposed in deep learning for every single task is probably time-consuming and complex; in its second part, the comparatively partial medical imaging modality dataset for training might hypothetically lead to overfit. A. Al-Masni et al. [11] propose a regional (R-CNN) for mass detection, followed by a fully connected CNN-based classifier for "benign versus malignant" prediction. Al-Masni et al. [11] proposed a R-CNN for a fully connected CNN framework for classifiers to predict mass detection in segmentation problem. The research anticipated by Alantari et al. [11] implements a completely automatic system which is designed for image classification, detection, and image segmentation by the deployment of possible deep learning models.

DefuQiu et al. [12] propose an end-to-end CNN built on a concatenated deep convolutional layer in reconstruction from lower resolution images to attain equivalent higher resolution images. In that progression of reconstruction of example based lower resolution osteoarthritis image as they routine the deep learning technique to improvise the peak to signal ratio after the process of image reconstruction. Kamnitsas et al. [13] proposed a DeepMedic three-dimensional construction for automatic segmentation that outdoes the state of the art on most crucial data available for training. The training structure so defined is efficient with respect to computational complexity and also partially improvises the intrinsic class imbalances of lesion segmentation.

15.3 ARCHITECTURE OF RESNET MEDICAL IMAGING MODALITIES

ResNet(n) is a n-layer residual network which has various parameters and attributes in it. It is the CNN model by Microsoft. In this architecture, the learning from residuals is deduction of features erudite from the input layer and it customs the skipped linking to promulgate information across all layers. The ResNet associates the nth layer's input straight to the [$n + x$]th layer which allows other layers to be stacked and built a deep neural network. The network uses a pretrained model in the testing and tuned it finely [14].

The principle for the ResNet approach preserves the spatial relationship among hidden layers of the input image and is an arrangement of the grid structure where layers preserve relationships through the operating procedures on previous layers in a probably small region as shown in Figure 15.1. It has several layers convolutions and likewise activation function which are highly recommended in improving efficiency of image-oriented tasks in three or higher dimension [15]. Through this notion, an input image is fed through the convolutional layers where the activation function from the previous layer is convolved into a set of parametrized filters in the first hidden layer (i, j), i is the filter number, and j is the represent layer [16]. The main role of the encoder is to condense the reconstruction error between the image input and network output, i.e., from input x to output $x\hat{}$ (i.e., $x \rightarrow x'$) which will decrease the dimensionality (specifically, latent space) from the dataset in supervised learning task. The middle frame between the image input and network output, i.e., bottleneck, can be considered as the main representative for the framework of this designed network. This framework creates tensor of the feature maps on applying convolution filters from the input layer to all hidden layers which ultimately fed by approximation of any nonlinear activation functions. These activation functions are "prerectified linear units" known as PreReLU, which is a function that will return 0 in case of a negative input and the value itself for a positive output. The function can be defined mathematically as $F(X) = \text{Max } [0, X]$; this can be simplified in the way that more tensors will be generated on feeding the feature maps through the given activation function. Each feature map is pooled into the pooling layer to produce a single numerical value by the input of a set of slight grid type regions [17]. The role of encoders plays a crucial part in this framework as it has the capability for learning the most complicated and nonlinear patters like in case of medical imaging. These layered network approaches acquire compressed illustration of the input through minimization of the reconstruction error amid the output and input values of the neural network. When numerous encoders are set in a formation, then it will

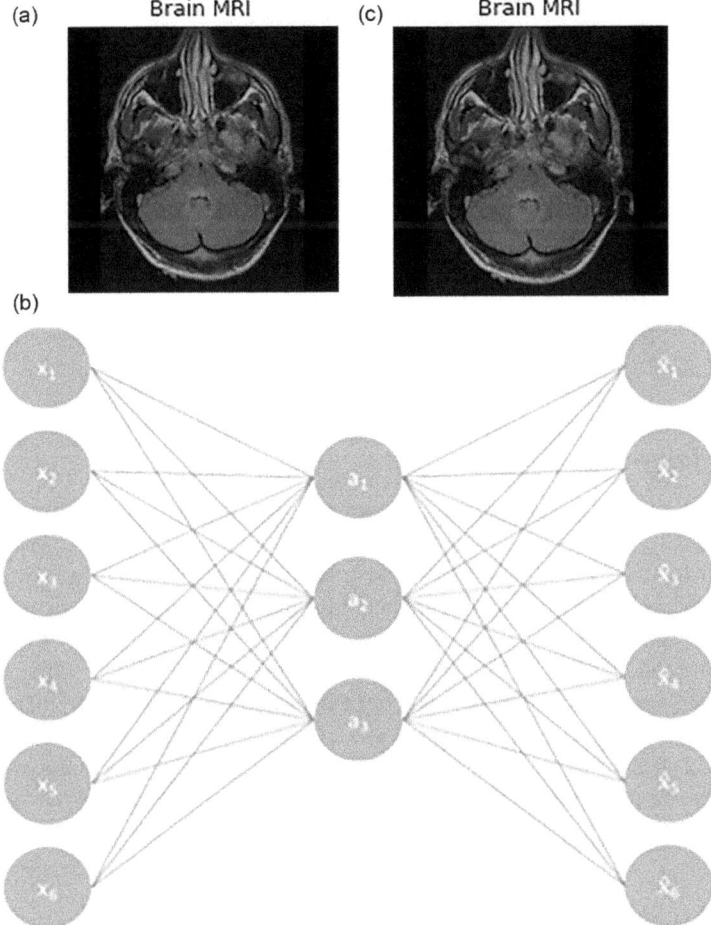

FIGURE 15.1   (a–c) Architecture of the ResNet with reconstruction for the output image.

suggestively enhance the visualization power with the usage of activation values for hidden units of the encoder as the input to the next high-level encoder [18]. Owing to this characteristic of the encoder, the lower layer handles the simplest patterns and as layer moves toward the higher level, the patterns are further complicate in the input vector. As discussed in the algorithm, the MRI image modalities undergo various problems such as optical blurring, noisy sequences, and poor contrast. The hidden layer will quote the structures from the input image and advanced resizing fine-tuned with reverence to the smaller image besides the fact that its size is steadily enlarged as the course of training progresses. The operative use of deep learning in this framework for medical imaging modalities gave upgrades to healthcare informatics as analysis of clinical diagnosis becomes more operative than before. Many establishments are increasing their research toward better visualization of medical imaging such as higher resolution, removal of noise, etc.

15.4 STAGES FOR IMPLEMENTATION OF THE RESNET FRAMEWORK

15.4.1 PREPROCESSING

The preprocessing medical input images range suggestively in relation to the size. Images are set defined to (128 × 128 × 3) pixel and the ResNet framework altered in two stages where in the first stage, newly added head of the untrained network is trained with the conservancy of the image-net assigned weights through the learning rate of approximately 1e − 3 for continuous three epochs. In the second phase, the entire untrained network got tuned via a discriminatory learning rate for five epochs [19]. Again, the network is fine-tuned with image-net assigned weights through the learning rate of approximately 1e − 4 for 25 epochs.

15.4.2 TRAINING THE NETWORK

Training the given model in multiple phases with diverse MRI image sizes using progressive resizing tends to accomplish a better oversimplification. This is also an example of transfer learning as one inputs the image size to the other one. One thing to notice is that as we proceed to the further stages of the training, we ensure to reduce the learning rates. The features which are erudite by the diverse layers of deep CNN approach are independent of the size of the input image. The features which are global in nature that are applicable to all layers are existing in a similar resized image through dissimilar pixel resolution. The training works in 3 states where each state matches to the image of dissimilar dimensions of the input image. Each enforces the cyclic learning rate scheme proposed by Smith [20] for the support of selection of the optimal rate for learning.

15.4.3 SEGMENTATION

Segmentation utilizes using a full-resolution CNN with the ResNet architecture [8]. It includes four stages of block and consists of two convolutional layers with PreReLU activation function, one maxpooling layer for the encoder and an upsampling layer in the decoder. Dependable on the ResNet framework, the layers from the encoder prance connections to the equivalent decoder. It is context with the gradient flow in the narrow layer during its phase of training. Segmentation aided as ground truth for model training for the segmentation carried out automatically. In the starting phase, two networks are available, where the first network is only fluid attenuated inversion recovery (FLAIR) and other network is trained and equipped with 3 available sequences which include precontrast, FLAIR, and postcontrast and secondly, only FLAIR. The FLAIR portion from either side of the region or slices of interest portion and channels were occupied with the adjacent neighboring slices of the tumor toward progression in the added information in the network. The corresponding slices which consist of the tumor significantly lower than the background session and functional oversampling with the augmented data image to process the training of the ResNet. To diminish the imbalances among pixels of tumor and nontumor, the slices, which are proved to be empty in nature, need to be discarded.

(a)   Brain MRI   (b)   Tumuor present

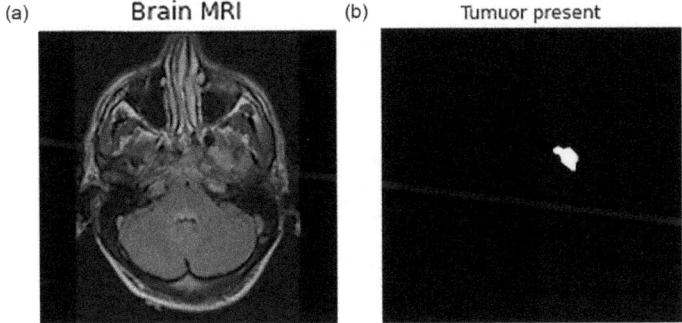

FIGURE 15.2   (a) Original brain MRI. (b) Mask to predict the location of the tumor in the MRI.

The aim of this conversion has various advantages in formation mask or the more density extent has dimension as compared to original image with three-dimensional channels as mentioned above in Figure 15.2. This will aid in reducing the dimension of an image, and as a result, the complexity level of the model will also reduce. This progression has been attempted since preparing a completely CNN with segmented images that does not consist of voxels which can be exceptionally adverse. The enormous portions of voxels in the irregular slices are standard, and in this manner, adequate negative information is accessible for training.

15.4.4 FOCAL LOSS FUNCTION

The main target of the focal loss function is to raise the issue of class imbalances between the background and forefront slices, or can say, classes while the training phase is working on the specified dataset. The objective is to detect the object which is imbalanced with respect to the background and the aforementioned slices showing the problem of classification in the class imbalances dataset available for the MRI of the brain.

The preliminary fact of loss function (focal) for the binary classification is defined as below:

$$x = \begin{cases} -(\log p) & y = 1 \\ -\log(1-p) & \text{otherwise} \end{cases}$$

Where

y belongs to $\{-1, 1\}$ and signifies the ground truth for the positive class and negative class, and

p belongs to the set of $[0, 1]$ and specifies the assessed probability for all class with label $y = 1$.

15.5 RESULTS AND DISCUSSIONS

The experiment on the MRI of the brain tumor on the self-learning method provides the slices of the FLAIR sequences in the input brain MRI image. The mask of the brain tumor is predicted sequences within the tumor slices while neglecting the empty slices.

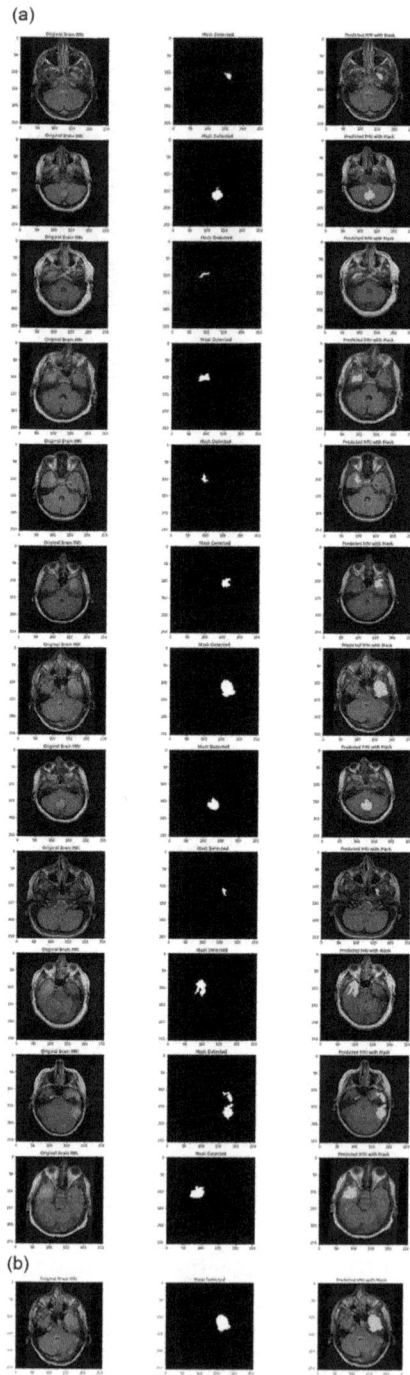

FIGURE 15.3 (a) Input MRI image. (b) Mask prediction of MRI generated through feature extraction. (c) Predicted brain MRI with segmented mask.

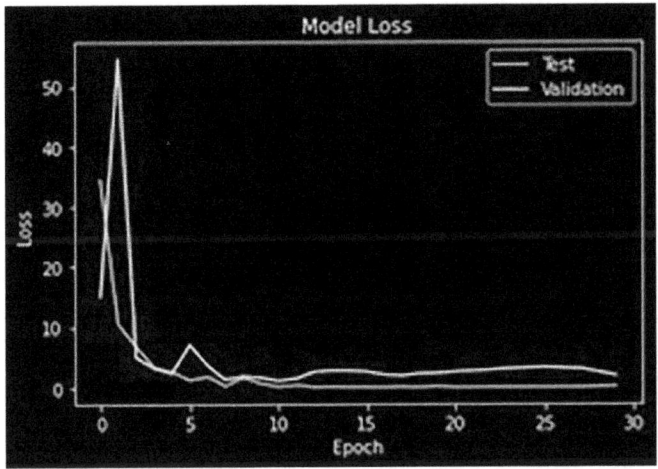

FIGURE 15.4 Graph for loss and epoch with respect to tested data and valid data.

As predicted in in Figure 15.3, the extraction of mask has to be situated to excerpt the area in the brain MRI, and slices are highlighted to extemporize the segmented mask in the trained image. The resultant output predicted mask image would be segmented mask in the higher dimensional view of the brain MRI through the ResNet framework.

The focal loss function would be impacted with epoch. The use of this function will be effective to measure the classification of the MRI image by taking up the tuning phase and configure it in the specified framework with network stabilizers of the given epochs. The function will be accomplished on the tested data, and then, the valid data is being differentiated from it. As shown in Figure 15.4, the graph is shown with measures of the focal loss and accuracy achieved with epochs by the aid of training parameters.

15.6 CONCLUSIONS AND FUTURE SCOPE

In this study, we propose a ResNet-based approach of a CNN which conveys a virtuous combination for the number of parameters in addition to the performance for faster training. The capability to provender images of sizes other than with which they are trained with. This is a perilous part of the training practice used for training a high routine network by means of few iteration epochs using the techniques introduced. In the first stage, it extracts the discriminatory features of slices of the tumor in neighboring slices. Mappings with the learning rate by the trained network are projected for segmented mask for specific subspace.

The scope of the CNN framework can further be used to compute possibly inferior computational overhead. Comprehensive valuations can be achieved on a big dataset of the brain MRI for the contrast enhancement. Medical MRI images which are evaluated can validate that the generative adversarial networks repossess higher resolution or fine quality MRI images with enhanced fine texture specifics related with predictable wavelet based compressed sensing schemes in the deep learning ResNet(n) framework to use pixel-based training scheme. Also, the ResNet(n) framework in further stage can suggest the value of reconstruction within milliseconds comprises briefings of magnitude closer to current state of the art for improved compressed sensing scheme.

REFERENCES

1. Elad, M., & Feuer, A. (1997). Restoration of a single super resolution image from several blurred, noisy, and undersampled measured images, *IEEE Trans. Image Process.* 6 (12), 1646–1658.
2. Mousa, A. H., Aldeen, Z. N., Mohammed, A. H., & Abboosh, M. G. K. (2020). A convolutional neural network-based framework for medical images analyzing in enhancing medical diagnosis, *Ann. Trop. Med. Public Health.* 23 (13), 4–10.
3. Peeters, R. R., et al. (2010). The use of super-resolution techniques to reduce slice thickness in functional MRI, *Int. J. Imaging Syst. Technol.* 14 (3), 131–138.
4. Razzak M.I., Naz S., Zaib A. (2018) Deep Learning for Medical Image Processing: Overview, Challenges and the Future. In: Dey N., Ashour A., Borra S. (eds) Classification in BioApps. Lecture Notes in Computational Vision and Biomechanics, vol 26. Springer, Cham. https://doi.org/10.1007/978-3-319-65981-7_12.
5. Choi, J. S., Base, S. H., & Kim, M. (2015). Single image super-resolution based on selfexamples using context-dependent subpatches, in: *Proceedings of the IEEE International Conference on Image Processing (ICIP)*, pp. 2835–2939, Quebec City, QC, Canada.
6. Freeman, W. T., Jones, T. R., & Pasztor, E. C. (2002). Example-based super-resolution, *IEEE Comput. Graph. Appl.* 22 (2), 56–65.
7. Yang, J. C., Wright, J., Huang, T. S., & Ma, Y. (2010). Image super-resolution via sparse representation, *IEEE Trans. Image Process.* 19 (11), 2861–2873.
8. Mane, V., Jadhav, S., & Lal, P. (2020). Image super-resolution for MRI Images using 3D faster super-resolution convolutional neural network architecture, in: *ITM Web of Conferences,* ICACC-2020, Vol. 32, No. 03044, pp. 1–7, Les Ulis.
9. Redmon, J., Divvala, S., Girshick, R., & Farhadi, A. (2016). You only look once: Unified, real-time object detection, in: *Proceedings of the IEEE Conference on Computer Vision and Pattern Recognition (CVPR)*, pp. 779–788, Las Vegas, NV.
10. Setio, A. A. A., et al. 2016. Pulmonary nodule detection in CT images: False positive reduction using multi-view convolutional networks, *IEEE Trans. Med. Imaging* 35, 1160–1169
11. Al-Masni, M. A., et al. (2018). Skin lesion segmentation in dermoscopy images via deep full resolution convolutional networks, *Comput. Meth. Prog. Biomed.* 162, 221–231.
12. Qiu, D., Zhang, S., Liu, Y., Zhu, J., & Zheng, L. (2020). Super-resolution reconstruction of knee magnetic resonance imaging based on deep learning, *Comput. Meth. Prog. Biomed.* 187, 105059. doi:10.1016/j.cmpb.2019.105059.
13. Kamnitsas, K., et al. (2017). Efficient multi-scale 3D CNN with fully connected CRF for accurate brain lesion segmentation, *Med. Image Anal.* 36, 61–78. doi:10.1016/j.media.2016.10.004.
14. Rizwan, I., Haque, I., & Neubert, J. (2020). Deep learning approaches to biomedical image segmentation. *Inform. Med. Unlocked* 18, 100297. doi:10.1016/j.imu.2020.100297.
15. Yang, C. Y., Huang, J. B., & Yang, M. H. (2010). Exploiting self-similarities for single frame super-resolution, in: *Proceedings of the Asian Conference on Computer Vision*, 2010, pp. 497–510, Springer, Berlin.

16. Khan, H. A., Jue, W., Mushtaq, M., & Mushtaq, M. U. (2020). Brain tumor classification in MRI image using convolutional neural network, *Math. Biosci. Eng.* 17 (10), 6203–6216.

17. He, K., Zhang, X., Ren, S., & Sun, J. (2016). Deep residual learning for image recognition, in: *Proceedings of the IEEE Conference on Computer Vision and Pattern Recognition*, Las Vegas, NV.

18. Farooq, M., & Hafeez, A. (2020). COVID-ResNet: A deep learning framework for screening of COVID19 from radiographs. arXiv preprint. arXiv:2003.14395.

19. Narayan Das, N., Kumar, N., Kaur, M., Kumar, V., & Singh, D. (2020). Automated deep transfer learning-based approach for detection of COVID-19 infection in chest X-rays. *Irbm* 1, 1–6. doi:10.1016/j.irbm.2020.07.001.

20. Zou, W. W., & Yuen, P. C. (2012). Very low resolution face recognition problem, *IEEE Trans. Image Processing*, 21 (1), 327–340.

Index

academic performance 230, 232
accelerometer 140, 141
activation function 206
advantageous condition 77
aggression 110
Airveda 193
amyotrophic lateral sclerosis 28
analytical techniques 244, 248
anxiety, depression and stress 253
AQI 196, 198
area under ROC 263
artificial intelligence 98, 141
artificial neural network 208
aspects/features 163
aspect weight estimation 170
assessment status 243, 244, 247
AUC 260
automatic segemantion 269
awareness status 244, 246–248

backpropagation neural network 214
Bargad wood 194
benign tumors 52
big data 96
binary array 91
binomial random variable 11
biological neural networks 205
bronchial cancer 101

CCR 75
central limit theorem 19
challenging geographic 73, 77
classification 230, 232, 235, 236–240
classification model 230, 236, 238–239
classifier 120
clustering 56
CNN *see* convolutional neural network (CNN)
CNN architecture 102
communicable disease 244
computer aided detection 140
conditional density 22
confusion matrix 264
context similarity 165
convolutional neural network (CNN) 67, 115, 146, 214, 268
cosine similarity 166
cow dung 190
cumulative distribution function 6

data analytic 243–245, 247, 249
database 244

data envelopment analysis 73, 75
decision tree 230–232, 236–237, 238, 239, 240
deep learning 66, 143, 252

efficiency 76
electrocardiogram 99
electromyograms 28
electromyography 141
embedded system 129
encoders 270
environment 199
error free reception 92
extreme gradient boosting 253
extreme learning machine 39

feature extraction 30
fire-fly optimization algorithm 34
FLAIR portion 272
frame structure 268
fuzzy 101

Gaussian random variable 8
Gayatri mantra 192
genetic algorithm 33
glioblastoma 100

hamming distance 91
Hawan 190
healthcare 97, 243, 244, 249
health sensor data management 99
heathcare 149
heterogeneous data fusion 99, 142
higher dimensional view 275
higher education 229–240

Indian culture 184
industrialization 186
infectious disease 243–245, 247–249
informatics 244, 248
input-oriented model 75
input parameters 77
input tragets 81
IoT 97
IoT sensor 140

joint moments of random variables 19

k-closest neighbors 262
k-nearest neighbor (KNN) 58, 114, 230, 236–240
KNN algorithm 91

LASSO 143
linear block code 89
local administration 150
logistic regression 56
loss function 273
lung adenocarcinoma 138
lung cancer 138

machine learning 54, 97, 141, 204, 230
Mahamrityunjay mantra 192
malignant tumor 53
mango wood 190, 194
mask prediction 273
mitigation 244, 248
model building 144
model evaluation 144
modulo-2 addition 90
moments of random variable 13
MRI 106
multiclass classification 252
multi layer feed forward neural network 212
multilingual feedback 160
multimodel data fusion 99
multinomial distribution 12
multiple random varibales 16

naïve Bayes 63, 264
nature inspired feature selection 32
neural network 115
neuromuscular impairment 27
Nigeria 243, 244, 247, 248
non-parametric 73, 75
NSM 75
Numpy 125

opinion mining 159
output parameters 77, 78

parity word 89
particle swarm optimization 32
particulate matter 197
pattern recognition 190
performance 72, 73, 76
point wise mutual information (PMI) 168
pollution 184
precision 263
prediction 230, 234, 236, 237, 238, 240
predictive model 231
prevalence 244, 245, 248
probability density function 6
public transport sector 71, 72
Python 130

questionnaire 156

random forest 230, 236, 238–240, 260
random variable 3
Raspberry Pi 120
recall 264
rectified linear unit 270
recursive neural network 212
reference set 76
reinforcement learning 210
resnet framework 272
risk factors 52
risk stratification 146
road network, in India 72
road transport , in Rajasthan 72

Sarcoma 100
science of mantra 188
science of yajna 188
segemented mask 275
sensors 97, 190
sentiment analysis 159
single hidden layer feed forward
          network 40
single layer feed forward neural network 211
slacks 75, 76
SMOTE 252
social inclination 159
Spyder IDE 105
students 229–240
subjectivity analyzer 172
supervised learning 55, 209
support vector machine (SVM) 60, 253
systematic block code 89

TF-IDF 166
tissue characterization 146
transformation of random variables 23
tuberculosis 243, 244, 248, 249
tumor grade 110

unsupervised learning 56, 210

vedic sciences 187

weight based code 90, 91
whale optimization algorithm 37
WHO 110
Wigner-Ville transform 30
wireless and wearable sensor 140

XGBoost 255
Xgboost tree 230, 236, 238–240

Yajna/Yagya 193
YOLO 116

Milton Keynes UK
Ingram Content Group UK Ltd.
UKHW022316190924
448439UK00018B/212